Hitler's Geographies

Hitler's Geographies

The Spatialities of the Third Reich

EDITED BY PAOLO GIACCARIA
AND CLAUDIO MINCA

The University of Chicago Press
Chicago and London

Paolo Giaccaria is assistant professor of political and economic geography at the University of Turin, Italy. **Claudio Minca** is professor and head of cultural geography at Wageningen University, the Netherlands.

The University of Chicago Press, Chicago 60637
The University of Chicago Press, Ltd., London

Printed in the United States of America

25 24 23 22 21 20 19 18 17 16   1 2 3 4 5

ISBN-13: 978-0-226-27442-3 (cloth)
ISBN-13: 978-0-226-27456-0 (e-book)
DOI: 10.7208/chicago/9780226274560.001.0001

Library of Congress Cataloging-in-Publication Data

Names: Giaccaria, Paolo, editor. | Minca, Claudio, editor.
Title: Hitler's geographies : the spatialities of the Third Reich / edited by Paolo Giaccaria and Claudio Minca.
Description: Chicago ; London : The University of Chicago Press, 2016. | Includes bibliographical references and index.
Identifiers: LCCN 2015045702 | ISBN 9780226274423 (cloth : alk. paper) | ISBN 9780226274560 (e-book)
Subjects: LCSH: National socialism. | Germany—History—1993–1945. | Geography—Political aspects.
Classification: LCC DD256.7 .H58 2016 | DDC 943.086—dc23 LC record available at http://lccn.loc.gov/2015045702

Chapter 3, "In Service of Empire: Geographers at Berlin's University between Colonial Studies and *Ostforschung* (Eastern Research)" by Jürgen Zimmerer, was originally published in German as "Im Dienste des Imperiums: Die Geographen der Berliner Universität zwischen Kolonialwissenschaften und Ostforschung," in *Jahrbuch für Universitätsgeschichte* 7 (2004): 73–100.

Chapter 5, "Race contra Space: The Conflict between German *Geopolitik* and National Socialism," by Mark Bassin, was originally published in *Political Geography Quarterly* 6, no. 2 (1987): 115–34. Reprinted by permission.

Chapter 7, "National Socialism and the Politics of Calculation," by Stuart Elden, was originally published in *Social and Cultural Geography* 7, no. 5 (2006): 753–69. Reprinted by permission.

Chapter 8, "Applied Geography and Area Research in Nazi Society: Central Place Theory and Planning, 1933 to 1945," by Mechtild Rössler, was originally published in *Environment and Planning* 7 (1989): 419–31. Reprinted by permission.

Chapter 11, "Nazi Biopolitics and the Dark Geographies of the *Selva*," by Paolo Giaccaria and Claudio Minca, was originally published in *Journal of Genocide Research* 13 (2011): 67–84. Reprinted by permission.

Chapter 14, "Hello Darkness: Envoi and Caveat," by Andrew Charlesworth, was originally published in *Common Knowledge* 9, no. 3 (2003): 508–19.

♾ This paper meets the requirements of ANSI/NISO Z39.48-1992 (Permanence of Paper).

Contents

Introduction

Hitler's Geographies, Nazi Spatialities

PAOLO GIACCARIA AND CLAUDIO MINCA

This book moves from the assumption that the Nazi project was, among other things, an eminently geographical project. Nazi ideology was in fact permeated by a broad spatial vision of the Reich and its territories, supported by a number of key geographical concepts, like those of *Lebensraum*, *Großraum*, Farther East, and *Geopolitik*, to name but a few. However, despite the popularity and widespread use of spatial concepts and metaphors in the Nazis' imperial discourse, including in policy pronouncements, and despite the fact that geographers and spatial planners played an important role in the Nazi project, a comprehensive examination of the relationship between geography, spatial theory, and the Third Reich remains to be developed. It is thus our contention that a geographical perspective on the spatialities of the Third Reich is much needed. Indeed, *Hitler's Geographies* aims to respond to the growing interest in the current academic literature in English—that is, the literature available to international debates—for a detailed investigation of the spatial imaginations of the Nazi regime and of the actual geographies it designed and implemented through its thirteen years of grand plans, colonization, exploitation, and genocide.

This volume provides a first overview of how recent research in English-speaking human geography and related disciplines has approached the spatialities of Hitlerism, and their relation to the geopolitical and, in some cases, biopolitical projections of the Nazi regime. While providing an analysis of "the spatial" in Nazi ideology from a multiplicity of theoretical perspectives, *Hitler's Geographies* also reflects on the entanglements between the Nazis' grand spatial plans and spatial practices "in place," something only marginally discussed in the key literature thus far. Furthermore, this collection is an attempt to introduce a geographical approach into most recent debates on the "cultural histories of the Third Reich" (see chapter 1 in this volume).

More specifically, we believe that *Hitler's Geographies* can contribute to broader debates on the spatialities of the Third Reich in two distinct ways. First, by providing an unprecedented collection of papers directly engaging with the specific relationship between spatial theory, Nazi ideology, and its geopolitical and genocidal practices. This is a theme that has recently gained momentum among scholars of National Socialism and the Holocaust and among geographers as well. This book intends to consolidate such interest by offering an ambitious lineup of chapters penned by geographers, alongside key interventions by prominent scholars in the field of Holocaust studies and historians of the Third Reich who have considered questions of space and spatial theory in their work. In addition, it brings together some of the key contributions on this topic in geography, which had previously only been available scattered across different journal issues. *Hitler's Geographies* represents therefore a first attempt to map the state of the art of geographers' and spatial thinkers' contribution to the literature on the Third Reich and the Holocaust available in English—that is, again, the literature available to international debates—but also an attempt to move the discussion a step further and propose this as a key area of future investigation for the field in the years to come.

The second objective of this project is more inherently theoretical, and it speaks to the increasing role geography and geographers play within interdisciplinary debates on "the political" and on the cultural histories of modernity. While this book clearly addresses the field of geography, its more general and significant ambition is to reflect on what the broader debate on Nazism and the Holocaust may learn from a deeper understanding of their spatialities and, more specifically, on how a geographical approach can contribute to such an analysis. But it is also an investigation of what geography may learn about the Third Reich—and its own (direct and indirect) relationship with it—by engaging with the work of other specialists preoccupied with the spatial dimension of Nazism and the Holocaust. Furthermore, we believe that this collection helps demonstrate how a closer look at the specificities of Hitler's geographies may draw attention to some undisclosed features of modernity and its spatialities. The growing interest in these issues in recent years on the part of non-geographers is a testament to the need for more interdisciplinary work on the geographical imaginations and on the implementation of the set of ideas, concepts, and practices that go under the label of "Hitlerism" (see, among others, Levinas 1990). The contributions from non-geographers are also key to the volume for this reason, and they confirm the wider purchase of our guiding argument. The entanglements between biopolitics and geopolitics, the pervasiveness of cartographic and calculative rationalities, and the endless search for new spatial orders and orderings are all geographical facets of modernity that can be fruitfully investigated with a closer interrogation of some of their manifestations in the regime, established by what Daniel Pick (2012) has called the "Nazi mind."

Through a direct and critical engagement with some of the most significant streams of the relevant literature, and with the aim of contributing more specifically to present debates on the cultural histories of Nazism and the Holocaust (for an overview of these, see chapters 1 [Giaccaria and Minca] and 2 [Stone] in this volume), *Hitler's Geographies* represents the first product of a broader project that intends to start formulating a tentative spatial theory of the Third Reich (see Giaccaria and Minca, this volume, chapter 1). The existing geographical literature, which has focused on Haushofer's *Geopolitik*, on the long trajectory of *Lebensraum* across the twentieth century, on the contribution of Walter Christaller and other geographers to the realization of the Nazi spatial imaginaries and broader ideology—and more recently on the "Holocaust Geographies"—has only partially completed its task. We argue instead that some of these geographical concerns must be scrutinized once again, especially in light of the perspectives on the spatial dimension/roots of the Third Reich made possible by the new lines of historical and philosophical research described above.

Hitler's Geographies sets out to complement the contemporary cultural histories of the Third Reich from a number of fresh geographical perspectives in order to offer new understandings on the spatialities that have characterized Nazi imaginaries and practices. The ambition of elaborating a tentative spatial theory of the Third Reich, starting with the present volume, is in fact grounded in our confidence that contemporary cultural and political geography might offer analytical tools capable of providing original insights into the *arcana* of the Nazi project, and a belief that the leadership of the Nazi regime—not only Hitler and Himmler, but also the "experts" who "worked towards" them (Kershaw 1993) in materializing their ideological imaginaries—were driven by a fundamentally spatial *Weltanschauung*, itself characterized by a racialized bio-geography of sorts. Accordingly, what we would like to suggest is that, while the specifically geographical *forma mentis* that has marked Hitlerism should be contextualized with reference to the geographical thought of the day, it should nonetheless be interpreted through the conceptual lenses that contemporary critical geographical theory provides.

The realization of a spatial theory of the Third Reich and the Holocaust—that is, an understanding of Nazism that adopts spatial and geographical theory as an interpretative framework—is a complex task, and *Hitler's Geographies* aims to represent just one first move in that direction. The essays collected here have in fact been deliberately conceived as moments/passages within a constellation of questions, intuitions, traces, archives, insights, investigations, and explorations that compose the fragments of the as-yet-incomplete patchwork produced by the *spatial cultural histories* of Hitlerism.

The Structure of the Book

In the chapter that follows and that opens the volume, "For a Tentative Spatial Theory of the Third Reich," we attempt to provide a brief critique of what we identify as the five main "streams of reflection" that have so far engaged with "Hitler's geographies," albeit in different ways and moments, and that mark the necessary starting point for this book project. Through a direct and critical engagement with these streams of literature, and with the aim of contributing to present debates on the cultural histories of Nazism and the Holocaust, this chapter tries to bridge the "Third Reich (academic) geography" with the (actual) "geographies of the Third Reich" in order to cast new light on three key aspects of the Nazi "spatial *Weltanschauung*": (1) between biopolitics and geopolitics, (2) between a topographical and topological approach to the government of people and space, and (3) between spatial ideology and spatial practice.

The second chapter, "Holocaust Spaces," written by Dan Stone, reflects on some of the issues discussed in this introduction, but from a different disciplinary perspective, that of historiography. In particular, Stone critically analyzes the ways in which historians of the Holocaust have dealt with the concept of space. In line with the book's overall architecture, he distinguishes two different "spatial" registers in the key literature: the first, focused on the examination of "actual spaces," that is, of the geographical and territorial dimensions of the Holocaust; the second, preoccupied with the "imagined spaces" or the figurative uses of space in, for example, Nazi propaganda or ideological geographical fantasies. The cultural histories of the Third Reich, he claims, can indeed help to better understand the spatial imaginations at the core of Nazi geopolitical projections and of the Final Solution: a combination of "rational"-sounding schemes that could be set out scientifically, in terms of planning, bureaucracy, and operational management, and of "nonrational" beliefs and fears—of Jewish world conspiracy, or racial decline, of history being driven by the clash of Aryan and non-Aryan forces—that these "rational" schemes were designed to overcome.

Stone's essay is deliberately complementary to the previous chapter, and the two essays together are intended to provide an introduction, from the perspectives of geography and history respectively, to the broader theoretical framework that makes up this book project. The remaining fifteen chapters are instead divided into two major parts, largely reflecting the two fields of inquiry that we describe as the "Third Reich Geographies" and the "Geographies of the Third Reich."

Third Reich Geographies

This first part investigates the "Third Reich Geographies," that is, the body of spatial theories and concepts that populated the Nazis' racialized imperial fantasies and animated their *Lebensraum* policies. It does so by engaging with the direct and indirect role geography and cognate disciplines played in the production of the Nazi spatial *Weltanschauung* and is articulated into two clusters of essays dedicated, respectively, to the tradition of academic geography in Germany, from the Wilhelmine Reich to the Weimar Republic and the Third Reich, and to the relationship between urban and regional planning and Nazi ideology and practice. The first cluster, "Biopolitics, Geopolitics, and *Lebensraum*," disentangles the complex genealogies of the Ratzelian legacy and assesses the involvement of geographers in German colonialism during the first half of the twentieth century. It is opened by a chapter penned by German historian Jürgen Zimmerer and titled "In Service of Empire: Geographers at Berlin's University between Colonial Studies and *Ostforschung*." Here Zimmerer diachronically reconstructs the relationship between German geography/geographers and the colonial theories and practices in both the Wilhelmine and Weimar periods. According to his reading, the sequence of chairs assigned by the Department of Geography at the Friedrich Wilhelm University in Berlin, from Ferdinand von Richthofen to Albrecht Penck, marked a clear continuity in the "colonial vocation" of German academic geography in those years. At the same time, there were important elements of discontinuity as well: while for Richthofen, German territorial expansion should have followed the tracks of other European colonial powers, in particular in Africa and Asia, Penck—influenced by his *völkisch* tendencies—argued that the geographical focus of German colonial ambitions should become Eastern Europe, a line of argument then incorporated by the Nazis in their search for *Lebensraum*.

In chapter 4, "The East as Historical Imagination and the Germanization Policies of the Third Reich," historian Gerhard Wolf delves into the relationship between geography, geographers, and Nazism by focusing on the work of Albrecht Penck and Wilhelm Volz and the role played by the institute

Mittelstelle für zwischeneuropäische Fragen, an *ante litteram* think tank of sorts for the new "discipline" of *Ostforschung* (broadly, Eastern research). In particular, Wolf suggests the existence of a clear continuity between the concepts elaborated by both Penck and Volz, and the spatial-geographical imagery proposed by Hitler and Himmler, but also the related practices of Germanization of Eastern Europe during the war period. The chapter shows how pivotal these perceptions were for the German occupiers in their attempts to segregate the native population in annexed Poland during World War II. In contrast to Hitler, who had unambiguously declared that Germany must "under no circumstances annex Poles with the intention to turn them into Germans," but "remove them and hand over the vacated territory to its own *Volksgenossen*," the German occupation regimes proved more flexible. When instructed to select the ethnic Germans from the native population, the chapter claims, the Nazis were guided less by Hitler's anthropological racism than by a more traditional *völkisch* definition of Germanness as promoted during the Prussian Germanization campaigns of the late nineteenth century and later refined by ideologues like Penck and Volz.

Chapter 5, Mark Bassin's "Race contra Space: The Conflict between German *Geopolitik* and National Socialism," first published in 1987 and here reprinted, represents a milestone article in past geographical debates on *Geopolitik* and Nazism. Despite the dialectic race/space here described to articulate Bassin's fundamental critique of *Geopolitik* may appear today as relatively outdated—especially in light of the literature of the past three decades—we suggest reading Bassin's article in the broader context of this section of the book. In fact, this article convincingly shows how the role of Haushofer had been excessively emphasized by the US media in the postwar period and, perhaps more importantly, how the environmental determinism that ideally connected—according to Bassin and many other Anglophone interpreters—the writings of Friedrich Ratzel and Karl Haushofer was fundamentally different from the biological racism that would connote Nazi ideology (see also Farinelli 1992; Minca 2006). At the same time, recent work has revealed clear analogies between German academic geography and Nazi mainstream spatial ideologies—like the genealogies discussed in the preceding two chapters. This allows a new light to be cast on the relationship between the discipline of geography and Nazism compared to the one emerging in Bassin's then pathbreaking article (see also Bassin's [2005] more recent account of the relationship between German geography and *völkisch* activism), perhaps offering a more articulate and complex picture of the contribution given by the discipline in the formulation of the spatial and colonial imagery permeating the Third Reich.

This first section is completed by Clemens Driessen and Jamie Lorimer's essay titled "Back Breeding the Aurochs: The Heck Brothers, National Socialism, and Imagined Geographies for Non-Human *Lebensraum*," where they investigate the bio-geographical imaginations behind the animal "back-breeding" programs carried out by the brothers Lutz and Heinz Heck—two influential German zoologists in charge of the Berlin and Munich zoos in the run-up to World War II. Favored by close connections to and patronage from the National Socialist elite, the Heck brothers sought to resurrect the wild cow (*aurochs*) and horse (*tarpan*) by breeding out the degeneration they associated with domestication. Drawing on archival material, this chapter situates these back-breeding initiatives in relation to the emerging field of geopolitics and demonstrates how the project to recreate extinct primordial wildlife functioned as part of particular discourses and practices of nature conservation emphasizing the ideal Germanic character of the European landscape, which would eventually require ethnic cleansing as a form of ecological restoration. While apparently differing from most conventional interpretations of German political geography, this approach brings evidence to our initial claim that Nazi geopolitics and biopolitics ended up merging in a sort of comprehensive project aimed at reorganizing both the human and animal spatial realm, in Germany, and in the whole continent of Europe.

The second cluster of chapters, labeled "Spatial Planning and Geography in the Third Reich," centers on another key element in the relationship between geographical tradition and the Third Reich, that is, the Nazis' aspiration to realize a spatial order coinciding with a political order, shaped in line with their *Weltanschauung*. This section opens with Stuart Elden's "National Socialism and the Politics of Calculation," originally published in 2006, which examines how mathematics and politics intertwined in National Socialist Germany, particularly in the period between 1933 and 1939. Elden, analyzing Martin Heidegger's key writings on Hitler's regime, argues that one of the philosopher's central questions revolved around the way in which what he calls "machination" and later "technology" depended upon a particular notion of metaphysics, a particular casting of being, that is, "to be is to be calculable." Elden suggests that Nazism sought control of the earth in ways that made possible and also exceeded its quest for world domination. Starting with a discussion of the notion of *Gleichschaltung*—that is, synchronization or political coordination—and its ontological underpinnings, the chapter discusses two key examples: the calculation of race in the Nuremberg laws, and the calculation of space in geopolitics. These come together in the racialized notion of *Lebensraum*. Elden's intervention hence bridges this cluster of essays to the previous one.

When one engages Nazi urban and regional spatial planning, two key figures who emerge are geographer Walter Christaller and economist and urban planner Gottfried Feder. The two central chapters of this section are dedicated to Christaller and his relation to geography and geographers of the day. They are followed by an essay on the realizations of the team led by Feder, a figure whose contribution to the spatial realizations of Nazism has been surprisingly overlooked by the geographical literature. Chapter 8 corresponds to what is perhaps the most often cited article on the relationship between geography and Nazism, Mechtild Rössler's "Applied Geography and Area Research in Nazi Society: Central Place Theory and Planning, 1933 to 1945," originally published in 1989. According to Rössler, under National Socialism, extraordinary research opportunities opened up for social scientists who wanted to implement their theories, especially if they could comply with the broader regime of truth that Hitlerism and its advocates imposed. The new legal context and central planning organizations, for example, provided the political and institutional basis for scientific research with an unprecedented amount of funding and opportunities for the actual implementation of theoretical and experimental work on space and society. Walter Christaller, who was at that time too old (and marginalized) for a university career, found unexpected recognition for his expertise after having been recruited by research institutions Himmler promoted. His personal and political biography is indeed imbued with paradoxes: a former member of the Social Democratic Party, he switched to the Nazi Party in 1940, to the Communist Party in 1945, and once again to the Social Democratic Party in 1959. However, these events merely hint at the complex nature of the political context in which Christaller and other scientists worked from 1933 to 1945, something critically revealed and examined by this fundamental article/chapter.

This issue is also discussed in detail, albeit from a different perspective, in chapter 9, Trevor Barnes's "A Morality Tale of Two Location Theorists in Hitler's Germany: Walter Christaller and August Lösch," where the contribution of central place theory to modern geography *and* to Nazi spatial planning in reference to the newly conquered Eastern territories is examined. Barnes further elaborates his engagement and his conversation "at a distance" with Rössler's work (see, for example, Barnes and Minca 2013) on the issue of "research opportunities" and the "Nazi choice" for geographers in all its complexity by comparing the lives and careers of Christaller and of the other recognized proponent of central place theory, August Lösch. Although central place theory is normally associated with both scholars in many reconstructions of the discipline's history—almost as if they worked shoulder to shoulder—their political (and academic) biographies in fact diverged dramatically. While they both supported a quantitative and calculative understanding of space, they were standing on two very different moral grounds. Christaller in fact decided to "work towards the *Führer*," to use Ian Kershaw's (1993) well-known description of the voluntary and oftentimes enthusiastic adherence/affiliation of scholars and academics to the Nazi cause. Lösch, on the contrary, always refused to be involved with the regime, to which he maintained a certain degree of distance. Barnes's reflections on *the possibility of not collaborating* are particularly important for this book's architecture and for the broader discussion on the role of intellectuals and academics under Hitler's rule.

The emphasis placed in this part of the book on the role that calculative conceptualizations of space played in the Nazis' "rule of experts"—and on the related aspiration to realize a carefully planned and fully ordered (*Lebens*) *Raum*—may appear in line with what the key literature normally defines as the "functionalist" interpretations of the Holocaust. These interpretations tend to highlight the "banality of evil" incorporated by the administrative routines and practices associated to the production of the conditions for systematic extermination. Zygmunt Bauman (1989), among others, famously goes as far as identifying Nazism as the extreme expression of the calculative rationalities characterizing modernity. However, as Barnes's chapter shows, despite the fact scholars like Lösch and Christaller shared the same "calculative" understanding of the relation between space and society, their individual trajectories in relation to Nazism have been profoundly different, possibly a demonstration of the limitations of a purely functionalist interpretation of the Third Reich and its complacent scholarship. We argue instead that these key elements of order and rationality always coexisted with other less rational elements, as largely illustrated by the so-called intentionalist interpretations of Nazism and the Holocaust (Stone 2010). This pervasive blending of myth and rational thought was indeed clearly present in the work of Gottfried Feder, a prominent academic (urban) theorist of the edification of the National Socialist state, at least until the first half of the 1930s.

In chapter 10, "Social Engineering, National Demography, and Political Economy in Nazi Germany: Gottfried Feder and His New Town Concept," Joshua Hagen examines in detail Feder's academic work and planning interventions. Feder's grand plan, in fact, aimed at creating a network composed of "one thousand new towns" each with a population of ten to fifteen thousand. This plan, drawn up by Feder in 1934 when he was the *Reichskommissar* for Settlement Affairs, would never be implemented by Hitler; however, it was at the origin of a putatively very important publication for the ways in which the urban hierarchy of the Reich and of the new territories was discussed in those years: *The New Town: An Attempt to Establish a New Art of Town Planning Out of the Social Structure of the Population* [orig.: *Die neue Stadt: Versuch der Begründung einer neuen Stadtplanungskunst aus der sozialen Struktur der Bevölkerung*]. What is particularly relevant for the main argument guiding *Hitler's Geographies* is the fact that Feder's work clearly shows an intent to establish a balanced combination between his claim of seeking "strictly scientific foundations" for his grand plan and a crowd of purely ideological tropes, deeply rooted into a *völkisch* tradition centered on a fundamental rejection of modern metropolitan developments and their presumed cosmopolitan culture. To these degenerate urban forms, Feder proposed to respond with plans focused on the realization of idealized and racially purified communities inhabiting an idyllic grand and pacified German *Lebensraum*.

Geographies of the Third Reich

The second part of the volume investigates the actual "Geographies of the Third Reich," and it is mainly preoccupied with how a geographical perspective on the cultural histories of Nazism may help to highlight some "spatial" aspects of the Third Reich and, particularly, of the Holocaust, that have been often overlooked or at least less scrutinized in previous work. The first cluster of chapters composing this part focuses on the "Spatialities of the Holocaust." In chapter 11, "Nazi Biopolitics and the Dark Geographies of the *Selva*," previously published in 2011, Giaccaria and Minca examine the spatialities of Nazi genocidal practices by engaging with the concepts of *selva* and *città*, as inspired by Italian philosopher Giorgio Agamben, and drawing upon a broader tradition in human geography. Although the historical events recalled here have been extensively discussed elsewhere, they are revisited through the lens of these two geographical metaphors in order to gain new insight into the spatial and philosophical dimensions of Nazi geopolitics and biopolitics. Giaccaria and Minca also reflect on how these concepts contributed to the merging of the "ideal" and the "factual" realms of the Nazi geopolitical project for the creation of new vital space for the German people. They suggest that much can be learned from an examination of the ways in which particular understandings of (imagined and material) space marked the genocidal plans and practices of the Nazi perpetrators, producing a specific geography of genocide, where (spatial) theory and the implementation of extermination came together into a true bio-geo-politics.

From a somewhat similar perspective, Tim Cole analyzes the spatialities of ghettoization within the Nazi imperial plans for the East by exploring the ways in which ghetto fences separated off Jewish and non-Jewish living spaces in his chapter "Geographies of Ghettoization: Absences, Presences, and Boundaries." Ghettoization functioned as a fundamental act of "clearing" and of removal of selected populations in order to create *judenfrei* territories for ethnic Germans, as part of the Nazis' search for new purified *Lebensraum* in the urban and rural landscapes of Eastern Europe. However, ghettoization also entailed the concentration of Jewish populations within segregated quarters in urban centers, with the consequence of actually increasing the Jewish presence in certain parts of the city (and not only by enforcing their absence in others). Once again, this interplay between presence and absence created an urban geography that fundamentally challenged—also because of its complicated and immediate practical implications (for example, an extraordinary density of Jewish residents in the core of the city)—Hitler's broader aspiration to create, in short time, a clear and stable (geo)political situation based on his radical acts of racial and spatial ordering.

While the Nazi "bio-geo-political" project generated endless fields of tensions between the city and the forest, inclusion and exclusion, pure and impure spatialities, presence and absence, it also established long-distance connections, which Michael Fleming describes in chapter 13 as distant "geographies of obligation." In "Spaces of Engagement and the Geographies of Obligation: Responses to the Holocaust," he suggests that while the Nazi destruction of Polish Jewry progressed incrementally from 1939 to 1945 through legal restrictions, confiscations, deportations, ghettoization, slave labor, death camps, and death marches, knowledge of the Nazi program increased in Poland and in the West over time. The "spaces of engagement" where knowledge of the unfolding Jewish tragedy was acquired, crucially influenced responses to it. To illustrate this point, the chapter focuses on the reactions of the British Foreign Office, the Polish government-in-exile (based in London), and the Polish Underground in Poland itself to information about the persecution and the systematic elimination of Jews. Fleming also argues that understandings of the Holocaust were heavily influenced by a specific set of "geographies of obligation"—that is, the hierarchies of responsibility and consideration between different groups at various spatial scales. This chapter explores how the notion of geographies of obligation can help deepen understandings of the way different institutions and key actors responded to the Holocaust while it was taking place.

This group of chapters is completed by "Hello Darkness: Envoi and Caveat," an article published in 2003 by Andrew Charlesworth and here reproposed. Charlesworth, one of the few geographers whose work has extensively engaged with Holocaust studies, investigates the complex relationship between place naming and the actual topographies of some key sites for the implementation of Nazi genocidal practices. Adopting a broadly phenomenological approach to his humanistic geographies of the Holocaust, he reflects on the power of names and places, and their ambivalent spatialities. On the one hand, the Nazi genocidal project seemed clearly indebted to the symbolic and material geographies of Eastern Europe, as expressed in their place naming and related topographies; on the other, this same geography is all too often overlooked and taken for granted by mainstream Holocaust studies. The geography of the Holocaust may have indeed many facets, but their ordinary landscapes can, precisely in their ordinariness and ambiguity, confront us with the need to take the appropriate step toward the understanding of what the British historian Alan Bullock once called "the geography of hell."

In shifting between past and present understandings of Auschwitz and other Holocaust sites, Charlesworth's intervention helps bridge this part of the book with the fourth and last section titled "Microgeographies of Memory, Witnessing, and Representation." A geographical understanding of the Third Reich and the Holocaust may indeed benefit from engaging with a more comprehensive spatiotemporal framework capable of capturing some of its cultural histories as they operated at different scales and with different temporal registers. While geographical methodologies and concepts may in fact help in highlighting some aspects of the past that often escape the grasp of historiography, at the same time incorporating the importance of the "sense of place" is in some cases key to the understanding of the work of memorializing, witnessing, and representing the Shoah. The three chapters closing the volume thus focus their attention on some specific microgeographies of memory and emotions related to the complicated presence, in place, of a traumatic past.

In chapter 15, "The Interruption of Witnessing: Relations of Distance and Proximity in Claude Lanzmann's *Shoah*," Richard Carter-White investigates the relationship between witnessing, Holocaust historiography, and spatiality by fully engaging with a geographical reading of Lanzmann's pathbreaking documentary film, *Shoah* (1985). According to Carter-White, Holocaust testimony is increasingly recognized as a complex and paradoxical mode of representation. Although imbued with a particularly geographical form of authority—the *proximity* of the witness to an occurrence—the evidential qualities of testimony are nonetheless placed in doubt by various forms of *distance* from the event: the interruptions of time, space, language, and subjectivity that hold the acts of witnessing and recounting apart and render the referential status of testimony uncertain. This chapter thus explores how Lanzmann's *Shoah* articulates this notion of witnessing as a *passage*, a disjuncture that nonetheless brings distant worlds into contact. *Shoah* is renowned for the emphasis Lanzmann placed on the sites of the Holocaust, but here a different geographical dynamic at work in the film is examined: the experiences of distance emerging from filmmaking strategies designed to gain proximity to the events recounted by witnesses. Through this analysis Carter-White argues that *Shoah* expresses the incalculable spatiality of witnessing, and thus the necessity of remaining open to the representational logic and demands of specific acts of witnessing.

The work of time on space is also at the center of chapter 16, "A Mobile Holocaust? Rethinking Testimony with Cultural Geography," by historian Simone Gigliotti, who reflects on the relationship between space, trauma, and memory during the Holocaust, as based on Jewish Holocaust survivor accounts of deportation train journeys from ghettos to concentration camps between 1941 and 1944. Historians have studied the Nazis' transportation of approximately three million Jews to concentration and extermination camps as evidence of the bureaucratic, administrative, clinical approach to mass death. This chapter proposes to move beyond that perspective by examining the effects of "cattle car" movement on deportees and their conceptions of mobility, time, and spatial dehumanization. Deportations were critical to the achievement of the "Final Solution," yet the experience of them remains an underexplored mobile topos of Nazi power and of the emotional geographies of the victims' suffering.

The final chapter, "What Remains? Sites of Deportation in Contemporary European Daily Life: The Case of Drancy," written by New York–based historian Katherine Fleming, engages with the interplay of memory and forgetting in Drancy, the site of the French deportation camp from which more than sixty thousand Jews were sent to their deaths, and where in 2009 construction workers renovating buildings discovered a large cache of graffiti, hitherto unseen, hiding on the buildings' walls underneath layers of drywall and other coverings. The camp at Drancy was first designed as a modernist experiment in urban living; a block of residential towers and interspersed open spaces, it had only just been nearing completion when the Nazis confiscated it in 1941. The Jews who were sent there were its first, temporary residents, and spent their internment in what was in effect a construction site. Its architects had named it La Cité de la Muette—the Silent City—as a nod to the peaceful yet urban environment it was designed to create. The name, of course, came to have deeper signification. With the discovery some sixty years after World War II of the wrenching last messages its internees wrote upon its walls, the Silent City spoke. Drancy today has been restored to its original purpose and is a housing complex that encompasses a Holocaust memorial. This chapter, starting from this exemplary case, considers various public sites of deportation and roundup, exploring the vast variation in the degree to which such sites have been banalized, sacralized, or otherwise set apart from or integrated into the ongoing daily life of the cities in which they are found. The author, closing the chapter—and the book—asks a series of key questions that are important for every scholar preoccupied with the memorialization of spaces of past violence today: is their integration a sign of forgetting, or, worse, of denial? Or can it be seen as something more affirming? Are the sites of atrocity and suffering better marked as sacral spaces, or given new life as something connected both to past and present?

Coda

Hitler's Geographies was conceived with several objectives in mind. First, we wanted to show how the Nazi high ranks were literally obsessed with spatial jargon, helped in this by the willing contribution of many academic geographers whose spatial theories and conceptualizations have been until recently largely underscrutinized, at least in Anglophone geography. Second, we intended to demonstrate that a geographical approach to Nazi ideology might help a great deal in understanding how this was indeed influenced by a specifically spatial *Weltanschauung*. Third, the combination of interventions included in this collection tried to bridge what we have defined as "Third Reich geographies" and the "geographies of the Third Reich." Fourth, influenced by the recent "regional turn" in Holocaust studies and by the work on the cultural history of the Third Reich, this book also aimed to illustrate how geography may contribute to realize a "cultural spatial history of the Third Reich," something of great relevance, we think, for the discipline of geography itself and also for all debates preoccupied with a more comprehensive incorporation of "the spatial" in the investigation of Hitlerism and its historical determinations. Finally, we believed that approaching Nazism and its spatialities with the novel perspectives proposed in this book inherently implied engaging with the complications of modernity and the fields of tensions permeating its conceptual body. Without leaning toward a functionalist interpretation of the Holocaust and the Third Reich by reading them merely as extreme expressions of the rationalities of modernity, we support the idea that the spatial manifestations of the calculative rationalities guiding Hitler's millennial project represented a catalyst and an unprecedented radical application of those same ambivalent rationalities, as many essays in this book show. We sincerely hope the reader will find in the pages of this book space and inspiration for new avenues of reflection on the deeply geographical nature of Hitlerism.

References

Barnes, Trevor, and Claudio Minca. 2013. "Nazi Spatial Theory: The Dark Geographies of Carl Schmitt and Walter Christaller." *Annals of the Association of American Geographers* 103 (3): 669–87.

Bassin, Mark. 2005. "Blood or Soil? The *Völkisch* Movement, the Nazis, and the Legacy of Geopolitik." In *How Green Were the Nazis? Nature, Environment, and Nation in the Third Reich*, edited by Franz-Josef Brüggemeier, Mark Cioc, and Thomas Zeller, 204–42. Athens: Ohio University Press.

Bauman, Zygmunt. 1989. *Modernity and the Holocaust*. London: Polity Press.

Farinelli, Franco. 1992. *I segni del mondo*. Florence: Nuova Italia Scientifica.

Kershaw, Ian. 1993. "'Working towards the Führer': Reflections on the Nature of the Hitler Dictatorship." *Contemporary European History* 2 (2): 103–18.

Levinas, Emmanuel. 1990. "Reflections on the Philosophy of Hitlerism." *Critical Inquiry* 17: 63–71.

Minca, Claudio. 2006. "Giorgio Agamben and the New Biopolitical Nomos." *Geografiska Annaler B* 88: 387–403.

Pick, Daniel. 2012. *The Pursuit of the Nazi Mind: Hitler, Hess, and the Analysts*. Oxford: Oxford University Press.

Stone, Dan. 2010. *Histories of the Holocaust*. Oxford: Oxford University Press.

Spatial Cultural Histories of Hitlerism

1

For a Tentative Spatial Theory of the Third Reich

PAOLO GIACCARIA AND CLAUDIO MINCA

In his account of modernity and barbed wire, Reviel Netz explicitly identifies "the geographical" at the core of National Socialism: "Here is how the Nazis saw it: people were divided into races, struggling for space . . . Germans were the landed people par excellence" (Netz 2004, 194). On many occasions, Hitler proclaimed the necessity of *Lebensraum* for the German people, a "*Volk ohne Raum*" (literally "people without space"—the title of a popular 1926 book by Hans Grimm), in Eastern Europe, which was depicted by the *Führer* in a 1937 secret meeting as an "empty space" waiting to be colonized. Again, from Netz:

> The Nazi aim for the future was . . . expansion, so that, in the plains stretching East of Germany, a living space—*Lebensraum*—would be created for present-day Germans. . . . Nazism was essentially a colonizing ideology. The goal—a typical twentieth-century goal—was to bring a certain space under complete control. (Netz 2004, 194)

If National Socialism was a revolutionary movement and the Third Reich a revolutionary government, as suggested, among others, by Martyn Housden in his biography of "*Lebensraum* manager" Hans Frank (2003, 71), they were both supported by a genuine "spatial revolution," to use the terms of Carl Schmitt, who was often described as the "Crown Jurist of the Third Reich" (see Minca and Rowan 2015a, 2015b). This "struggle for space," at the core of the Nazis' geographical imaginations, reached its peak with the war between what they saw as two antithetical races—the Germans and the Jews—in competition for resources, not least among them space itself. The related "spatial revolution" began with urban pogroms in the 1930s, which violently disputed the right of the Jews to inhabit certain parts of the city and their very living spatialities, but by the 1940s, with the advent of the war, the spatial revolution took the form of ghettoization, deportation, detention, death marches, and, finally, systematic extermination.

In the process of producing Nazi Europe—a *judenfrei* and Germanized continent—millions of people were forced to leave their homes and were relocated, with the millennial aim of creating a revolutionary new cultural and political geography of Europe, where the Germans and the other (surviving) peoples were supposed to find "their proper place" (see, among others, Aly and Heim 2002; Browning 2004).

> The entire geography of Jewish Europe now revolved around the death camps . . . As killing institutions, the geographical reach of the death camps (in particular that of Auschwitz) was remarkable. The death camps killed people coming from the entire continent—all the way from Greece to Norway, from France to the Soviet Union. This was based on a geography of concentration and transportation spread across the continent. (Netz 2004, 219)

As noted in the introduction to this volume, despite the popularity and widespread use of spatial concepts and metaphors in the Nazis' imperial discourse, including in policy pronouncements, a comprehensive examination of the relationship between geography, spatial theory, and the Third Reich remains to be developed. However, the topic has not been entirely neglected. In the following section, we therefore attempt to provide a brief outline of what we tentatively identify as five main "streams of reflection" that have so far engaged with "Hitler's geographies," albeit in different ways and moments, and that mark the necessary starting point for this book project. While being entirely aware of how imprecise and incomplete such an operation may prove, we would like to qualify these five "moments" as: (1) the *Lebensraum* debate in the 1980s, mainly in the field of geography; (2) the (rather scattered) literature on the dark geographies of genocide of the following decades; (3) the "regional turn" in Third Reich historiography; (4) the recent work on the "cultural history" of the Third Reich; and, finally, (5) the intersections between political philosophy and geography that in recent years have focused on questions of space, biopolitics, and calculation in relation to National Socialism.

Hitler's Geographies

THE *LEBENSRAUM* DEBATE

Karl Haushofer's popularity in the United States during the war period, when the German political geographer was depicted by many—even outside of academia—as the evil inspiration of the Third Reich's fantasies of territorial expansion, was followed in the wake of World War II by a sort of *damnatio memoriae*—at least in English speaking geography—around this figure, and the whole *Geopolitik* project (Troll 1949); an implicit politics of forgetting that, in our view, has delayed if not obstructed any broader critical reflection on the relationship between geography, spatial theory, and Nazism (see Barnes and Minca 2013; also Abrahamsson 2013). In fact, it was not until 1987 that a potential debate on this topic emerged among geographers, when a special issue dedicated to the theme was published by the journal *Political Geography Quarterly*, with the contribution of both German (Heske 1987) and Anglo-American geographers (Bassin 1987; Parker 1987; Paterson 1987). This debate, largely centered on the history of *Geopolitik* and the *Lebensraum* concept ("living/vital space"), was then continued by another special issue in the same journal, two years later, this time entirely composed of articles penned by German geographers (Fahlbusch, Rössler, and Siegrist 1989; Herb 1989; Kost 1989; Ossenbrügge 1989; Sandner 1989; Schultz 1989; see also Sandner 1988; Sandner and Rössler 1994; Kost 1998). The discussion on that occasion went beyond the confines of a critical reflection on *Geopolitik* in relation to Hitler's imperialism, at least to some degree, including for example an analysis of the role played by Walter Christaller—a key international figure in the fields of economic geography and regional planning in the first postwar decades in the United States and many European countries—who was "employed on the staff of the Main Office for Planning and the Soil, an agency within Himmler's Reich Commissariat for the Strengthening of Ethnic Germandom" (Rössler 1989, 427; reprinted in this volume). Arguably, Rössler's denunciation of Christaller's controversial implication in the Nazi project went substantially unnoticed in the broader geographical community, including those still engaged with Christaller's urban geometries and the related literature and applications until recently. This aspect alone would deserve a close investigation in order to understand why Christaller—and his work—was never held accountable for his direct scholarly implication with Himmler's grand plans for the realization of German *Lebensraum* in Eastern Europe (see Barnes, this volume, chapter 9).

However, while geographers only began to engage with these complicated legacies in the late 1980s, scholars in other disciplines had already conducted work on these topics. For instance, an analysis of the concept of *Lebensraum* in relation to Haushofer and *Geopolitik* had been at the core of historian Woodruff Smith's scholarly preoccupation since the early 1980s (see Smith 1980, 1986). During the same period the planner John Mullin had illustrated the importance of urban and regional planning within the Nazi "rule of experts" (1981, 1982a, 1982b). In human geography, however, the debate on Nazi geography and spatial thought lost momentum quite rapidly after the two special issues highlighted above. The rather isolated interventions on similar areas that appeared in the following decades moved away from historical analysis to focus on other aspects, for example, the legacy of the *Geopolitik* tradition in relation to contemporary geopolitics (see O'Loughlin and van der Wusten 1990; O'Loughlin and Heske 1991), something that would crucially contribute to the emergence of the field of "critical geopolitics" in the 1990s (Ó Tuathail and Agnew 1992; Ó Tuathail 1996, 16–43). The broader reflection on the relationship between Nazism and geography—on "the geography and the geographers of the Third Reich"—proposed by those two special issues in *Political Geography Quarterly* was thus not incorporated by the discipline's mainstream and survived only in a somehow mutated form. Whenever questions of *Geopolitik* and *Lebensraum* reemerged during the following decades, they normally appeared outside of geography, mostly but not exclusively in the field of history (Murphy 1997, 1999; Diner 1999; Herwig 1999; Natter 2003, 2005; Baum 2006; Danielsson 2009; Neumann 2009) with the exception of Kenneth Olwig's article on the concept of *Raum* (Olwig 2002) and Guntram Herb's work on German cartography and *Geopolitik* (1997, 2002; also Hagen 2009b). Another field of inquiry, geographical in nature, is that preoccupied with the relationship between the Nazis and the environment/environmentalism. The literature with this focus is important, and it has been rapidly growing in recent years (see Lekan 2004; Brüggemeier, Cioc, and Zeller 2005; Uekoetter 2006; Nyhart 2009). The question of nature was indeed central to Nazi ideology, also because it was often articulated via connections to ideas of blood, soil, and regional planning. Two chapters of the present book are dedicated to this issue, one in relation to animal breeding and zoology (Driessen and Lorimer, chapter 6, this volume), the other on conceptualizations of the *selva*, the forest, as a counter-spatiality compared to the one produced by the city, and the *civitas* (Giaccaria and Minca, chapter 11, this volume).

DARK GEOGRAPHIES OF GENOCIDE

Under the label of what we have broadly defined above as the "geographies *of* the Third Reich" we include a vast array of very different and relatively isolated interventions that have appeared in past decades. The debate hosted by *Political Geography Quarterly* in the late 1980s (focused on the "Third Reich geographies/geographers") was in fact followed by important but rather fragmented attempts to study the spatialities of Nazism and the Holocaust using analytical tools associated with geography. These interventions have spanned from the electoral geographies of Nazi Germany (O'Loughlin, Flint, and

Anselin 1994; O'Loughlin 2000, 2002), to the postmodern topologies of the Holocaust (Clarke, Doel, and McDonough 1996), to even the most recent experimental attempts to incorporate GIS-led technology within historiographical accounts of the Shoah (Beorn et al. 2009; Knowles, Cole, and Giordano 2014). While some of this work was undoubtedly original and highly illustrative, the debate initiated by *Political Geography Quarterly* was fundamentally over. Not only were there fewer and fewer articles focused on these topics in geography journals, but also those that were published were largely unrelated to each other, and very different from one another and the existing literature in terms of the theoretical and methodological approach adopted. These include work on some emblematic spatial manifestation of Nazism—the city (Hagen 2004, 2009a; Till 2005; Hagen and Ostergren 2006), the ghetto (Cole 2003, 2013), the camp (Charlesworth 1992, 2003, 2004b; Minca 2006, 2007, 2015), specific sites and landscape making the geographies of the Holocaust (Giaccaria and Minca, chapter 11 this volume; Knowles, Cole, and Giordano 2014)—or on questions of memory and heritage of the genocidal geographies of the Third Reich, and in particular of Auschwitz-Birkenau and other extermination camps (Charlesworth 1994, 2004a; Charlesworth and Addis 2002; Charlesworth et al. 2006; Carter-White 2009, 2012, 2013; Giaccaria and Minca 2011). Arguably, these approaches, while of great interest if taken individually, remained largely limited to present-day investigations of the spaces and the spatial practices of the Nazis; these "geographies of the Third Reich" were in fact rarely linked to the broader production of the "Third Reich geographies," that is, the body of spatial theories and concepts that populated the Nazis' racialized imperial fantasies and animated their *Lebensraum* policies (perhaps the exceptions being Clarke, Doel, and McDonough 1996; Rössler 2001; Barnes and Minca 2013).

The aim of the present chapter, however, is that of envisaging, if not a general "spatial theory" of Nazism as yet, at least an embryonic interpretative framework capable of bridging precisely these "Third Reich geographies" and the actual "geographies of the Third Reich." We will return to this aspect later in the chapter, where we lay out the guiding theoretical assumptions behind this project. However, before doing that, it is important to briefly introduce the remaining three fields of inquiry that have inspired this book and our overall project. Two of these "streams of literature" are historiographical in nature. While an exhaustive review of historical work on the spaces of Nazism and the Holocaust is beyond the scope of this volume, and certainly this introduction (see Stone 2010; also Stone's chapter 2 in this volume), it may nonetheless be useful to recall a few recent developments in this literature, since they represent in many ways the condition of possibility for offering a reading of Hitler's geographies consistent with the guiding spirit of this volume.

THE "REGIONAL TURN" IN THIRD REICH HISTORIOGRAPHY

In the previous pages we mentioned a few historians who have investigated in depth the concept of *Lebensraum* and the contribution of German academic geography to the production of a specific Nazi spatial ideology. Historical work on the Holocaust has also dedicated considerable attention to specific sites of extermination—particularly to the ghettos (see Tim Cole's pathbreaking contributions at the intersection of history and geography [2003, 2013]; see also Horwitz 2008; Engelking-Boni and Leociak 2009; Michman and Schramm 2011) and, above all, to Auschwitz, presented as the true "capital of the Holocaust" (see Gutman and Berembaum 1994; Hayes 2003; but especially Wolfgang Sofski's detailed work on the topographies of the camp [1997] and Debórah Dwork and Robert Jan van Pelt's [1996] monumental analysis of Auschwitz and the relationships among the camp, the city, and the surrounding region). However, this interest for "the spatial" has benefited in the last decade or so from a growing number of innovative interventions and the availability of new material coming from the former Soviet Union and Eastern European archives after the end of the Cold War. This material and the related focus on questions of space have contributed to producing a new stream of research on Nazism and the Holocaust that British historian Dan Stone has defined as "regional studies" (Stone 2010, 90–95; see also Herbert 2000):

> Remarkably, until a decade or so ago, historians actually knew very little about the ways in which genocidal policies developed in the *Generalgouvernement*, in the Wartheland, in Ukraine, Belarus, the Baltic States, or Romania. Now each of these places has at least one major study devoted to it. (Stone 2010, 91)

Despite having been sometimes criticized in the broader field of history for their emphasis on the role of peripheral decisions in the actual production of the conditions that made the Holocaust and its practices possible, these works are particularly relevant for our project for at least two main reasons. First, the study of "local" decisions—together with the appearance of an increasing number of biographies dedicated to influential high-ranking Nazi officials (see Housden 2003; Gerwarth 2011; Longerich 2011; Epstein 2012)—has helped to highlight how complex and often how fundamentally contradictory the practices related to Nazi geopolitics and racial policies were when implemented "on the ground."

Second, these "regional studies" have shown the key role spatial categories played in the formulation of the Nazi project of purifying German territories and realizing a new *Lebensraum*, but also the clear overlap between colonial imaginations and geopolitical practices that defined Hitler's imperialism, and with it, the production of the conditions for the extermination of the European Jews. In particular, the acquisition of more detailed documents concerning the campaigns of expansion in Eastern Europe has allowed a sort of postcolonial reading of Nazism to emerge (see Lower 2005; Baranowski 2011; also Burleigh 1988; Mazower 2008). On the one hand, very detailed and innovative studies on the relationship between colonialism and genocide have appeared in the last decade or so (Moses and Stone 2006; Moses 2008; Kühne 2013b; Pergher et al. 2013), with a particular focus given to the so-called "settler colonialism" (Wolfe 1999, 2006; Moses 2004; Finzsch 2008). On the other, recent historiographical work has provided in-depth analyses of German colonialism in Namibia, and of the related genocidal practices put in place by colonists and the military against the Herero populations (Stone 2001; Conrad 2013). This work has recognized in the genocidal strategies implemented by General Lothar von Trotha a precedent of the Nazi strategies and related practices in the occupied territories of Eastern Europe (Madley 2005; Zimmerer 2005, 2008; Fitzpatrick 2008). Particularly important is also the work of cultural theorist Paul Gilroy in connecting the experience of Hitler's Europe and its complicated legacies to questions of race, empire, biopolitics, and geopolitics in a truly postcolonial perspective (see Gilroy 2004). Finally, other authors have investigated the influence of the American "Far West" on Nazi colonial imagination and practices (Guettel 2012; Kakel 2013).

THE "CULTURAL HISTORIES" OF THE THIRD REICH

The second "historiographical" stream of research may be defined as a "cultural history of the Third Reich," that is, the result of a multiplicity of approaches and methodologies coming from a broader array of disciplines and merging into a debate that has opened the study of Nazism and the Holocaust to an entirely novel and, to some extent, more comprehensive set of perspectives (see, for example, the special issue appeared in *Dapim* in 2009, and in particular Goldberg's and Stone's interventions). If it is true that "the real contribution of cultural history to Holocaust history is only just beginning to be made" (Stone 2009, 62), it is possibly equally true that the incorporation in this field of studies of new approaches related to the "cultural turn" (Burke 1997, 2008) and the "linguistic turn" (Clark 2004) has contributed to lay the foundations for what, again, Dan Stone has described as an "integrated historiography of the Holocaust" (2010, 5). While on the conditions and overall motivations that have produced the Final Solution there persists a significant degree of opaqueness (Stone 2009, 62)—and related disagreement in contemporary scholarly interpretations—the inclusion of new theoretical perspectives on the nature and the development of the genocidal violence the Nazis perpetrated has nonetheless offered the space for broader and alternative readings (Confino 2012, 2014; Stone 2013; Aly 2014). What is important for our main argument is that these readings are capable of moving beyond the limits imposed by the use of archival evidence in historiographical work to include analyses reliant on interpretative tools coming from the fields of philosophy, anthropology, and the social sciences and the humanities in general.

This "theoretically inclined historiography on the Holocaust" (Finchelstein 2009)—and the Third Reich more broadly—has been nourished by new research directions of great interest and originality. Key contributions to the perspectives brought in by these cultural histories include Alon Confino's analysis of the relationship between national identity and memory in modern Germany (1997), David Blackbourn's study of the concepts of landscape and nature and their related transformations throughout twentieth-century Germany (2007), and Vejas Liulevicius' examination of the perception of Eastern Europe in modern German culture (2000, 2010). Further, Peter Fritzsche's work on *Life and Death in the Third Reich* (2008), Claudia Koonz's on Nazi ethics (2003), Daniel Pick's on the psychology of the "Nazi mind" (2012), and Shelley Baranowski's on the role and the practices of tourism under Hitler (2007) have also opened new territories of investigation, together with Dominick LaCapra's (1996, 1998, 2004) reflections on the role of memory and representation, Moishe Postone and Eric Santner's (2003) on "catastrophe and meaning," both in relation to the Holocaust, and Christopher Hutton's (2005) and Andreas Musolff's (2010) linguistic analysis of the Nazi jargon, to name just a few. More recently, Thomas Kühne (2013a) and Martina Steber and Bernhard Gotto (2014) have analyzed in detail the role that the concept of *Volksgemeinschaft* (people's community) played in the making of the Nazi cultural imagination. This bourgeoning literature has also created a favorable context for pathbreaking investigations of all forms of calculative rationalities the Nazis implemented in order to shape and govern the German population (see, for example, Götz Aly and Karl Roth's work on the Nazi census [2004]; also Elden, this volume, chapter 11). What is more, according to Stone (2009, 59–65), these cultural histories of the Third Reich and the Holocaust have ignited a renewed historiographic interest in Nazi ideology and the conditions that contributed to generating it. This interest, however, does not merely represent a return to more classic "intentionalist" interpretations, fundamentally based on the analysis of the role of anti-Semitism in determining the Final Solution, or on discussions concerning the existence of an original genocidal plan at the roots of the Nazi conceptualization of the "Jewish problem." What we are witnessing is instead a growing attention for the multiple and complex genealogies of the Nazi *Weltanschauung*, that is, a "worldview" based on the coming together of biopolitics and geopolitics (see Neumann 2009), the "world of meanings that allowed the idea of the 'Final Solution' to emerge as an option for the Third Reich's leaders" (Stone 2009, 60).

BIO-GEO-POLITICAL PHILOSOPHIES OF THE THIRD REICH

In the context of debates focused on these cultural histories, the study of the "Third Reich geographies" and the "geographies of the Third Reich" may play an important role in the ongoing redefinition of research on Nazism and the Holocaust. On the one hand, modern German geography has made a significant contribution to producing a specifically spatial *Weltanschauung*, providing inspiration and legitimacy to both Nazi ideology and genocidal practice (see Wolf [chapter 4] and Zimmerer [chapter 3] in this volume). On the other, Nazi conceptualizations of space have been key to Hitler's imperial projections and plans and therefore represent a crucial field of inquiry to enrich and complement the present work on the cultural histories of the Third Reich. Within this very productive widening of the historiographical debate, political philosophy has also recently begun to play a greater role. Indeed, its contribution is now so great that it can be considered an important fifth "stream of literature" converging in the project that stands behind this book. While recent work in geography on Martin Heidegger (Elden 2003, 2006b) and Carl Schmitt (Elden 2010; Legg 2011; Minca and Rowan 2014, 2015a, 2015b) has discussed their respective relationships with Nazism, the broader engagement with the ideology and practice of Nazism that is growing in geography is arguably related to the influence of Italian political philosopher Giorgio Agamben. Agamben's reflection on modern sovereign power (1998; see also Minca 2007), largely (albeit critically) draws on Schmitt's conceptualizations of the state of exception, but also on the deeper biopolitical nature of what Levinas has described as "the philosophy of Hitlerism" (Levinas 1990); this has in fact inspired work of geographers on the spaces of exception (Gregory 2006; Minca 2006), the structure of the ban (Minca 2007; Giaccaria and Minca, chapter 11, this volume), and the spatialities of camp (Minca 2005, 2006, 2015; Ek 2006; Ramadan 2009; Giaccaria and Minca 2011). The return of the "bios" at the core of Italian political philosophy (see Esposito 2008, 2011) and the role given in this context to questions of space has certainly helped to create the conditions for further advancements in the analysis of the relationship between geography, biopolitics, and the Third Reich (see, for example, Carter-White 2009, 2012; Romanillos 2011; Minca and Rowan 2015a).

Although geographers such as David Atkinson (2011), Stuart Elden (2006a), and Mike Heffernan (2011), among others, have recently carried out work that is critically sensitive to the historical complexity of the relationship between Nazi spatial thought and the intellectual trajectories that passed through it, a degree of implicit resistance to a fuller engagement with Nazism and its legacy on the part of (English-speaking) geography seems to persist, something underlined in recent interventions by Minca and Rowan (2014, 2015a). Perhaps this resistance also reflects the discipline's relative overall unease with the historic relationship between Nazism and German geopolitics, and German geography more broadly, as demonstrated for instance by the abovementioned amnesia concerning Christaller's implication with the SS. For this reason, in addition to the recent trends in the interdisciplinary literature related to the Third Reich, we believe that the time for a full engagement with this legacy has arrived. If Nazi spatial theory represented a truly dark hour for the discipline of geography, this should not only be fully acknowledged but also presented as an opportunity to reflect on urgent contemporary concerns—as has happened, for example, in recent debates about the biopolitical dimensions of contemporary politics that have placed the analysis of the Nazi biocracy once again at their center (see Agamben 1998, 2002, 2005; Esposito 2011).

For a Spatial Theory of the Third Reich?

Through a direct and critical engagement with these five streams of literature, and with the objective of contributing to present debates on the cultural histories of Nazism and the Holocaust, this chapter, as noted above, represents an early product of a broader project that intends to start formulating a tentative spatial theory of the Third Reich. Accordingly, we argue that the bridging of the "Third Reich (academic) geography" and of the (actual) "geographies of the Third Reich" should be attempted. We believe in fact that this bridging may cast new light on what we consider to be three fundamental aspects of the Nazi spatial *Weltanschauung*, three fields of dialectical tension within the Nazi project: (1) between biopolitics and geopolitics; (2) between a topographical and topological approach to the government of people and space, in other words, between "space" and "place"; and (3) between spatial ideology and spatial practice.

BIOPOLITICS/GEOPOLITICS

The first field of tension reflects one of the fundamental theses guiding our overall project; that is, the fact that the Nazi spatialities were the result of a "worldview" based on the coming together of biopolitics and geopolitics. In the philosophy of Hitlerism, we claim, space and race were neither separate nor counterpoised. Rather, they coexisted in a permanent tension, given that a specific bio-geo-political dispositive lay at the core of the entire Nazi project. All racial and biopolitical practices of the Third Reich were conceived as inextricably entangled with spatial and geopolitical practices. The space-race link was pivotal for Nazi expansionism and genocide, but also for the production of a new biopolitically perfected German Man and German community (see Aly and Heim 2002 on this). Space and race were connected *ab origine* in a quasi-ontological relationship in Hitlerism. Biology and race, as manifest in Nazi ideology, were fundamentally conceived in spatial terms, for example, in the projected reorganization of territories in order to accommodate a progressively purified and homogeneous Aryan population. At the same time, space was theorized and spatial practices implemented according to bio-geo-political principles. Spatial theory, spatial practice, and spatial imaginations, in the Third Reich, were intertwined and necessarily conflated with the biopolitical (see Giaccaria and Minca, chapter 11, this volume; also Minca and Rowan 2015a, in particular chapters 5 and 6). The outcome of this process was thus that Nazism was a truly *bio-geo-political* project. However, the bio-, the geo-, and -politics, were not brought together into a coherent unity via a smooth synthesis. Rather, as with many other aspects of Hitlerism and its implementation, it was all too often characterized by improvisation, impulsive decisions, ad hoc adjustments, and controversy (Stone 2010, in particular chapter 3). Accordingly, the dialectical tension between the biopolitical and the geopolitical was endlessly negotiated, reimagined, and translated into new practices and decisions, even as they often conflicted with those that had come before.

SPACE/PLACE

The second aspect we consider fundamental to understanding the Nazi spatial *Weltanschauung* is related to the tension between conceptualizations of space and place, in some cases conceivable as a tension between the topographical and the topological in the genocidal imaginations (and implementations) of Hitlerism. This is not the place for a full theoretical engagement with the dialectic between topography and topology, as discussed in contemporary cultural studies and cultural geography in particular (see, for example, Martin and Secor 2013). Suffice it to say that, within our project, the topographical is intended as a "measure of space" and/or spatial measure, but also as the attempt to "spatialize" a set of calculative rationalities into a projected reduction of reality to the coordinates of that very same measure. The "topological" is instead considered a broad category that comprises, to speak with Giorgio Agamben, that "which remains" beyond/despite the imposition of such a strict and violently implemented regime of rational spatiality (see Agamben 2002; also Giaccaria and Minca 2011). We would thus like to incorporate the idea that the spatialities the Nazi regime produced were in many ways an extreme expression of a specifically modern understanding of space, an understanding primarily based on a set of calculative rationalities, as convincingly demonstrated, among others, by Stuart Elden (see Elden 2006a, reprinted in this volume, chapter 11; also Elden 2013).

At the same time, as Carl Schmitt famously declared (see, among others, Minca and Rowan 2015a, 2015b), the political and the spatial were to be understood as inevitably intertwined and entangled, and the appropriation of land and the marking of its space to be considered foundational to every order and process of ordering. This was expressed, according to this controversial spatial thinker, in the necessary coincidence between spatial order (*Ortung*) and political ordering (*Ordnung*). Read through these lenses the Nazi project consisted of a violent attempt to merge space and place, to make "the spatial" perfectly coincide with "the political," the cultural, and the racial (even if Schmitt would likely not have come to the same conclusion, despite his complicity with the regime).

Conceptualizations of *Raum* ("space" in Germanic languages) were indeed central to political debates in German academia and in popular discourse throughout the 1930s, including the *Lebensraum* ideologies popularized by Haushofer and incorporated into Hitler's imperial fantasies, or theorizations of *Großraum* (greater space) Carl Schmitt formulated during the years of Nazi expansion toward the East (see Minca and Rowan 2015a, in particular chapter 6). A spatial jargon populated the speeches and the writings of many influential ideologues, from Himmler to hardcore ideologues like Alfred Rosenberg and Walter Darré, among many others. However, as noted by Kenneth Olwig, in German the term *Raum* does not merely refer to calculative and measurable space. According to Olwig, *Raum* incorporates a constitutive ambiguity, referring to space not only as *res extensa* but also as a place, as lived space; it comprises also a qualitative idea of cultural space entirely devoid of the calculative rationalities described above (Olwig 2002). What we would like to argue is that this tension inherent to the concept of

FIGURE 1.1. Map of the planned transfer of *Volksdeutsche* from the Dobrudja (*Bodenständiges und rückgeführtes Bauernvolkstum in Ost und West*, from Konrad Mayer, *Landvolk im Werden*, 1941). Source: Collectie Steegh-Teunissen, Leiden.

Raum found its epitome in the spatialities of the Third Reich. On the one hand, the Nazi broader "bio-geo-political" ideology was founded on a sort of rational geographical vision, according to which spatial order and racial order were supposed to coincide. On the other, however, the Nazi *Raum* was also marked by a mystical and irrational vein, based on a fundamental "search for origins," for an authentic and uncontaminated past spatial order. This order was accordingly identified, at times with some idealized German medieval past (Arnold 1990), and at other times with a mythical Orient from where the pure Aryan population that made up the core of the German *Volk* was presumed to have its origins (Goodrick-Clarke 1985).

In line with the first interpretation, space was then carefully theorized and planned with the support of qualified scientific expertise, ranging from geography to planning, from economics to anthropology, from biology to demography, from agronomy to genetics. At the same time, this real and imagined geography was strongly permeated with culturalist interpretations of those same spaces (Schenk and Bromley 2003). Jeffrey Herf (1984) describes this peculiar trait of modernity and reaction that permeated many intellectual figures and public debates during that period and the preceding decades in Germany as a "Reactionary Modernism."

This tension characterizing the concept of *Raum* in particular was recomposed and even exploited by the Nazis via, for example, the extremely racialized pseudoscientific expressions animating the discourse of *Blut und Boden*, "blood and soil," so dear to some of the most radical Nazi advocates (Bramwell 1985). This inherent tension between space and place in the Nazi projections nevertheless allowed recognized academics to be part of plans supported by political leaders driven by fanaticism and often nonsensical interpretations of the realities they were supposed to theorize and appropriate in their (scientific) calculations/projections. This also explains perhaps how people like geographer Walter Christaller and agronomist Konrad Meyer could work in the same ideological climate together with characters of the kind of Feder, Darré, Rosenberg, and even Himmler. More widely, the tension between space and place inhabiting this interpretation of *Raum*, in the cultural and political context that Hitlerism produced, became a field in which the rational and functional dimensions of the Nazi spatial plans, and their millennial, mythical, and irrational elements could be merged and fused in what Sheila Faith Weiss (2010) has described as the "Nazi Symbiosis," especially given the fact that "the description of biological racism as represented in sober and objective language seems insufficient to account for a topic whose essence was inner demons and hallucinations" (Confino 2005, 309).

SPATIAL IDEOLOGY/SPATIAL PRACTICE

Finally, we argue that these two fields of tension—between space and race *and* between space and place—ended up being materialized in a third field of tension, that between spatial ideology and spatial practice, where metaphors and acts, projections and decisions inevitably came together. What we have defined as the Nazi broader bio-geo-political ideology was in fact characterized by a perpetual tension between a "spatial" *Weltanschauung* and a set of actual "territorial" practices and policies. Indeed, for Boaz Neumann, "National Socialism did not present and experience itself as an ideology but as a world-view. Whereas ideology was conceived as articulated through the intellectual faculty and associated with ideas concerning life, *Weltanschauung* was coupled with life experience and the senses. . . . In the case of Nazism, there was no gap between the idea/ideology and practice" (Neumann 2002, 109). (*Lebens*)*Raum* was thus the locus where concepts and practices had to become one and the same, with no remnants, no gap. The occupation of new territories in Eastern Europe represented the actual field in which Nazi worldview/ideology and its related conceptual apparatus were translated into actual daily decisions and acts with real effects on the subjected lands and people.

Perhaps the most important aspects for our argument is the fact that these three fields of tension did not end up merging into the desired synthesis when the Nazi spatial *Weltanschauung* was translated into action and practice. The Nazi aspiration of making race, space, and place coincide in fact never materialized. Despite the presence of an army of experts willing to contribute to its practical realizations, Nazi ideology was based on poor spatial theory and a good dose of fanaticism/fantasy-thinking. Therefore, all attempts to translate the related spatial imaginations into actual practice—into real spaces/places—were doomed to fail from their very inception. This of course does not mean that they did not have effects, quite the contrary. Nonetheless, the Nazi spatial plans, as much as it was centered on the ambition of establishing a definitive bio-geo-political order for the foundation of a millennial Reich, was constitutionally unable to realize the desired symbiosis among its concepts and objectives.

It is worthwhile dwelling a little more on this dimension. The need to govern the fields of tension inhabiting the Nazi spatial *Weltanschauung* in fact implied the parallel need to decide who should be living and in which space—a decision that necessarily required a value assessment of the subjected populations and their related (living/vital) spaces. The obvious result was a hierarchy of people and spaces, not only in the territories of the Reich but also in the whole of Europe. The Nazi bio-geo-political grand plans clearly identified their condition of possibility in their capacity for drawing and protecting new borders—between the racial and the spatial categories envisaged—especially after Operation Barbarossa was translated into a new, expanded, application of the principles of the Nazi *Weltanschauung*. However, these fields of tension, when materialized in the practices of the German troops and the SS "in place" and in the realities of the Eastern territories under Nazi domination, had to be constantly renegotiated, which in fact rendered those very same borders—and related categorizations—unstable and uncertain. Indeed, on the ground, when confronted with unexpected regional and local specificities, the broader armchair racial and spatial categories the Nazi ideologues produced had to be endlessly reformulated as new territories were progressively included in the march of Nazi boots, first in the subjugation of Poland and later in the immense territories of the Soviet Union. The main consequence was that those who were supposed to implement, case by case, these same categories, had to constantly recalibrate/reformulate their practices, in their attempt to adjust the individual context to the Nazi grand racial and spatial theories (and to changes in Hitler's will). This explains in many ways why, in practice, this bio-geo-political millennial plans often resulted in acts of arbitrary pure violence, provoked not only by the fanaticism that pervaded many Nazis in the hubris of their war of conquest but also in part by their inability to manage the tensions described above when on the ground (Giaccaria and Minca, chapter 11, this volume).

UNTENABLE *ENDREICH* (FINAL REICH)

With the continuation of the war and the first signs, already in late 1941, that the grand Nazi spatial plans may not be so easily accomplished, their internal tensions became even more accentuated, and the contradictions inherent to the Third Reich's spatial *Weltanschauung* began to show their genocidal effects. The coincidence between the Nazis racial and geopolitical maps proved impossible to realize in practice, since the ordered and fully planned spatialities the Nazi "experts" envisioned could not be translated into a return to an idealized *Heimat*, and the colonial fantasies of a new German *Lebensraum* could not be materialized on the ground when applied to the occupied territories of Eastern Europe (Kühne 2013a). Between plans and realizations, imaginations and possibilities, theories of race and people, and the complex realities to be governed in the new territories there always existed an unbreachable gap, a remnant, a leftover—the genuine product of those unsolved tensions inherent in Hitler's geographies.

The main thesis of our project concerning the bio-geo-political nature of Hitler's geographies is therefore that these gaps, these remnants, were the threshold, the "third spaces" where both the Holocaust and the collapse of the Third Reich literally *took place*. If it is true that the so-called territorial solution to the "Jewish problem"—that is, the deportation of the European Jews "somewhere"—seemed to represent, at a certain moment in time, the necessary condition and consequence of the grand bio-geo-political plans to realize a *judenfrei Reich*, it is possibly equally true that these very plans could not be accomplished in the first place for their intrinsic limitations and contradictions, for the existence of the abovementioned unsolvable fields of tension (Giaccaria and Minca, chapter 11, this volume). As the eastward expansion pushed onward, the early plans to deport the German Jews to Nisko, the experimental reservation near Lublin managed by infamous SS Odilo Globocnik, were thus quickly replaced by the more ambitious plans to "Germanize" increasingly larger and larger portions of Eastern Europe, including the General Government administrated by Hans Frank and, eventually, vast former Soviet territories in areas like Ukraine and Belarus (Lower 2005). Accordingly, the *judenfrei Reich* project was soon reconceptualized as *judenfrei*

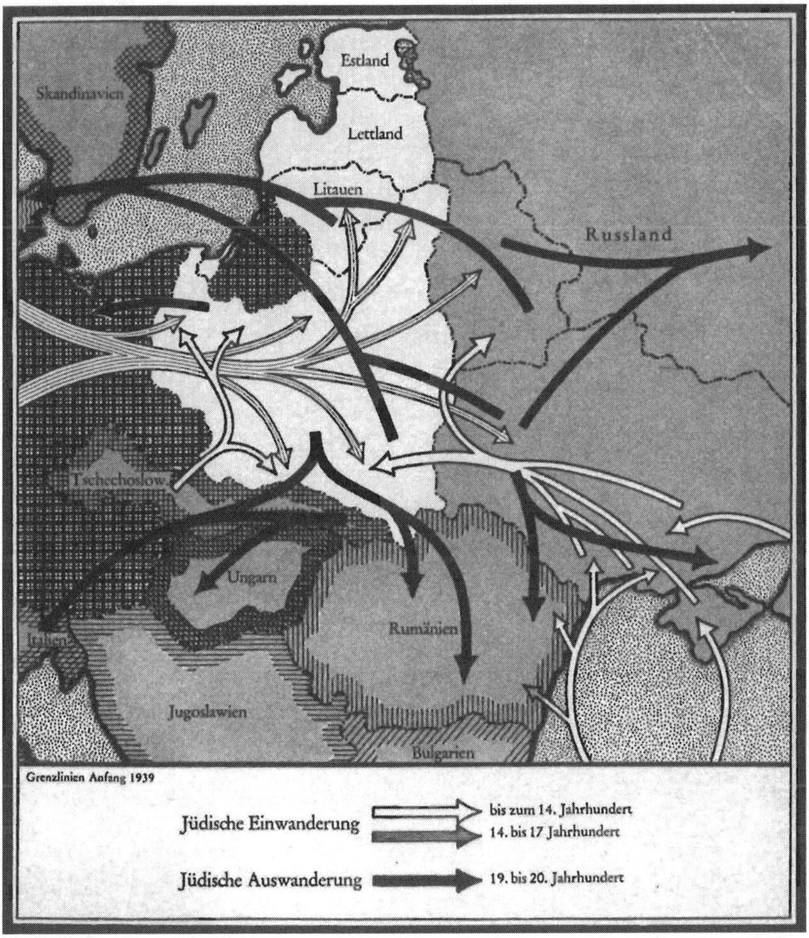

FIGURE 1.2. Jewish migrations toward the German East (*Jüdische Wanderungsbewegung im Vorfeld des deutschen Ostens*, from Walther Jantzen, *Geopolitik im Kartenbild*, 1940). Source: Collectie Steegh-Teunissen, Leiden.

Europe, with the result that the Nazis were suddenly confronted with the presence of—and the related need to manage—unwanted and rapidly increasing subjugated populations, composed not only of Jews but also Poles, Russians, Roma people, and with "shrinking" space available to (re)locate these "human materials" in order to make room for the dreamed German *Lebensraum*. As Aly notes, "just as the failure of the 'Jewish reservation of Lublin' project had led to the ghettoization of the Lodz Jews, the failure of the Madagascar Plan led to the ghettoization of the Warsaw Jews in late autumn 1940" (Aly 1999, 81).

At the same time, the practical impossibility of stabilizing the ghettoization of the Jews within the newly established—and constantly changing—*Ortung/Ordnung* of the millennial-Reich-in-the-making, made the elimination of the ghettos all the more necessary. This somewhat "geographical" decision is what had possibly helped to render the *Endlösung*, the extermination of European Jewry, not only conceivable but also something to be presented as a "spatial problem," as confirmed by the Nazi jargon adopted in the secret protocols of the Wannsee Conference, held on January 20, 1942: "emigration has now been replaced by evacuation of the Jews to the East. . . . The evacuated Jews will first be taken, group by group, to so-called transit ghettos, in order to be transported *farther East* from there" (Arad, Gutman, and Margaliot 1999, 253, 256). This obsession with "spatial" discourse among the Nazi high ranks thus appears as one of the "inner demons and hallucinations" populating the zeitgeist of the Third Reich and its imperial ambitions, an obsession with which present debates on the cultural histories of the Holocaust should engage in full, especially if we accept, as argued by Peter Longerich (2010), that the territorial solution was genocidal at its very inception.

What is more, the desperate attempts to match race and *Raum*, space and place, *Weltanschauung* and practice on the ground, toward the end seemed to get entirely out of hand and to contribute, paradoxically, to the very collapse of the Third Reich. Vejas Liulevicius convincingly expresses this when describing the role of the concept of *Raum* in the failure of the bio-geo-political Nazi ideology and plans:

> The regime used modern techniques for the goal of a terrible future utopia which classical modernity would not recognize, seeking space, rather than development. While the Soviets retreated, "trading space for time," the Nazis gave up time to gain space—seeking an everlasting, timeless present of destructive expansion in their vision of the *Ostland*. As the tide of events turned in the East, Hitler refused to give up the spaces conquered and forbade withdrawal again and again, producing military disasters. The ideological primacy of *Raum* was fatal in its consequences in the East. At long last, this was brought home to Germans as the Red Army invaded their territory by 1945,

> turning the utopia of *Raum* into a nightmare of the advancing East. (Liulevicius 2000, 272)

The violent interventions to tame the fields of tension permeating the Nazi bio-geo-political ideology therefore finally resulted in the production of infinite genocidal black holes, an abyss in which the European Jews were forced to enter facing annihilation—but also, paradoxically, in self-destruction for the regime itself.

To be sure, it is far from our intention to compare in any possible way the condition of the perpetrators to that of the victims, and we certainly do not contend that the German population had been victimized by the Nazi project in ways similar to European Jews or the other subjected peoples during the war. Rather, what we are arguing is that in the foolish attempts to realize a new German purified *Lebensraum*, German and non-German individuals and populations, victims and perpetrators, ended up being somehow linked—in their very existence, as individual and collective subjects—by the red thread that made Nazi bio-geo-politics the core intellectual engine of a genocidal machinery. In other words, the same spatial *Weltanschauung*, because of its bio-geo-political nature, was at the foundation of both the Germanification and the related purification of the *Volksgemeinschaft*, of the implementation of the "Final Solution" to the *Judenfrage* (Jewish question), but *also* of the very collapse of the *Endreich*. In particular, we suggest that Nazi spatial ideology was essentially a bio-geo-political ideology, based on an overall violent attempt to realize an ordered set of spatialities in which the typically modern tensions between ideas of race *and* space, of place *and* space were supposed to dissolve into a newly founded Nazi and German *Lebensraum*, literally, "living" but also "vital" space. This very thesis rests at the core of our attempt to start drafting a spatial theory of the Third Reich, of which this chapter is a first experimental and tentative step.

Acknowledgments

We are particularly grateful to John O'Loughlin, James Sidaway, and Dan Stone for their helpful comments and suggestions on this chapter.

References

Abrahamsson, Christian. 2013. "On the Genealogy of *Lebensraum*." *Geographica Helvetica* 68: 37–44.

Agamben, Giorgio. 1998. *Homo Sacer: Sovereign Power and Bare Life*. Stanford, CA: Stanford University Press.

———. 2002. *Remnants of Auschwitz: The Witness and the Archive*. New York: Zone Books.

———. 2005. *State of Exception*. Chicago: University of Chicago Press.

Aly, Götz. 1999. *Final Solution: Nazi Population Policy and the Murder of the European Jews*. London: Arnold.

———. 2014. *Why the Germans? Why the Jews?: Envy, Race Hatred, and the Prehistory of the Holocaust*. New York: Henry Holt.

Aly, Götz, and Susanne Heim. 2002. *Architects of Annihilation: Auschwitz and the Logic of Destruction*. Princeton, NJ: Princeton University Press.

Aly, Götz, and Karl H. Roth. 2004. *The Nazi Census: Identification and Control in the Third Reich*. Philadelphia: Temple University Press.

Arad, Yitzhak, Yisrael Gutman, and Abraham Margaliot. 1999. *Documents on the Holocaust*. Lincoln: University of Nebraska Press.

Arnold, Bettina. 1990. "The Past as Propaganda: Totalitarian Archaeology in Nazi Germany." *Antiquity* 64: 464–578.

Atkinson, David 2011. "Remembering Nazi Intellectuals." In *Spatiality, Sovereignty, and Carl Schmitt*, edited by Steve Legg, 201–10. London: Routledge.

Baranowski, Shelley. 2007. *Strength through Joy: Consumerism and Mass Tourism in the Third Reich*. Cambridge: Cambridge University Press.

———. 2011. *Nazi Empire: German Colonialism and Imperialism from Bismarck to Hitler*. Cambridge: Cambridge University Press.

Barnes, Trevor, and Claudio Minca. 2013. "Nazi Spatial Theory: The Dark Geographies of Carl Schmitt and Walter Christaller." *Annals of the Association of American Geographers* 103 (3): 669–87.

Bassin, Mark. 1987. "Race contra Space: The Conflict between German Geopolitik and National Socialism." *Political Geography Quarterly* 6 (2): 115–34.

———. 2005. "Blood or Soil? The *Völkisch* Movement, the Nazis, and the Legacy of Geopolitik." In *How Green Were the Nazis? Nature, Environment, and Nation in the Third Reich*, edited by Franz-Josef Brüggemeier, Mark Cioc, and Thomas Zeller, 204–42. Athens: Ohio University Press.

Baum, Rob K. 2006. "Deconstruction of National Identity in the Third Reich: Nazisprache und Geopolitik." *National Identities* 8 (2): 95–112.

Bauman, Zygmunt. 1989. *Modernity and the Holocaust*. London: Polity Press.

Beorn, Waitman, Tim Cole, Simone Gigliotti, Alberto Giordano, Anna Holian, Paul B. Jaskot, Anne Kelly Knowles, Marc Masurovsky, and Erik B. Steiner. 2009. "Geographies of the Holocaust." *Geographical Review* 99 (4): 563–74.

Blackbourn, David. 2007. *The Conquest of Nature: Water, Landscape, and the Making of Modern Germany*. New York: Norton.

Bramwell, Anna. 1985. *Blood and Soil: Richard Walther Darré and Hitler's "Green Party."* London: Kensal Press.

Browning, Christopher. 2004. *The Origins of the Final Solution*. Lincoln: University of Nevada Press.

Brüggemeier, Franz-Josef, Mark Cioc, and Thomas Zeller, eds. 2005. *How Green Were the Nazis? Nature, Environment, and Nation in the Third Reich*. Athens: Ohio University Press.

Burke, Peter. 1997. *Varieties of Cultural History*. Cambridge: Polity Press.

———. 2008. *What Is Cultural History?* Cambridge: Polity Press.

Burleigh, Michael 1988. *Germany Turns Eastwards*. Cambridge: Cambridge University Press.

Carter-White, Richard. 2009. "Auschwitz, Ethics, and Testimony: Exposure to the Disaster." *Environment and Planning D: Society and Space* 27 (4): 682–99.

———. 2012. "Primo Levi and the Genre of Testimony." *Transactions of the Institute of British Geographers* 37: 287–300.

———. 2013. "Towards a Spatial Historiography of the Holocaust: Resistance, Film, and the Prisoner Uprising at Sobibor Death Camp." *Political Geography* 33: 21–30.

Charlesworth, Andrew. 1992. "Towards a Geography of the Shoah." *Journal of Historical Geography* 18 (4): 464–69.

———. 1994. "Contesting Places of Memory: The Case of Auschwitz." *Environment and Planning D: Society and Space* 12: 579–93.

———. 2003. "Hello Darkness: Envoi and Caveat." *Common Knowledge* 9 (3): 508–19.

———. 2004a. "A Corner of a Foreign Field That Is Forever Spielberg's: Understanding the Moral Landscapes of the Site of the Former KL Płaszów, Kraków, Poland." *Cultural Geographies* 11: 291–312.

———. 2004b. "The Topography of Genocide." In *The Historiography of the Holocaust*, edited by Dan Stone, 216–51. Basingstoke: Palgrave Macmillan.

Charlesworth, Andrew, and Mike Addis. 2002. "Memorialization and the Ecological Landscapes of Holocaust Sites." *Landscape Research* 27 (3): 229–51.

Charlesworth, Andrew, Alison Stenning, Robert Guzik, and Michal Paszkowski. 2006. "'Out of Place' in Auschwitz? Contested Development in Post-War and Post-Socialist Oświęcim." *Ethics, Place, and Environment* 9 (2): 149–72.

Clark, Elizabeth A. 2004. *History, Theory, Text: Historians and the Linguistic Turn*. Cambridge, MA: Harvard University Press.

Clarke David B., Marcus A. Doel, and Francis X. McDonough. 1996. "Holocaust Topologies: Singularity, Politics, Space." *Political Geography* 15 (6–7): 457–89.

Cole, Tim. 2003. *Holocaust City: The Making of a Jewish Ghetto*. New York: Routledge.

———. 2013. *Traces of the Holocaust: Journeying in and out of the Ghettos*. London: Bloomsbury.

Confino, Alon. 1997. *The Nation as a Local Metaphor: Württemberg, Imperial Germany, and National Memory, 1871–1918*. Chapel Hill: University of North Carolina Press.

———. 2005. "Fantasies about the Jews: Cultural Reflections on the Holocaust." *History and Memory* 17 (1–2): 296–332.

———. 2012. *Foundational Pasts: The Holocaust as Historical Understanding*. Cambridge: Cambridge University Press.

———. 2014. *A World without Jews: The Nazi Imagination from Persecution to Genocide*. New Haven, CT: Yale University Press.

Conrad, Sebastian. 2013. "Rethinking German Colonialism in a Global Age." *Journal of Imperial and Commonwealth History* 41 (4): 543–66.

Danielsson, Sarah K. 2009. "Creating Genocidal Space: Geographers and the Discourse of Annihilation, 1880–1933." *Space and Polity* 13 (1): 55–68.

Diner, Dan. 1999. "Knowledge of Expansion on the Geopolitics of Karl Haushofer." *Geopolitics* 4 (3): 161–88.

Dwork, Debórah, and Robert Jan van Pelt. 2002. *Auschwitz*. New York: Norton.

Ek, Richard. 2006. "Giorgio Agamben and the Spatialities of the Camp: An Introduction." *Geografiska Annaler B* 88 (4): 363–86.

Elden, Stuart. 2003. "Taking the Measure of the Beiträge." *European Journal of Political Theory* 2 (1): 35–56.

———. 2006a. "National Socialism and the Politics of Calculation." *Social and Cultural Geography* 7 (5): 753–69.

———. 2006b. *Speaking against Number: Heidegger, Language, and the Politics of Calculation.* Edinburgh: Edinburgh University Press.

———. 2010. "Reading Schmitt Geopolitically: Nomos, Territory, and Großraum." *Radical Philosophy* 161: 18–26.

———. 2013. *The Birth of Territory*. Chicago: University of Chicago Press.

Engelking-Boni, Barbara, and Jacek Leociak. 2009. *The Warsaw Ghetto: A Guide to the Perished City*. New Haven, CT: Yale University Press.

Epstein, Catherine. 2012. *Model Nazi: Arthur Greiser and the Occupation of Western Poland.* Oxford: Oxford University Press.

Esposito, Roberto. 2008. *Bíos: Biopolitics and Philosophy*. Translated by Timothy Campbell. Minneapolis: University of Minnesota Press.

———. 2011. *Immunitas: The Protection and Negations of Life*. Cambridge: Polity Press.

Fahlbusch, Michael, Mechtild Rössler, and Dominik Siegrist. 1989. "Conservatism, Ideology, and Geography in Germany, 1920–1950." *Political Geography Quarterly* 8 (4): 353–67.

Finchelstein, Federico. 2009. "Regarding History, Holocaust, and Culture." *Dapim: Studies on the Holocaust* 23 (1): 87–88.

Finzsch, Norbert. 2008. "'[. . .] Extirpate or Remove that Vermine': Genocide, Biological Warfare, and Settler Imperialism in the Eighteenth and Early Nineteenth Century." *Journal of Genocide Research* 10 (2): 215–32.

Fitzpatrick, Matthew P. 2008. "The Pre-History of the Holocaust? The *Sonderweg* and *Historikerstreit* Debates and the Abject Colonial Past." *Central European History* 41 (3): 477–503.

Fritzsche, Peter. 2008. *Life and Death in the Third Reich*. Cambridge, MA: Belknap Press of Harvard University Press.

Gerwarth, Robert. 2011. *Hitler's Hangman*. New Haven, CT: Yale University Press.

Giaccaria, Paolo, and Claudio Minca. 2011. "Topographies/Topologies of the Camp: Auschwitz as a Spatial Threshold." *Political Geography* 30: 3–12.

Gilroy, Paul. 2004. *Between Camps*. London: Routledge.

Goldberg, Amos. 2009. "Introduction." *Dapim: Studies on the Holocaust* 23 (1): 45–51.

Goodrick-Clarke, Nicholas. 1985. *The Occult Roots of Nazism: Secret Aryan Cults and Their Influence on Nazi Ideology*. New York: New York University Press.

Gregory, Derek. 2006. "The Black Flag: Guantanamo Bay and the Space of Exception." *Geografiska Annaler B* 88 (4): 405–27.

Guettel, Jens-Uwe. 2012. "The American West and the Nazi East: A Comparative and Interpretative Perspective." *Journal of Genocide Research* 14 (3–4): 514–19.

Gutman, Yisrael, and Michael Berembaum. 1994. *Anatomy of the Auschwitz Death Camp.* Bloomington: Indiana University Press.

Hagen, Joshua. 2004. "The Most German of Towns: Creating an Ideal Nazi Community in Rothenburg ob der Tauber." *Annals of the Association of American Geographers* 94 (1): 207–27.

———. 2009a. "Historic Preservation in Nazi Germany: Place, Memory, and Nationalism." *Journal of Historical Geography* 35: 690–715.

———. 2009b. "Mapping the Polish Corridor: Ethnicity, Economics, and Geopolitics." *Imago Mundi* 62 (1): 63–82.

Hagen, Joshua, and Robert Ostergren. 2006. "Spectacle, Architecture, and Place at the Nuremberg Party Rallies: Projecting a Nazi Vision of Past, Present, and Future." *Cultural Geographies* 13: 157–81.

Hayes, Peter. 2003. "Auschwitz: Capital of the Holocaust." *Holocaust and Genocide Studies* 71 (2): 330–50.

Heffernan, Michael. 2011. "Mapping Schmitt." In *Spatiality, Sovereignty, and Carl Schmitt Geographies of the Nomos*, edited by Steve Legg, 234–43. London: Routledge.

Herb, Guntram H. 1989. "Persuasive Cartography in Geopolitik and National Socialism." *Political Geography Quarterly* 8 (4): 289–303.

———. 1997. *Under the Map of Germany: Nationalism and Propaganda, 1918–1945*. London: Routledge.

———. 2002. "A Journey into the Ticket of German Geopolitik." *Geopolitics* 7 (3): 175–82.

Herbert, Ulrich, ed. 2000. *National Socialist Extermination Policies: Contemporary German Perspectives and Controversies*. Oxford: Berghahn.

Herf, Jeffrey. 1984. *Reactionary Modernism: Technology, Culture, and Politics in Weimar and the Third Reich*. Cambridge: Cambridge University Press.

Herwig, Holger. 1999. "Geopolitik: Haushofer, Hitler, and *Lebensraum*." *Journal of Strategic Studies* 22 (2–3): 218–41.

Heske, Henning. 1987. "Karl Haushofer: His Role in German Geopolitics and in Nazi Politics." *Political Geography Quarterly* 6 (2): 135–44.

Horwitz, Gordon J. 2008. *Ghettostadt: Łódź and the Making of a Nazi City*. Cambridge, MA: Belknap Press of Harvard University Press.

Housden, Martyn. 2003. *Hans Frank: Lebensraum and the Holocaust*. London: Palgrave Macmillan.

Hutton, Christopher. 2005. *Race and the Third Reich*. Cambridge: Polity Press.

Kakel, Carroll. 2013. *The Holocaust as Colonial Genocide: Hitler's "Indian Wars" in the "Wild East."* Basingstoke: Palgrave Macmillan.

Knowles, Anne, Tim Cole, and Alberto Giordano, eds. 2014. *Geographies of the Holocaust*. Bloomington: Indiana University Press.

Koonz, Claudia. 2003. *The Nazi Conscience*. Cambridge, MA: Harvard University Press.

Kost, Klaus. 1989. "The Conception of Politics in Political Geography and Geopolitics in Germany until 1945." *Political Geography Quarterly* 8 (4): 369–85.

———. 1998. "Anti-Semitism in German Geography, 1900–1945." *GeoJournal* 46: 285–91.

Kühne, Thomas. 2013a. *Belonging and Genocide: Hitler's Community, 1918–1945*. New Haven, CT: Yale University Press.

———. 2013b. "Colonialism and the Holocaust: Continuities, Causations, and Complexities." *Journal of Genocide Research* 15 (3): 339–62.

LaCapra, Dominick. 1996. *Representing the Holocaust: History, Theory, Trauma*. Ithaca, NY: Cornell University Press.

———. 1998. *History and Memory after Auschwitz*. Ithaca, NY: Cornell University Press.

———. 2004. *History in Transit: Experience, Identity, Critical Theory*. Ithaca, NY: Cornell University Press.

Legg, Stephen, ed. 2011. *Spatiality, Sovereignty, and Carl Schmitt: Geographies of the Nomos*. London: Routledge.

Lekan, Thomas M. 2004. *Imagining the Nation in Nature Landscape Preservation and German Identity, 1885–1945*. Cambridge, MA: Harvard University Press.

Liulevicius, Vejas. G. 2000. *War Land on the Eastern Front*. Cambridge: Cambridge University Press.

———. 2010. *The German Myth of the East: 1800 to the Present*. Oxford: Oxford University Press.

Longerich, Peter. 2010. *Holocaust: The Nazi Persecution and Murder of the Jews*. Oxford: Oxford University Press.

———. 2011. *Heinrich Himmler: A Life*. Oxford: Oxford University Press.

Lower, Wendy. 2005. *Nazi Empire-Building and the Holocaust in Ukraine*. Chapel Hill: University of North Carolina Press.

Madley, Benjamin. 2005. "From Africa to Auschwitz: How German South West Africa Incubated Ideas and Methods Adopted and Developed by the Nazis in Eastern Europe." *European History Quarterly* 35 (3): 429–64.

Martin, Lauren, and Anna Secor. 2013. "Towards a Post-Mathematical Topology." *Progress in Human Geography* 38 (3): 420–38.

Mazower, Mark 2008. *Hitler's Empire*. London: Penguin.

Michman, Dan, and Lenn J. Schramm. 2011. *The Emergence of Jewish Ghettos during the Holocaust*. Cambridge: Cambridge University Press.

Minca, Claudio. 2005. "The Return of the Camp." *Progress in Human Geography* 29 (4): 405–12.

———. 2006. "Giorgio Agamben and the New Biopolitical Nomos." *Geografiska Annaler B* 88: 387–403.

———. 2007. "Agamben's Geographies of Modernity." *Political Geography* 26 (1): 78–97.

———. 2015. "Geographies of the Camp." *Political Geography* 49 (1): 74–83.

Minca, Claudio, and Rory Rowan. 2014. "The Trouble with Carl Schmitt." *Political Geography* 4: A1–A3.

———. 2015a. *On Schmitt and Space*. London: Routledge.

———. 2015b. "The Question of Space in Carl Schmitt." *Progress in Human Geography* 39: 268–89.

Moses, Dirk. 2004. *Genocide and Settler Society: Frontier Violence and Stolen Indigenous Children in Australian History*. Oxford: Berghahn.

———, ed. 2008. *Empire, Colony, Genocide: Conquest, Occupation, and Subaltern Resistance in World History*. Oxford: Berghahn.

Moses, Dirk, and Dan Stone, eds. 2006. *Colonialism and Genocide*. London: Palgrave Macmillan.

Mullin, John Robert. 1981. "The Impact of National Socialist Policies upon Local City Planning in Pre-War Germany (1933–1939): The Rhetoric and The Reality." *Journal of the American Planning Association* 47 (1): 35–47.

———. 1982a. "Ideology, Planning Theory, and the German City in the Inter-War Years: Part One." *Town Planning Review* 52 (2): 115–30.

———. 1982b. "Ideology, Planning Theory, and the German City in the Inter-War Years: Part Two." *Town Planning Review* 52 (3): 257–72.

Murphy, David T. 1997. *The Heroic Earth: Geopolitical Thought in Weimar Germany, 1918–1933*. Kent, OH: Kent State University Press.

———. 1999 "'A Sum of the Most Wonderful Things': *Raum*, Geopolitics, and the German Tradition of Environmental Determinism, 1900–1933." *History of European Ideas* 25: 121–33.

Musolff, Andreas. 2010. *Metaphor, Nation, and the Holocaust: The Concept of the Body Politic*. London: Routledge.

Natter, Wolfgang. 2003 "Geopolitics in Germany, 1919–45." In *A Companion to Political Geography*, edited by John Agnew, Katharyne Mitchell, and Gerard Toal, 187–203. London: Blackwell.

———. 2005. "Friedrich Ratzel's Spatial Turn." In *B/ordering Space*, edited by Henk Van Houtum, Olivier Kramsch, and Wolfgang Zierhofer, 171–87. Aldershot: Ashgate.

Netz, Reviel. 2004. *Barbed Wire: An Ecology of Modernity*. Middletown, CT: Wesleyan University Press.

Neumann, Boaz. 2002. "The National Socialist Politics of Life." *New German Critique* 85: 107–30.

———. 2009. "The Phenomenology of the German People's Body (*Volkskörper*) and the Extermination of the Jewish Body." *New German Critique* 36 (1): 149–81.

Nyhart, Lynn K. 2009. *Modern Nature*. Chicago: University of Chicago Press.

O'Loughlin, John. 2000. "Can King's Ecological Inference Method Answer a Social Scientific Puzzle: Who Voted for the Nazi Party in Weimar Germany?" *Annals of the Association of American Geographers* 90 (3): 592–601.

———. 2002. "The Electoral Geography of Weimar Germany: Exploratory Spatial Data Analyses (ESDA) of Protestant Support for the Nazi Party." *Political Analysis* 10 (3): 217–43.

O'Loughlin, John, Colin Flint, and Luc Anselin. 1994. "The Geography of the Nazi Vote: Context, Confession, and Class in the Reichstag Election of 1930." *Annals of the Association of American Geographers* 84 (3): 351–80.

O'Loughlin, John, and Herman Heske. 1991. "From 'Geopolitik' to 'Geopolitique': Converting a Discipline for War to a Discipline for Peace." In *The Political Geography of Conflict and Peace*, edited by Nurit Kliot and Stanley Waterman, 37–59. London: Belhaven Press.

O'Loughlin, John, and Herman van der Wusten. 1990. "Political Geography of Panregions." *Geographical Review* 80 (1): 1–20.

Ó Tuathail, Gearóid. 1996. *Critical Geopolitics: The Politics of Writing Global Space*. London: Routledge.

Ó Tuathail, Gearóid, and John Agnew. 1992. "Geopolitics and Discourse: Practical Geopolitical Reasoning in American Foreign Policy." *Political Geography* 11 (2): 190–204.

Olwig, Kenneth R. 2002. "The Duplicity of Space: Germanic 'Raum' and Swedish 'Rum' in English Language Geographical Discourse." *Geografiska Annaler* 84 (1): 1–17.

Ossenbrügge, Jürgen. 1989. "Territorial Ideologies in West Germany, 1945–1985: Between Geopolitics and Regionalist Attitudes." *Political Geography Quarterly* 8 (4): 387–99.

Parker, Geoffrey. 1987. "French Geopolitical Thought in the Interwar Years and the Emergence of the European Idea." *Political Geography Quarterly* 6 (2): 145–50.

Paterson. J. H. 1987. "German Geopolitics Reassessed." *Political Geography Quarterly* 6 (2): 107–14.

Pergher, Roberta, Mark Roseman, Jürgen Zimmerer, Shelley Baranowski, Doris L. Bergen, and Zygmunt Bauman. 2013. "The Holocaust: a Colonial Genocide? A Scholars' Forum." *Dapim: Studies on the Holocaust* 27 (1): 40–73.

Pick, Daniel. 2012. *The Pursuit of the Nazi Mind: Hitler, Hess, and the Analysts*. Oxford: Oxford University Press.

Postone, Moishe, and Eric Santner, eds. 2003. *Catastrophe and Meaning: The Holocaust and the Twentieth Century*. Chicago: University of Chicago Press.

Ramadan, Adam. 2009. "Destroying Nahr el-Bared: Sovereignty and Urbicide in the Space of Exception." *Political Geography* 28 (3): 153–63.

Romanillos, José Louis. 2011. "Geography, Death, and Finitude." *Environment and Planning A* 43 (11): 2533–53.

Rössler, Mechtild. 1989. "Applied Geography and Area Research in Nazi Society: Central Place Theory and Planning, 1933 to 1945." *Environment and Planning D: Society and Space* 7: 419–31.

———. 2001. "Geography and Area Planning under National Socialism." In *Science in the Third Reich*, edited by Margit Szöllösi-Janze, 59–78. New York: Berg.

Sandner, Gerhard. 1988. "Recent Advances in the History of German Geography, 1918–1945: A Progress Report for the Federal Republic of Germany." *Geographische Zeitschrift* 76: 120–33.

———. 1989. "The Germania Triumphans Syndrome and Passarge's Erdkundliche Weltanschauung: The Roots and Affects of German Political Geography beyond Geopolitik." *Political Geography Quarterly* 8 (4): 341–51.

Sandner, Gerhard, and Mechtild Rössler. 1994. "Geography and Empire in Germany, 1871–1945." In *Geography and Empire*, edited by Anne Godlewska and Neil Smith, 115–27. Oxford: Blackwell.

Schenk, Tilman A., and Ray Bromley. 2003. "Mass-Producing Traditional Small Cities: Gottfried Feder's Vision for a Greater Nazi Germany." *Journal of Planning History* 2 (2): 107–39.

Schultz, Hans D. 1989. "Fantasies of *Mitte: Mittellage* and *Mitteleuropa* in German Geographical Discussion in the 19th and 20th Centuries." *Political Geography Quarterly* 8 (4): 325–39.

Smith, Woodruff S. 1980. "Friedrich Ratzel and the Origins of Lebensraum." *German Studies Review* 3 (1): 51–68.

———. 1986. *The Ideological Origins of Nazi Imperialism*. Oxford: Oxford University Press.

Sofsky, Wolfgang. 1997. *The Order of Terror: The Concentration Camp*. Princeton, NJ: Princeton University Press.

Steber, Martina, and Bernhard Gotto, eds. 2014. *Visions of Community in Nazi Germany: Social Engineering and Private Lives*. Oxford: Oxford University Press.

Stone, Dan. 2001. "White Men with Low Moral Standards? German Anthropology and the Herero Genocide." *Patterns of Prejudice* 35 (2): 33–45.

———. 2009. "Holocaust Historiography and Cultural History." *Dapim: Studies on the Holocaust* 23 (1): 52–68.

———. 2010. *Histories of the Holocaust*. Oxford: Oxford University Press.

———. 2013. *The Holocaust, Fascism, and Memory: Essays in the History of Ideas*. Basingstoke: Palgrave Macmillan.

Szejnmann, Claus-Christian, and Maiken Umbach, eds. 2012. *Heimat, Region, and Empire. Spatial Identities under National Socialism*. Basingstoke: Palgrave Macmillan.

Till, Karen. 2005. *The New Berlin: Memory, Politics, Place*. Minneapolis: University of Minnesota Press.

Troll, Carl. 1949. "Geographic Science in Germany during the Period 1933–1945: A Critique and Justification." *Annals of the Association of American Geographers* 39 (2): 99–137.

Uekoetter, Frank. 2006. *The Green and the Brown: A History of Conservation in Nazi Germany*. Cambridge: Cambridge University Press.

Weiss, Sheila Faith. 2010. *The Nazi Symbiosis: Human Genetics and Politics in the Third Reich*. Chicago: University of Chicago Press.

Wolfe, Patrick. 1999. *Settler Colonialism and the Transformation of Anthropology*. London: Cassell.

———. 2006. "Settler Colonialism and the Elimination of the Native." *Journal of Genocide Research* 8 (4): 387–409.

Zimmerer, Jürgen. 2005. "The Birth of the Ostland out of the Spirit of Colonialism: A Postcolonial Perspective on the Nazi Policy of Conquest and Extermination." *Patterns of Prejudice* 39 (2): 197–219.

———. 2008. "Colonialism and the Holocaust: Towards an Archeology of Genocide." *Development Dialogue* 50: 95–124.

2

Holocaust Spaces

DAN STONE

> The Nazis were out to eliminate not only groups of people like the Jews, but to destroy all the inhabitants of an area, along with all their cultural manifestations, in order to create "space" for their own people.
>
> RAPHAEL LEMKIN (1992, 168)

> Geopolitics, or applied geography, is synonymous with National Socialist dogma.
>
> ROBERT E. DICKINSON (1943, 209)

Introduction

It is hard to avoid geography when thinking about the Holocaust. Almost any book on the subject deals with spatial, geographical, or territorial concepts, whether they are explicitly theorized as such or not. To take an example pretty much at random, consider this passage from Kenneth Helphand's fascinating study of ghetto gardens:

> The prevailing landscape of the Holocaust was a man-made one. Claude Lanzmann, the director of *Shoah* . . . said it was "a film about topography, about geography." That geography included the creation of ghettos, places that W. A. Douglas Jackson has called "pariah-scapes"—or places to which people have been banished. Because they were existing urban areas that were walled in, each ghetto had a unique character and topography . . . the ghettos were essentially holding stations where the inhabitants died of hunger and disease or were deported to extermination camps. Everyone there shared a common fate; the ghettos were death traps.[1] (Helphand 2006, 63–64)

Yet although geographers have been thinking for a while now about the ways in which their discipline can contribute to understanding Holocaust spaces, whether via GIS or through more interpretive methods, historians are only just catching up.[2] This chapter builds on the earlier historical geographical research and the other chapters in this book dealing with Nazi *Raumplanung*, geographical theory, and projects for colonial settlement to show how Nazism's greatest crime, the Holocaust, was not only a geographical undertaking but also cannot be thought outside of geographical categories. Scholars, especially historians who write about the Nazi perpetrators, have long used these categories, but the recent, fruitful conjunction of cultural geography, history, and memory studies means that these concepts and terms can now be explicitly theorized and their meanings for Holocaust studies brought out to the full. Whether one thinks about the Nazi perpetrators' fantasies and visions of a *judenrein* Europe, of the victims' experiences of ghettoization and massive physical restraint in hiding, deportation, or in camps, or finally of Holocaust memory, which is deeply rooted in place, and where specific sites are intimately connected to triggering certain memories, we shall see in this chapter that spatial concepts are not just unavoidable but, once conceived as spatial terms in the strict sense (and not just, for example, Nazi rhetoric), fundamental for understanding the nature of the Holocaust as a genocidal project. In other words, whether one speaks of "physical space," that is, the territorial, especially transcontinental dimensions of the Holocaust, the actual places where the Holocaust happened, or "cognitive space," that is, figurative or imaginative uses of space, in, for example, propaganda, mapping, or ideological metaphors, understanding the Holocaust is inseparable from geography, because "geographic representations undergird political practices" (Tyner 2008, 11). I will back up this claim with reference to the perpetrators of the Holocaust, the victims and, especially, post-Holocaust memories bound up with sites, and, finally, the idea of genocide per se as a spatial phenomenon.

Perpetrators

Cultural geographers regularly discuss the significance of cognitive space, in particular by examining the creation of symbolically resonant political landscapes which work as dominant representations of nations, dynasties, religious institutions, or powerful families (see Graham 1998; Till 2004).[3] With respect to Nazism, it is easy to pinpoint the geographical concepts and terms which quickly dominated Nazi thought and then, after 1933, the Third Reich. Friedrich Ratzel's bio-geographical theories, which included the concept of *Lebensraum*, and Karl Haushofer's *Geopolitik* were, along with a well-known combination of *völkisch* ideas, race science, and eugenics, among the key components of Nazi thought, which approved of the notion that "blood" (racial belonging) was necessarily tied to territory and that a people's characteristics were linked to the landscape in which they lived and the uses to which they put the land. One does not therefore have to exaggerate the importance of Richard Walther Darré and the "blood and soil" wing of Nazism in order to grasp the fact that "land" and "territory" were basic building blocks of Nazi thought.[4] All geography involves the relationship of people and place; Nazism, among the other ways it can be defined, can be regarded as a radical geographical project to reorder the relationships of peoples to their places in Europe—and, in theory, beyond Europe—which would radically intervene in deciding who was and who was not permitted to have "space" at all. This fact was recognized very early on by wartime commentators on the Third Reich; Frank Munk, for example, wrote about "the mission of space politics" in his 1943 book *The Legacy of Nazism* (Munk 1943, 48), and in 1940 Francis Aldor wrote that "the fate of Poland since the German occupation shows only too clearly that what the Germans mean by living space is at the same time the dying space of other peoples" (Aldor 1940, 141, cited in Cole 2003, 252n22). In a nice example of English understatement, one academic geographer wrote, at the conclusion of a detailed survey of German territory in terms of trade, population distribution, and cultural patterns, that "the concepts of the *Kulturboden*, *Volksboden*, *Lebensraum*, and, above all, *Deutschland*, though all containing germs of truth and the makings of valuable conceptions, have been brought into evil repute by the exaggeration of scholars fired with an overcharge of emotionalism" (Dickinson 1943, 206).[5]

Nazi terms such as *Lebensraum* (living space), *Großraumwirtschaft* (large economic space), *Generalplan Ost* (General Plan for the East), and *Entfernung* (removal) are all geographical concepts. Nazi practices, whether involving the "resettlement" (*Aussiedlung*) of its victims or the tourist and leisure practices of the *Volk*, sought to create new senses of who had access to mobility and space in Europe.[6] And the Holocaust's nature as a "continent-wide" genocide is more and more widely recognized. The significance of that designation is that it reveals the extent of collaboration in the murder of the Jews, whether national (regimes such as Vichy France), institutional (for example, Ukrainian auxiliary police forces), or individual (looting Jews' apartments, for example). It also removes the Holocaust from an a-topian "nowhere" space, rooting it more concretely in actual sites of destruction. For where the death camps can be thought of philosophically as places of exception, as liminal sites, as thresholds and the like (themselves spatial metaphors worthy of investigation), historical and geographical research into the Holocaust on the ground across occupied Europe makes us confront the actual sites of destruction, including not just pits, forests, and forts where Jews were shot, ghettos and camps where they died or were murdered en masse, but also the warehouses, collection points, rail termini, and ordinary streets, roads, and fields where Jews were held prior to deportation or along which they were forced to drag themselves on the so-called death marches.

Modern Europe, in other words, is dotted with Holocaust-related sites because the genocide of the Jews was a continent-wide affair; it cannot be hidden away in exceptional sites such as Treblinka or Auschwitz—just as those places cannot be taken as topoi or logical consequence of modern European culture, for this move too occults the vast range of sites, institutions, and power relations that differentiated the daily practices of genocide in this massive area. Recent historical research breaks the Holocaust down into microhistories of violence in different locales, all connected by the wider Nazi mania for destruction, but all aided (and otherwise far harder to realize) by local and regional social, political, and cultural norms and intergroup tensions. This is a necessarily geographical way of conceiving the Holocaust, in which we are dealing not with a master plan implemented from Berlin such that there is nothing of interest to say about the implementation of the process on the ground (as it is assumed to be uniform and smooth functioning—a view which replicates how the Nazis themselves portrayed it); historical geography helps to understand that the overall figure of six million murdered Jews was arrived at not only mechanically by "modern," factory-line killing in a few key installations. Rather, detailed examination shows how many of these deaths were facilitated according to different local circumstances—including constraints of physical space (places of hiding, the nature of terrain, flat or hilly, forested or open) and of cognitive space (notions of national identity, the perceived power or otherwise of the state, fear of occupying forces or indigenous collaborators, and so on). The chronological notion of the "unfolding" of the Final Solution—still common in historiography—runs up against a geographical viewpoint that sees the Holocaust in terms of multiple spatial acts of displacement, relocation, concentration, and segregation being carried out simultaneously across Europe.[7]

For the Nazis, *Lebensraum* meant "the great reward of a vigorous and pure living space; for their victims, it is the denial of any space at all" (Wood 2001, 63; cf. Jureit 2012). One can understand the Holocaust as a process of gradually restricting the Jews' spaces, until the point at which they are denied any space at all. From the perpetrators' perspective, this gradual diminution of living space for the Jews was a necessary correlate of the creation of the expanding *Volksgemeinschaft*.[8] "Space" in this view should be understood both physically and cognitively. In the case of the former, the killing of Jews was part of the process whereby the German Empire was being created through territorial expansion and colonial settlement (although we should recognize that only the former process, killing Jews, was ever really achieved; the "Nazi empire" barely existed before it was overthrown).[9] And with respect to the latter, killing Jews followed from the Nazi belief that Aryan prosperity required the elimination of the threat posed by the "international Jew," who therefore had to be removed in order that the Germans could have living space—"living space" understood now not as physical territory but as the possibility of cultural activity (in the sense that we commonly talk of "breathing space"). Clearly, in this latter scenario, space can only be understood imaginatively: the few million Jews in Europe were not literally "in the way" of German territorial expansion; however, in the Nazis' minds, the Jews were hindrances to German expansion and racial rejuvenation. If one thinks of some of the most famous scenes of Nazi propaganda, such as Jewish international finance or the Jews depicted as rats spreading across the map of Europe in *The Eternal Jew*, one immediately understands the Holocaust as a spatial project, as long as one means by that a combination of physical and cognitive space, the two necessarily bound together in the Nazi worldview. Indeed, we might here follow Boaz Neumann's phenomenological rendering of the historians' "intentionalist" argument, and think of the gradualism of the Jews' ever-decreasing space as the perpetrators brought that about as the working through of a process whereby Jews were always already ontologically dead as soon as the Nazis came to power (Neumann 2010). As Neumann shows, in the Nazi worldview, each race received the environment it deserved: for the Germans, the new settlements of the East; for the Jews, the *Judenreservat* (ghetto) or extermination camp. "The Nazi ethnic-biological-racist space of the Lebensraum in the East," he writes, "could not have existed without the extermination of the Jew in the space of the camp—the extermination camp" (Neumann 2012, 108; see also Neumann 2011).

Victims

A decade ago Andrew Charlesworth complained that historians too readily overlooked the topographical aspects of genocide, focusing on documents but neglecting the sites to which they referred. Noting approvingly Raul Hilberg's insistence on the importance of written documents for providing historical proof of the Holocaust after all the survivors die, Charlesworth adds: "The place is still there. There will still be a place called Treblinka after all the survivors are dead" (Charlesworth 2004, 217). Similarly, Tim Cole, following Robert Jan van Pelt, has stressed the need to think of the architectural planning and the physical structure of Holocaust sites, which otherwise exist in a "no place" in our imagination, shrouded in mist and fog (Cole 2003; Pelt and Westfall 1991). Historians have to learn how to read landscapes as well as written texts. This is so not only so that historians can show how the Nazi genocide was underpinned by certain spatial concepts, but also so that they can explain how the Holocaust's victims experienced what was happening to them as a physical reality—the shrinking of space—and as a cognitive process—the removal from the world of meaning.

The victims of the Holocaust were spatially separated from their normal social context. In the first phase of the Holocaust, the Einsatzgruppen shootings in the western Soviet Union, Jews were shot in pits and trenches dug out of the Eastern European sand or in forests on the edges of towns. Such sites were not always hidden—Babi Yar is on the outskirts of Kiev, and the enormity of what happened there could hardly be kept secret—but they were rendered anonymous by being covered and left unmarked (Charlesworth 2004, 222; Desbois 2008).[10] When the ghettos of occupied Poland were established, the Jews were physically isolated from the rest of society, usually under the pretext of preventing the spread of diseases such as plague or typhus, which Jews were accused of carrying. The designation of the ghettos as "death boxes" is startling, but does not do justice to what happened to their inhabitants' understanding of their environment, as for the victims their world shrank to infinitesimal proportions—this physical restriction reflecting their inability to act in ways they chose—and for the Polish gentiles outside them their cities suddenly included not just a no-go zone but also a place that became primarily a zone of the imagination onto which fears and fantasies were projected. For the German authorities, in cases such as Łódź, which were to become places of German settlement (in this case, renamed Litzmannstadt), the presence of the ghettos confirmed the Nazi script of German superiority based on the need to control the Jewish threat, but did so with unsettling proximity, as this *unheimlich* presence existed, in the case of Łódź until summer 1944, in the midst of the new *Heimat* (see Horwitz 2008).

The ghettos are perhaps the best examples of how victimization during the Holocaust could be understood spatially, if only because they existed for a relatively long period of time and their inhabitants could experience the shrinking of their world on a daily basis, as so many surviving diaries agonizingly record. For many victims, ghettos formed their longest experience of "Nazi space." Tim Cole shows how spatial concepts can shed light on the ghettos with reference to Budapest and, in more recent work, smaller Hungarian ghettos, including Veszprém, Szeged, and Körmend. Cole argues that the Holocaust was not a monolithic, unitary phenomenon, but that the ways in which it unfolded on the ground in different locations varied considerably from place to place. In Hungary, the state removed some Jews it considered fit for forced labor from the deportation process with the result that the inhabitants of some of Hungary's ghettos were almost exclusively female. And it made an enormous difference to those people whether they had to move to the area designated as the ghetto or whether, as Cole puts it, "the ghetto came to them." Those who could stay put fared far better than those for whom ghettoization meant dislocation as well as concentration and segregation (Cole 2011, 30–31).

This focus on the actuality of physical space is becoming more common. In their remarkable reconstruction of the Warsaw ghetto (published in Polish in 2001, English in 2009), Polish historians Barbara Engelking and Jacek Leociak provide detailed maps and enough information to recreate the ghetto on a street-by-street and even house-by-house basis. Their aim is to reconstruct the ghetto in as microscopic a way as possible, straining the limits of the historical text's ability to *re*-present the past down to its tiniest detail—an almost "sacred" project whereby no scrap of information, no matter how small, can be left out. This Herculean act of recovery is only really possible on the basis of a spatial reconstruction of the ghetto, for the book is almost a replica of the ghetto in word and image, and perfectly illustrates what Frank Ankersmit means when he argues that historical texts operate not as windows onto the past but as substitutes for it (Ankersmit 1990, 291).[11] The book's function as a kind of memorial or *matzeva* in its own right is especially noteworthy given the authors' observation that the area where the ghetto stood is "today almost entirely deprived of physical traces that might aid our memory." Although the ghetto is no more, the authors point out, the "here" remains hidden behind today's district of Muranów: "The framework remains within which a different reality is located; a topographical point remains, a cartographical abstraction" (Engelking and Leociak 2009, 810).

But the ghettos are by no means the only ways into understanding the spatial experience of the Holocaust. By way of example, Simone Gigliotti writes powerfully of the deportation trains, explaining the train journey as a kind of transition between life and death, while never losing sight of the terrible physical reality of the process (see also Gigliotti, this volume, chapter 16). She contrasts the attitudes of those responsible for the deportations, who regarded the Jews as "cargo" (in Franz Stangl's words) or Auschwitz as "only a train station" (Franz Novak, Eichmann's transport officer in the RSHA [Reich Security Main Office]), with the experiences of the victims, whose more visceral descriptions convey a sense of a process unfolding in time and space: "temporal moments of increasing danger and diminished spatial expression," as Gigliotti puts it (Gigliotti 2009, 50, 51, 66). Elsewhere, the so-called death marches have belatedly received scholarly attention, with their main analyst partly employing geographical terminology and methods to make some sense of an otherwise more or less incomprehensible phenomenon (Blatman 2011). Jews hiding in the huge, primeval forests of eastern Poland have been examined with a view to showing how their environment changed the nature of their social structures (Weber 2008). And, in the rare cases where surviving victims' documents do reveal an understanding of events—contrary to the claim that they could not grasp what was happening at the time it occurred—the use of spatial and territorial (that is, cognitive and physical space respectively) terminology is striking. I have in mind here the so-called Scrolls of Auschwitz, the texts written by men of the Auschwitz *Sonderkommando* and buried in the earth by the gas chambers and crematorium, some of which were dug up after the war. Zalman Gradowski's text, especially, is offered to the reader as the chance to take a "journey," accompanying the soon-to-be-dead: "Take with you, O member of mankind, all the supplies you need to keep you from cold and damp, hunger and thirst. For in the frosty nights we will stand in desolate fields, escorting my tortured brothers on their last journey, their last death march" (Gradowski 1985, 174–75).[12] Gradowski's text was written for the future reader and reminds us that the journeys made by the Holocaust's victims are in some senses being repeated in the post-Holocaust context with every reading.

Spatial Aftermaths

Holocaust survivor testimony can thus be thought of as a "spatial trajectory" in the sense Simone Gigliotti propounded, following cultural theorist Michel de Certeau. "By means of a whole panoply of codes, ordered ways of proceeding and constraints," de Certeau writes, narrative structures "regulate changes in space . . . made by stories in the form of places put in linear or interlaced series: from here one goes there" (Certeau 1984, 115, cited in Gigliotti 2009, 66–67). This claim acquires powerful meaning in the context of Holocaust testimonies, for as well as their narrative structures being spatial metaphors their content usually mirrors this spatial trajectory as the author recounts fleeing or being forcibly moved from one place to another, or both. *Yizker-Bikher*, or memorial books, function similarly; these texts are rooted in place—usually they are publications of the *landsmanshaft* of a particular locale—and are in some senses guidebooks to a no-longer-existing place and can therefore only be reconstructed in the imagination. *Yizker-Bikher* and testimonies are aids in the creation of geographical collective memories, "substitutes for the lost town chronicle," as the historians of memorial books put it (Kugelmass and Boyarin 1998, 2; see also Roskies 1984).

But a concern with space in "aftermath issues" is by no means restricted to testimonies or other texts. Memory and place are intertwined (see Hoelscher and Alderman 2004). In fact, memorial sites, especially former concentration and death camps, are always changing as they are musealized and reshaped under different political circumstances. And the material remains of prewar Jewish life, particularly as they work in tandem with cultural memories, both anchor memories in space and give post-Holocaust memories an irreducibly spatial dimension. If this sounds circular, it is because we see here how social structures, including cultural memories, are constituted by spatiality, and vice-versa. "Jewish ruins in postwar Europe," writes Michael Meng, "reflected the results of an enormous genocide. Shards of broken stone—abandoned streets, homes, synagogues, cemeteries, and communal spaces—became the main traces of Jewish life after the Holocaust" (Meng 2011, 3). How could such devastation be cognized other than through the imagination, juxtaposing such sites as they existed after the war with images of them as they were beforehand, in reality or in idealized form?

Take, for example, Jonathan Webber and Chris Schwarz's work on traces of Jewish memory in Galicia (*Photographing Traces of Memory*). Schwarz's photographs are attempts to bring to life something that was only spoken about in the abstract ("Jewish life in Poland"); at the same time their simplicity jolts us into recognizing the enormity of what befell the centuries-old Jewish civilization in Poland. Likewise, the Polish debates over Jan Tomasz Gross's *Neighbors* have been so bitter partly because the book removed the Holocaust from "nowhere" to "somewhere specific," a real and very small place with real people involved who could be identified by name. The hand-drawn maps in such microhistories are as powerful as many written testimonies, for they provide a shock of recognition: this could be anywhere in Eastern Europe (and there are other such cases), but it is not, it is Jedwabne. The reality of what this means breaks through the patina of scholarly detachment when facts and figures return to actual sites: in Paweł Łoziński's short film *Birthplace* (1992), the writer Henryk Grynberg comes back from the United States to the village of his birth, and of his younger brother's death. His attempts to engage the locals in discussion of the "open secret" as to who killed him is familiar to viewers of Lanzmann's *Shoah* but more open, less aggressive and accusatory. At the end, Grynberg digs up the skeletal remains of his brother; the scene not only is almost emotionally unbearable but also shows how, ultimately, the Holocaust needs to be understood as deeply tied to the soil of Europe in hundreds of thousands of similar sites. Auschwitz and Treblinka are the biggest Jewish cemeteries in Europe; but the geography of the Holocaust links thousands of villages and fields across the continent.

The shock that is engendered by the specificity of place is the reason behind the rise to prominence of microhistory in Holocaust historiography. Where the "functionalist" consensus of the 1980s made us think—very productively, we must say—in terms of "modernity," bureaucracy, and factory-line genocide, recent work on the face-to-face killing in the "Wild East" has forced us to encounter the brutality of the killing process, which is hidden by inaccurate depictions of faceless and emotionless murder carried out by distant "desk-killers."[13] Structures of genocide, including leadership patterns and chains of command, remain central to understanding the occurrence of the Holocaust and other cases of genocide (see below); however, scholars increasingly recognize that the specific shape that genocide takes is determined by orders from above as well as by interactions between many different individuals, groups, and institutions on the ground. In fact, when one examines the way in which the Holocaust took place across Europe, the dynamics of destruction depend considerably not only on the degree of SS penetration of a place but also on the extent of police collaboration, the absence or presence of state institutions, and, most important for microhistories, inter-ethnic and group tensions, rivalries, and friendships at the very local level. Local history (like *Alltagsgeschichte* or "history of everyday life" before it) is thus playing a vital role first in reconstructing the sequence of events on the ground—which is providing a more meaningful context for the notion of "industrial genocide"—and, second, in showing how memories and places are so indissolubly linked. The former is clear in Yehuda Bauer's work on the *shtetl* (Bauer 2009) or in Omer Bartov's on his mother's birthplace, Buczacz, now in Ukraine (Bartov 2007). The latter is apparent in the attention being increasingly paid not to memorials in the United States or Israel, but to the changing patterns of life and layers of memory in evidence in formerly vibrant centers of Jewish life, such as Czernowitz (today Chernivtsi in Ukraine).[14] This applies to German cities too, or formerly German cities such as Breslau (today Wrocław). Scholars have unveiled the memories of different eras to show how Nazism has been dealt with (or not) in different periods since the end of the war, through analyses of architecture, heritage, and photography.[15]

Apart from what remains of specific sites, certain places (or what Paul Connerton [2009, chap. 4] identifies as *loci*[16]) have been invested with what one might call a "memory duty." That is to say, they have become not just places that people can visit as sites of "dark tourism" but places which are actively put to work as interactive *lieux de mémoire* (see, for example, Lennon and Foley 2000). For example, in the former Warsaw ghetto, the "Route of Memory, Martyrdom, and Struggle of the Jews" was opened on April 19, 1988, incorporating the Monument to the Heroes of the Ghetto unveiled originally in 1948 (famous as the site of Willy Brandt's "Kniefall" in 1969) and following the route taken by the deported Jews to the *Umschlagplatz* (collection point), where Hanna Szmalenberg and Władysław Klamerus's monument marks the place from which so many were sent off to Treblinka (Engelking and Leociak 2009, 809). In a more commercial vein, in the wake of the 1993 film version of *Schindler's List*, parts of Kraków's former Jewish quarter have turned into a kind of network of place-dependent memory.

Genocide

Territory is central to genocide studies. Questions of "territorial integrity" and "ethnic cleansing" concern the "correct" relationship between land and population, and the relevance of land and the "removal" of population groups are obvious in the context of settler colonialism.[17] Although some definitions of genocide include a territorial dimension, and despite the existence of a few books dealing with "the geography of genocide," the spatial aspects of genocide are rarely brought out explicitly.[18] Yet genocide always involves "the brutal rending of the deep ties between a targeted community and its environment" (Wood 2001, 57). Indeed, across otherwise very different cases of genocide—for all are unique in their own ways—one sees a striking continuity of spatial practices. The disposal of bodies or ash in rivers, for example, occurred during the Holocaust (in the Danube, Neris, and Soła in Budapest, Kaunas, and Auschwitz respectively) as well as during the Rwandan genocide (when the dead Tutsis were sent on their way "back to Ethiopia"); Charlesworth notes with reference to the Holocaust that this riverine expulsion "must have constituted a cleansing of these unwanted elements from the national body politic" (Charlesworth 2004, 221; see also Blackbourn 2006, chap. 5). That was certainly also the case in Rwanda.

All cases of genocide involve—so obviously, by definition, that the phenomenon is rarely analyzed—the physical separation of one group from another. Sometimes, as in Rwanda, victims are killed on the spot, at roadblocks, in their fields and houses, or trying to flee across the hills. But in most cases of genocide, the victim group has been separated, often placed in inhospitable territory as a way (perhaps unconsciously from the perpetrators' point of view) of bringing about their deaths "naturally." From Andrew Jackson's "trail of tears," which forced the "five civilized tribes" to leave their ancestral lands and "settle" (a euphemistic term for being dumped) in barren land in what is today Oklahoma, to Tasmania's surviving Aborigines being dumped on Flinder's Island, where many died of exposure, to the imprisonment and large-scale deaths on Shark Island of those Hereros who had escaped being expelled to the desert, the deportation of Armenians to the Syrian desert, and the expulsion of Cambodia's city dwellers to the countryside and forced labor—the "killing fields"—genocide is inherently a phenomenon that depends on control of territory and the ability to move population groups.

Such forced removals—for all the theoretical distinctions between "ethnic cleansing" and "genocide"—are often in practice genocidal.[19] The deportation of Armenians outside of Anatolia was not just to the desert but, as Donald Bloxham points out, was to an area with a majority Arab population beyond the Ottoman heartland which was subject to Great Power pressures. Sending the Armenians "to these inhospitable areas was akin to a death sentence for many in the medium term at most, and given general wartime shortage even that term was liable to be sharply reduced" (Bloxham 2012, 239). The Nazis' Madagascar Plan—known to historians, following Heydrich's terminology, as "the territorial final solution"—also illustrates this point; although the plan was not realized, because the British controlled the sea lanes in the Indian Ocean, thereby preventing the Nazis from deporting Jews to the former French colony, the scheme itself, historians now widely recognize, was genocidal since there were no plans to provide the expelled Jews with infrastructure, food supply, or shelter. The Jews, as one Nazi memorandum put it, would "be able to run their own administration in this territory." What this actually meant, as another memorandum by the same author (Franz Rademacher of the Foreign Office) confirmed, was clear: "From a German perspective, the Madagascar solution means the creation of a huge ghetto."[20] As Peter Longerich notes, with the failure of the plan, "Madagascar" came, for a few months until policy moved on again, to stand for "'anywhere' that might permit the execution of a 'final solution,' or in other words for the option of initiating a slow and painful end for the Jews of Europe in conditions inimical to life." This "anywhere" was soon replaced by a "nowhere" as explicit killing became the order of the day (Longerich 2010, 164). In this sense, what happened to the Jews of Europe has many similarities with what has happened to victims of other genocides. While some scholars worry that genocide studies might somehow diminish the significance of the Holocaust by thinking in broader conceptual or comparative terms, taking a "spatial perspective" suggests that the genocide studies mandate is not unattainable, that is, making meaningful comparisons without sacrificing awareness of the specificities of each case.

Conclusion

As a graduate student, I had the good fortune of being driven around western Galicia (today's southeastern Poland), with Jonathan Webber as guide, as part of a group putting together an exhibition called *Representations of Auschwitz*

(Doosry 1995). Working on an exhibition whose aim was to critique the ways in which certain images of the Holocaust have become iconic, it was too easy to separate out such images from their original context—which was the point the exhibition was trying to make but also the net effect of working on such material. As Rick Crownshaw points out, "victims, specifically Jewish ones, become remythologized through the current iconicity of images of them" (Crownshaw 2010, 119). It was therefore salutary to spend hours on the road traveling to remote places to see where Jews were burned in synagogues or shot in pits on the outskirts of small towns such as Rzeszów, Jarosław, or Dąmbrowa Tarnowska. The sheer physicality of the Holocaust, as a "territorial" phenomenon—the effort of going from place to place over large distances to isolate and murder the Jews—and as a "spatial one"—the rendering of Europe *judenrein*—becomes especially clear as one travels, albeit in rather more comfortable conditions, along some of the paths trodden by the perpetrators and victims. If, as theorists of space claim, space is constitutive of social structure, then the spatial dimension of the Holocaust is unavoidably a central aspect of its occurrence.

"Space" is central to Holocaust historiography whether one examines perpetrators, victims, "gray zones," post-Holocaust memory, or even the wider context of the Holocaust, including European history, colonialism, and genocide. With respect to the perpetrators, their spatial dreams and fantasies illustrate something familiar about Nazism in general and the Holocaust in particular: a combination of "rational"-sounding schemes that can be set out scientifically, in terms of planning, bureaucracy, and operational management, and the "nonrational" beliefs and fears—of Jewish world conspiracy, or racial decline, of History being driven by the clash of Aryan and non-Aryan forces—that these rational schemes were designed to overcome. With respect to the victims, theories of space provide immediacy and concreteness to the history of suffering that otherwise all too easily disappears into a "no place," and it helps to understand how the victims experienced persecution as an unfolding process that removed their room for action. Although I have not discussed "bystanders" or "gray zones" in this chapter, it is clear that these concepts are as much spatial as they are moral constructs and judgments. There were people who were caught in between different groups and therefore in between different places, and their histories can also be rendered spatially. As Primo Levi noted, in the Lagers (camps), people could not be divided into "'we' inside and the enemy outside, separated by a sharply defined geographic frontier" (Levi 1989, 23). Post-Holocaust memory is fundamentally rooted in place, particularly the loss of place and, often, the "fixing" of place in the memory which is sometimes confronted with the harsh reality of what a place has become, as in Henryk Grynberg's return to the site of his little brother's murder. But all memory is always also a forgetting, and while the recent turn to loci of memory might make us aware of the necessary selection process involved in memorials' and museums' constructions of memory, we should bear in mind that a similar process is at work when we deal with spatial memories. Finally, with respect to debates concerning how to contextualize the Holocaust—which currently center on questions of European history, global history, colonialism, and genocide—all of these are topics which only make sense spatially, whether one thinks of the inseparability of territoriality from the diplomatic history of the Great Game, theories of ethnic homogeneity in Europe in the age of nationalism, or the relationship between economic rationality and fears/fantasies of savages in the process of European overseas colonial settlement—and the question of how that history might relate to that of Nazism and the Holocaust. As this introductory survey indicates, thinking about Holocaust history through "space" is not only a way of getting to the heart of the problems and debates underpinning contemporary thought on the origins and execution of the Holocaust. It also presents a challenge to historians who think in terms of chronology and temporality to consider how geography might adjust, and, helpfully, strengthen their interpretations of how the Holocaust took place.

Acknowledgments

I am very grateful to Simone Gigliotti, Becky Jinks, Jessica Rapson, and the editors of this book for their comments on an earlier draft of this chapter.

Notes

1. On Douglas Jackson's concept of "pariah landscapes," referring to spaces of separation and banishment, see Cole (2003, 21). On Lanzmann's comments on *Shoah*, see also Charlesworth (2004, 216).

2. For starting points, see Clarke, Doel, and McDonough (1996); Beorn et al. (2009); Giordano and Cole (2011); Giaccaria and Minca, chapter 11 this volume; Giaccaria and Minca (2011); the activities of the USHMM Holocaust Geographies Project; and the chapters in this volume.

3. For a good example of how spatial imaginaries change over time and are subjected to political and cultural shifts, see Bialasiewicz (2003). For the *locus classicus* of theorizing "space," see Tuan (1977); Lefebvre (1991).

4. See Bramwell (1985). Cf. my comments in Stone (2013). And on the reasons why geopolitics was not always looked on favorably by the Nazis, see Bassin (1987 and 2005).

5. Dickinson was reader in geography at University College London.

6. See, for example, Baranowski (2004).

7. I'm grateful here to Simone Gigliotti for this wording.

8. See Kühne (2010) for a radical version of this argument.

9. On Nazi "colonialism," see Lower (2005); Baranowski (2011).

10. See also Waitman W. Beorn, "A Geography of Complicity," part of the Holocaust Geographies Project, online at http://www.ushmm.org/maps/projects/holocaust-geographies/?content=einsatz (accessed February 11, 2013).

11. Cf. my discussion in Stone (2003, 231).

12. For further discussion of the "Scrolls of Auschwitz," see the essays in Chare and Williams (2013).

13. For more detailed discussion, see Lawson (2010) and Stone (2010).

14. Hirsch and Spitzer (2010). See in general Brauch, Lipphardt, and Nocke (2008).

15. For example: Baer (2002); Gregor (2008); Macdonald (2008); Rosenfeld and Jaskot (2008); Rosenfeld (2011); for a moving example, see also Paulus, Raim, and Zelger (1995). See also Confino (2006); Szejnmann (2009); Kühne and Lawson (2011); Szejnmann and Umbach (2012).

16. See, for example, Yale's impressive Cambodian Genocide Program GIS sites at: http://www.yale.edu/cgp/maps.html (accessed February 21, 2013).

17. See, inter alia, Wolfe (2001); Moses (2004); Barta (2005); Moses and Stone (2007); Moses (2008).

18. See Kiernan (2007) and Cooper (2008) for a definition of genocide that includes territory.

19. See Lieberman (2010) and, for a stronger defense of the distinction between ethnic cleansing as a phenomenon of forced removal from territory and genocide as mass killing, see Ther (2012).

20. Franz Rademacher, memorandums of July 3, 1940, and July 2, 1940, respectively; cited in Longerich (2010, 163).

References

Aldor, Francis. 1940. *Germany's "Death Space": The Polish Tragedy*. London: Francis Aldor.

Ankersmit, F. R. 1990. "Reply to Professor Zagorin." *History and Theory* 29 (3): 275–96.

Baer, Ulrich 2002. *Spectral Evidence: The Photography of Trauma*. Cambridge, MA: MIT Press.

Baranowski, Shelley. 2004. *Strength through Joy: Consumerism and Mass Tourism in the Third Reich*. Cambridge: Cambridge University Press.

———. 2011. *Nazi Empire: German Colonialism and Imperialism from Bismarck to Hitler*. Cambridge: Cambridge University Press.

Barta, Tony. 2005. "Mr Darwin's Shooters: On Natural Selection and the Naturalizing of Genocide." *Patterns of Prejudice* 39 (2): 116–37.

Bartov, Omer. 2007. *Erased: Vanishing Traces of Jewish Galicia in Present-Day Ukraine*. Princeton, NJ: Princeton University Press.

Bassin, Mark. 1987. "Race contra Space: The Conflict between German *Geopolitik* and National Socialism." *Political Geography Quarterly* 6 (2): 115–34.

———. 2005. "Blood or Soil? The *Völkisch* Movement, the Nazis, and the Legacy of Geopolitik." In *How Green Were the Nazis? Nature, Environment, and Nation in the Third Reich*, edited by Franz-Josef Brüggemeier, Mark Cioc, and Thomas Zeller, 204–42. Athens: Ohio University Press.

Bauer, Yehuda. 2009. *The Death of the Shtetl*. New Haven, CT: Yale University Press.

Beorn, Waitman, Tim Cole, Simone Gigliotti, Alberto Giordano, Anna Holian, Paul B. Jaskot, Anne Kelly Knowles, Marc Masurovsky, and Erik B. Steiner. 2009. "Geographies of the Holocaust." *Geographical Review* 99 (4): 563–74.

Bialasiewicz, Luiza. 2003. "Another Europe: Remembering Habsburg Galicja." *Cultural Geographies* 10 (1): 21–44.
Blackbourn, David. 2006. *The Conquest of Nature: Water, Landscape, and the Making of Modern Germany*. New York: W. W. Norton.
Blatman, Daniel. 2011. *The Death Marches: The Final Phase of Nazi Genocide*. Cambridge, MA: Belknap Press of Harvard University Press.
Bloxham, Donald. 2012. "The Holocaust and European History." In *The Holocaust and Historical Methodology*, edited by Dan Stone, 233–54. New York: Berghahn.
Bramwell, Anna. 1985. *Blood and Soil: Richard Walther Darré and Hitler's "Green Party."* Bourne End: Kensal Press.
Brauch, Julia, Anna Lipphardt, and Alexandra Nocke, eds. 2008. *Jewish Topographies: Visions of Space, Traditions of Place*. Aldershot: Ashgate.
Certeau, Michel de. 1984. *The Practice of Everyday Life*. Berkeley: University of California Press.
Chare, Nicholas, and Dominic Paul Williams, eds. 2013. *Representing Auschwitz: At the Margins of Testimony*. Houndmills: Palgrave Macmillan.
Charlesworth, Andrew. 2004. "The Topography of Genocide." In *The Historiography of the Holocaust*, edited by Dan Stone, 216–52. Houndmills: Palgrave Macmillan.
Clarke, David B., Marcus A. Doel, and Francis X. McDonough. 1996. "Holocaust Topologies: Singularity, Politics, Space." *Political Geography* 15 (6–7): 457–89.
Cole, Tim. 2003. *Holocaust City: The Making of a Jewish Ghetto*. London: Routledge.
———. 2011. *Traces of the Holocaust: Journeying in and out of the Ghettos*. London: Continuum.
Confino, Alon. 2006. *Germany as a Culture of Remembrance: Promises and Limits of Writing History*. Chapel Hill: University of North Carolina Press.
Connerton, Paul. 2009. *How Modernity Forgets*. Cambridge: Cambridge University Press.
Cooper, Allan D. 2008. *The Geography of Genocide*. Lanham, MD: University Press of America.
Crownshaw, Richard. 2010. *The Afterlife of Holocaust Memory in Contemporary Literature and Culture*. Houndmills: Palgrave Macmillan.
Desbois, Patrick. 2008. *The Holocaust by Bullets: A Priest's Journey to Uncover the Truth behind the Murder of 1.5 Million Jews*. New York: Palgrave Macmillan.
Dickinson, Robert E. 1943. *The German Lebensraum*. Harmondsworth: Penguin.
Doosry, Yasmin, ed. 1995. *Representations of Auschwitz*. Oświęcim: Auschwitz-Birkenau State Museum.
Engelking, Barbara, and Jacek Leociak. 2009. *The Warsaw Ghetto: A Guide to the Perished City*. Translated by Emma Harris. New Haven, CT: Yale University Press.
Giaccaria, Paolo, and Claudio Minca. 2011. "Topographies/Topologies of the Camp: Auschwitz as a Spatial Threshold." *Political Geography* 30 (1): 3–12.
Gigliotti, Simone. 2009. *The Train Journey: Transit, Captivity, and Witnessing in the Holocaust*. New York: Berghahn.
Giordano, Albert, and Tim Cole. 2011. "On Place and Space: Calculating Social and Spatial Networks in the Budapest Ghetto." *Transactions in GIS* 15: 143–70.
Gradowski, Zalman. 1985. "Writings." In *The Scrolls of Auschwitz*, edited by Ber Mark, 173–205. Tel Aviv: Am Oved.
Graham, Brian, ed. 1998. *Modern Europe: Place, Culture, and Identity*. London: Routledge.
Gregor, Neil. 2008. *Haunted City: Nuremberg and the Nazi Past*. New Haven, CT: Yale University Press.

Helphand, Kenneth I. 2006. *Defiant Gardens: Making Gardens in Wartime*. San Antonio, TX: Trinity University Press.

Hirsch, Marianne, and Leo Spitzer. 2010. *Ghosts of Home: The Afterlife of Czernowitz in Jewish Memory*. Berkeley: University of California Press.

Hoelscher, Steven, and Derek H. Alderman. 2004. "Memory and Place: Geographies of a Critical Relationship." *Social and Cultural Geography* 5 (3): 347–55.

Horwitz, Gordon J. 2008. *Ghettostadt: Łódź and the Making of a Nazi City*. Cambridge, MA: Belknap Press of Harvard University Press.

Jureit, Ulrike. 2012. *Das Ordnen von Räumen: Territorien und Lebensraum im 19. und 20. Jahrhundert*. Hamburg: Hamburger Edition.

Kiernan, Ben. 2007. *Blood and Soil: A World History of Genocide and Extermination from Sparta to Darfur*. New Haven, CT: Yale University Press.

Kugelmass, Jack, and Jonathan Boyarin. 1998. *From a Ruined Garden: The Memorial Books of Polish Jewry*. Bloomington: Indiana University Press.

Kühne, Thomas. 2010. *Belonging and Genocide: Hitler's Community, 1918–1945*. New Haven, CT: Yale University Press.

Kühne, Thomas, and Tom Lawson, eds. 2011. *The Holocaust and Local History*. London: Vallentine Mitchell.

Lawson, Tom. 2010. *Debates on the Holocaust*. Manchester: Manchester University Press.

Lefebvre, Henri. 1991. *The Production of Space*. Oxford: Blackwell.

Lemkin, Raphael. 1992. *Thoughts on Nazi Genocide: Not Guilty?* Edited by Steven L. Jacobs. Lewiston: Edwin Mellen Press.

Lennon, J. John, and Malcolm Foley. 2000. *Dark Tourism: The Attraction of Death and Disaster*. London: Continuum.

Levi, Primo. 1989. "The Grey Zone." In *The Drowned and the Saved*, 22–51. London: Abacus.

Lieberman, Benjamin. 2010. "'Ethnic Cleansing' versus Genocide?" In *The Oxford Handbook of Genocide Studies*, edited by Donald Bloxham, and A. Dirk Moses, 42–60. Oxford: Oxford University Press.

Longerich, Peter. 2010. *Holocaust: The Nazi Persecution and Murder of the Jews*. Oxford: Oxford University Press.

Lower, Wendy. 2005. *Nazi Empire-Building and the Holocaust in Ukraine*. Chapel Hill: University of North Carolina Press.

Macdonald, Sharon. 2008. *Difficult Heritage: Negotiating the Nazi Past in Nuremberg and Beyond*. London: Routledge.

Meng, Michael. 2011. *Shattered Spaces: Encountering Jewish Ruins in Postwar Germany and Poland*. Cambridge, MA: Harvard University Press.

Moses, A. Dirk, ed. 2004. *Genocide and Settler Society: Frontier Violence and Stolen Indigenous Children in Australian History*. New York: Berghahn.

———. 2008. *Empire, Colony, Genocide: Conquest, Occupation, and Subaltern Resistance in World History*. New York: Berghahn.

Moses, A. Dirk, and Dan Stone, eds. 2007. *Colonialism and Genocide*. London: Routledge.

Munk, Frank. 1943. *The Legacy of Nazism: The Economic and Social Consequences of Totalitarianism*. New York: Macmillan.

Neumann, Boaz. 2010. *Die Weltanschauung des Nazismus: Raum, Körper, Sprache*. Göttingen: Wallstein.

———. 2011. *Land and Desire in Early Zionism*. Waltham, MA: Brandeis University Press.

———. 2012. "National Socialism, Holocaust, and Ecology." In *The Holocaust and Historical Methodology*, edited by Dan Stone, 101–23. New York: Berghahn.

Paulus, Martin, Edith Raim, and Gerhard Zelger, eds. 1995. *Ein Ort wie jeder andere: Bilder aus einer deutschen Kleinstadt. Landsberg 1923–1958*. Reinbek: Rowohlt.

Pelt, Robert Jan van, and Carroll William Westfall. 1991. *Architectural Principles in the Age of Historicism*. New Haven, CT: Yale University Press.

Rosenfeld, Gavriel D. 2011. *Building after Auschwitz: Jewish Architecture and the Memory of the Holocaust*. New Haven, CT: Yale University Press.

Rosenfeld, Gavriel D., and Paul B. Jaskot, eds. 2008. *Beyond Berlin: Twelve German Cities Confront the Nazi Past*. Ann Arbor: University of Michigan Press.

Roskies, David G. 1984. *Against the Apocalypse: Responses to Catastrophe in Modern Jewish Culture*. Cambridge, MA: Harvard University Press.

Stone, Dan. 2003. *Constructing the Holocaust: A Study in Historiography*. London: Vallentine Mitchell.

———. 2010. *Histories of the Holocaust*. Oxford: Oxford University Press.

———. 2013. "Rural Revivalism and the Radical Right in France and Britain between the Wars." In *The Holocaust, Fascism, and Memory: Essays in the History of Ideas*, 110–22. Houndmills: Palgrave Macmillan.

Szejnmann, Claus-Christian W. 2009. "Regional History, Nazism, and the Holocaust." In *Rethinking History, Dictatorship, and War: New Approaches and Interpretations*, edited by Claus-Christian W. Szejnmann, 55–67. London: Continuum.

Szejnmann, Claus-Christian W., and Maiken Umbach, eds. 2012. *Heimat, Region, and Empire: Spatial Identities under National Socialism*. Houndmills: Palgrave Macmillan.

Ther, Philipp. 2012. "Ethnic Cleansing." In *The Oxford Handbook of Postwar European History*, edited by Dan Stone, 141–62. Oxford: Oxford University Press.

Till, Karen E. 2004. "Political Landscapes." In *A Companion to Cultural Geography*, edited by James S. Duncan, Nuala C. Johnson, and Richard H. Schein, 347–64. Oxford: Blackwell.

Tuan, Yi-Fu. 1977. *Space and Place: The Perspective of Experience*. Minneapolis: University of Minnesota Press.

Tyner, James A. 2008. *The Killing of Cambodia: Geography, Genocide, and the Unmaking of Space*. Aldershot: Ashgate.

Weber, Suzanne Weiner. 2008. "The Forest as a Liminal Space: A Transformation of Culture and Norms during the Holocaust." *Holocaust Studies* 14 (1): 35–60.

Wolfe, Patrick. 2001. "Land, Labor, and Difference: Elementary Structures of Race." *American Historical Review* 106 (3): 866–905.

Wood, William B. 2001. "Geographic Aspects of Genocide: A Comparison of Bosnia and Rwanda." *Transactions of the Institute of British Geographers* 26 (1): 57–75.

PART I

Third Reich Geographies

SECTION 1

Biopolitics, Geopolitics, and *Lebensraum*

3

In Service of Empire: Geographers at Berlin's University between Colonial Studies and *Ostforschung* (Eastern Research)

JÜRGEN ZIMMERER

> Whoever perceives the increasing importance of German trade and German shipping along the coasts [= China] and gives consideration to how much both still have the ability to grow painfully touches upon the small political importance that Prussia has on the lands of Eastern Asia. . . . The acquisition of such a tremendously convenient sea harbor [on the island of Chusan] would be of great importance at this critical moment; the immanent opening of the large coal mines and the long-term construction of railways in China means a certain considerable upswing in international relations, especially regarding the initiation of the development of mercantilist interests, the size of which is yet to be determined. . . . I have hazarded to approach your Excellency as a stranger; I beg of you . . . to apply to Eulenburg for further references to whose former mission to Eastern Asia I was attached as Geologist. I am presently carrying out scientific research in China that has been ongoing since Eulenburg's mission ended. Should your Excellency command further reports from me, I will eagerly make myself available. I would count myself fortunate if I could use my small powers to make myself useful to my Fatherland.
>
> FERDINAND VON RICHTHOFEN (1869), in Gründer (1999)

With these words, the still largely unknown explorer and geographer Ferdinand von Richthofen introduced himself to the chancellor of the then North German Confederation, Otto von Bismarck, in 1869.[1] Fifteen years would pass before the newly unified German Empire acquired its first colony and another eighteen before it actually established a stronghold in East Asia with the occupation of Kiautschou. By this time, Richthofen had long been one of the most influential figures in German geography. As a professor, he was in fact appointed head of the Department of Geography at Friedrich Wilhelm University in Berlin, one of the most prestigious teaching and research institutes of geography in Germany.[2] Richthofen's career and the development of geographic research at Friedrich Wilhelm University are symptomatic of the dramatic rise in importance geography experienced as an academic discipline in the nineteenth century. In 1871, there were only two geography professorships in the newly established Reich. Geography was not even an independent school subject, but was mostly instructed as a supplementary discipline to other subjects. Three years later, the Prussian government decided to establish chairs for geography at all universities; within a decade, geography was taught in all schools (Schulte-Althoff 1971, 14, 120).

This rise in popularity was closely connected to broader political developments in Germany, especially the growing interest and enthusiasm for colonial expansion that eventually resulted in the formal announcement throughout 1884–85 that German Southwest Africa, German East Africa, Cameroon, Togo, and some islands in the Pacific would serve as German protectorates.[3] Geographers used this new interest in faraway lands in order to secure the establishment of their subject as an academic discipline (see Schulte-Althoff 1971; Osterhammel 1994). A downright symbiotic relationship between the subject's acceptance and the growing colonial enthusiasm in Germany[4] offered geographers the opportunity to demonstrate the discipline's practical value, resulting in a growth of social relevance, the creation of permanent university posts, and easier access to research grants. Although some geographers were critical of colonialism, the large majority of them shared, together with other academics and possibly most of the public opinion, a strong support for Germany's newly acquired prestige in international politics obtained via the realization of an overseas empire.[5] Additionally, they used Germany's growing foreign trade to thrust their subject out of the shadow of historical studies and to ensure its independent institutionalization. In the aftermath of colonial acquisitions, these institutional changes were formalized. Geographers thus became members of the Colonial Council, conducted political consultations, trained colonial personnel, and provided much of the necessary funding for practical colonial work.

The history of this codependency is not only of interest for the identity of geography as a discipline but also for the study of German colonialism and, more generally, European imperialism. On the one hand, it shows the wide popularity of colonial fantasies among educated citizens and, therefore, how personal interests and national inferiority complexes can be combined into collective feelings and a search for the nation's "place in the sun" (Bülow in Hildebrand 1995, 193). On the other, it provides evidence of the presence of a rational and scientific core in the reproduction of the colonial euphoria marking those years. The colonial undertakings were not just the play area of adventure-seeking rogues, failing aristocrats, and disowned sons of the bourgeoisie searching for wealth, power, adventure, and sexual gratification or of unscrupulous trading companies with their greed for economic profits; it was also a place for sober operating magistrates, explorers, and researchers who were seen as the epitome of European rationality.[6] As the field of postcolonial studies has widely demonstrated, the colonies have always served as research topics and exercise fields for new, rational, and scientific methods of exploitation and "development."

The colonial project of the Wilhelmine era, as it evolved most distinctly in German Southwest Africa, was an attempt to implement a modern utopian state and economic concept, in which the "undesirable developments" of modern society were to be avoided; the administration in fact strove to create a colonial state that did not have to shy away from a comparison with the German homeland. Development and expansion, order and efficiency: these were the basic principles of government to be followed in those new overseas endeavors. Characteristically then, there was heavy emphasis on bureaucratic rule and supposedly "scientific method" in "indigenous policy." The goal was the construction of an efficient economic system and a racially organized society where the local population provided the labor force through which the economic development of the colony was accelerated and the extraction of raw materials warranted.[7] However, for this to occur, key information about land and people, traffic routes, raw materials, settlement possibilities, and detailed knowledge concerning "useful" areas worth considering for European cultivation were necessary. The colonial administrative machine needed universities to act as training centers and academics to collect and elaborate crucially strategic data.

Geography was in that respect the colonial discipline par excellence; this does not mean that all its research was geared toward the exploration or development of overseas territories, but that, out of all sciences, geography provided the closest links to the projects of European expansionism. Like medicine (Eckart 1997) and engineering (Van Laak 2001), geographers trained specialized staff for service overseas; like eugenics (Grosse 2000), they delivered ideological and (pseudo)scientific justifications for colonial exploitation and the racial hierarchy of the political and legal system. Through categories like "Colonial Population," "Colonial Economic Production," "Colonial Settlement and Infrastructure," "Colonial Typology," "Colonial States," and "Comparative Colonial Empires," political geography was busy with racial research, geo- and tropical medicine, commodity exchange, colonial production and economic exchange processes between metropolis and the colony, and with questions concerning cultural intermixing, colonial settlement, and local development (Kost 1988). From there it took up the insights of other disciplines and transformed them into knowledge valuable to the administrations at home and in the colonies.

Arguably, German historians have dealt with this complex problem only in a cursory manner.[8] This neglect becomes noticeable when contrasted with the vast literature related to Germany's second attempt at building an empire, the Nazi policies of *Lebensraum* in the "East." In those debates, the involvement of academics in conquest and occupation is one of the most intensely analyzed topics of the last decades. First the doctors, hospitals, and "population experts" actively involved in euthanasia programs entered into view, but soon the occupation policies in Eastern Europe and the murder of the Eastern European Jews followed. Thus it became clear to all that scientists in disciplines as diverse as eugenics, medicine, biology, the social sciences, geography, economics, engineering, law, and even the arts and humanities supported the planned reordering of Europe by the National Socialists. Many academics in fact willingly provided the legitimization for racial and expansion politics, developed some of the key related concepts, and functioned as reviewers of racial hygienic measures or as experts for the military and civil administrations.

Although these studies closely examine the contribution of academics in occupation and war, they mostly tend to overlook their middle- and long-term historical context. While historical research has exposed the relationship between academics and other "experts" and Nazi extermination policy, the same research has done little to bring to light the academic traditions belonging to scholars providing "solutions" to Hitler's regime, the exception perhaps being racial hygiene.[9] This is particularly true for the relationship between colonial science and colonialism, the playing field by definition for "population politics" before the outbreak of World War II.[10] This was also important for geography, since the concept of space—especially in relation to race—was key to both European colonialism and Nazi imperial strategies. European colonialism and, later, Nazi politics of occupation in the East, were operating according to logic based on macrospatial economies (*Großraumwirtschaften*), trademarked by their endeavor to tap into the resources of extended dependent territories (see Zimmerer 2008). For this reason alone it is key to analyze the relationship between academic geography and German politics diachronically, from the foundation of the colonial empire to the eastward expansion under the National Socialists. There are, naturally, other connections aside from this one in the field of geography, for example, geopolitics, which blossomed during the Third Reich,[11] or military geography research,[12] but they will not be treated here. This chapter instead attempts to historically trace the manifold connections between universities, politics, and the public based on the research developed by geographers, which contributed in an important way to the rise of colonial enthusiasm long before the founding of the Third Reich.

Armchair Explorers

To understand the development of geography in the nineteenth century, it is necessary to look above and beyond universities into the enhanced interest in foreign lands, the production of knowledge about them, the biographies of the men who obtained it, and, perhaps most importantly, the role played by those institutions created to organize the engagement with foreign continents: the geographical societies.

In a unique way, these societies combined the different groups of researchers responsible for gathering geographical data, developing regional spaces, and organizing and supporting the adventurers, explorers, scholars, bourgeois and aristocratic dignitaries, and "armchair explorers."[13] Further, they took a key position in formulating peculiar colonial fantasies in Germany.[14] During a time when the boundaries between academic researchers and explorers were still liquid, and geography as a discipline was still in the process of being developed, geographical societies formed sanctuaries for geographic research. They also created a public climate in which the rise of geography was made possible.

The first geographical society was founded in London in 1788 under the title African Association for Promoting the Discovery of the Interior Parts of Africa. In 1830, this title developed into the Royal Geographical Society of London. On the continent, the geographical societies of Paris (Societé de Géographie) and Berlin (Gesellschaft für Erdkunde zu Berlin) were established in 1821 and 1828 respectively. Other geographical societies soon followed, such as those in Frankfurt am Main (1836), Bombay (1838), Mexico City (1839), Petersburg (1845), Darmstadt (1845), Delft (1851), New York (1852), Vienna (1856), Buenos Aires (1856), Geneva (1858), Leipzig (1861), Dresden (1863), Munich (1869), and Bremen (1870). In their early days, these geographical societies were primarily concerned with a relatively small circle of explorers and expeditions; there was little interaction with the public sphere and with related academic branches. Academics, in fact, were hardly the only explorers; soldiers, diplomats, missionaries, and adventurers all published reports on foreign lands. Since driven by completely different principles and purposes, the writings of these groups were not usable for scientific purposes. Only a shift toward professional training in this area transformed later travels into true geographical expeditions. Quite a few prominent university professors, for example the aforementioned Richthofen, began their careers as explorers (Schulte-Althoff 1971, 17–19).

A general problem for expeditions of all kinds was the costs since very little financial support was available in the German Confederation. Consequently,

German explorers increasingly signed on to the expeditions of other countries. While this helped international exchange, many considered it a thorn in the flesh. This was a critical time in the emergence of German nationalism and, especially in light of the absence of a German Empire, achievements in the area of "discovery" served as sources of national pride in a colonial stage where Germany otherwise considered itself weak. Scientific achievements thus became a tool to be involved in the battle for the international partition of the planet. In relation to Africa, geographer Adolf Bastian put this credo into words in 1875:

> Guided by the conviction that the long row of African discoveries, with which German (!) travellers have been actively involved in the most excellent way and with the upmost enthusiasm, were gradually beginning to approach their end, Germany cannot neglect to take part in the competition so that the history of geography can, in future, be attributed to the German *Volk*, when the names of the discoverers are written into its annals. (Bastian 1875, 6, in Schulte-Althoff 1971, 55)

Many critics, however, believed that expressing this ideological credo was not enough and that Germans should officially become active in creating a long-term colonial empire. Behind this new proposed strategy were not only nationalism and colonial romance but also opportunities for academic disciplines, like geography, to prove their practical use, gain heightened public respect, be awarded research money, and obtain institutional recognition via the establishment of permanent departments in schools and universities.

A good example of the interaction among international academic achievements, attempts to further German geographical research, and strategies to secure the discipline's institutional position can be found in the life and work of cartographer August Petermann. At age twenty-three, he went to London and became a member of the Royal Geographic Society, eventually taking the role of secretary (Schulte-Althoff 1971, 23–38). From London, he networked with leading German explorers and linked them with international expeditions. In 1854, he returned to Germany and founded the famous *Petermanns Geographische Mitteilungen*. There, as an initiator and tireless seeker of funding to support geographical research, he gave a decisive contribution to the possibility for German expeditions to take part in the global race for scientific exploration.

After the establishment of its colonial empire, Germany was in fact expected to further participate in international discoveries. Tellingly, the 1873 founding statement of the Afrikanischen Gesellschaft in Deutschland

(African Society of Germany) formulated the necessity for geography to undertake such an endeavor:

> The political validity of a nation is measured by the amount of obligations incumbent upon it to find solutions to the challenges of civilization. Since Germany once again resumed its rightful seat on the council of nations, it must, more than ever before, dedicate itself to the cultivation of science. It behooves Germany to cultivate the management of geographic ventures where new areas of excellence will win notice, to be at the forefront because such determinably decided the course. (*Petermanns Geographische Mitteilungen* 1873, 72, in Schulte-Althoff 1971, 50–51)

Thus the road from a dedication to geographical discovery to a dedication to the colonial empire extolled by many in 1848 was not a long one. That this positioning is detailed in the founding statement of a geographical society demonstrates the close relationship and the mutual benefit between the general public enthusiasm for expeditions, geographic research, and the institutionalization of geography. The development of academic and nonacademic interest in geography continued in parallel, as both contributed to debates associated with the founding of empire. Between 1828 and 1871, only seven geographic societies existed in Germany. By 1884, nine more were present; at the outbreak of the World War I, an additional five were active. The first German Geographers Day was held in 1881 and was celebrated through visits by state representatives, explorers, generals, and members of the educated middle class. Along with curious onlookers and adventure seekers, and animated by the reports of Stanley's Africa crossing, also many economists expressed fantasies about German expanding colonialism; many economists were in fact members of geographical societies since interested in practical knowledge concerning economic structures, traffic connections, and information regarding the presence of raw materials in foreign countries (*Petermanns Geographische Mitteilungen* 1873, 72, in Schulte-Althoff 1971, 47–48).

This general growth in interest and public visibility of geographical research finally gained financial support, as clearly stated in an 1873 article that appeared in *Correspondenzblatt der Afrikanischen Gesellschaft* (in Schulte-Althoff 1971, 60):

> Such things are capable of great efforts, being science on the one hand and, on the other, useful for trade and industry; geography stands on a line of exchange between the theoretical and practical life. The ways established by pioneering geographers lead, sooner or later, to new exchange markets followed by established mercantilism upon which new sources of income based on ac-

tive exchange develop. Strong global trade grows out of the prudent utilization of resources offered by geography.

Opportunities the newly established empire offered were used to great advantage by academic geography. As noted above, in the year the empire was founded there were only two chairs for geography, one in Berlin and one in Göttingen, and the subject was not taught independently in schools and universities. In 1871, however, a chair for geography was established in Leipzig and, two years later, also in Munich, Halle, Strasburg, and Dresden. In 1874, the Prussian government further attempted to establish chairs for geography in Prussian universities, a measure not immediately put into practice due to a significant lack of qualified personnel (in Schulte-Althoff 1971, 41). In high school syllabi of 1882 and 1892, geography was noted as an independent subject and was thus "freed" from its relationship with history (ibid., 120).

When the benefits of the German colonial empire began being perceived as increasingly important for the country's economic and political global role after 1884, the specialized knowledge geographers offered became even more requested. The colonies needed to be surveyed, their economic potential analyzed, and the colonial management personnel educated. Geography, because of its capacity to respond to these needs by providing key information about foreign lands and because of the enthusiasm in the general public for colonial adventures, managed to establish itself as an independent scientific discipline. For this reason, academic geographers were, in overwhelming numbers, supporters of German colonialism.

In the Service of Colonialism

Ferdinand von Richthofen was—as the opening quote shows—from the very beginning among the most vehement supporters of the establishment of a German colonial empire. On the grounds of his position as a professor at Friedrich Wilhelm University and his reputation as an explorer, his voice carried great weight within the discipline. He accordingly endeavored to create a close connection with the colonial movement and made the Geographical Institute in Berlin the most influential in Germany on questions concerning colonialism.

Born in the upper Silesian Karlsruhe, the young Richthofen studied geology in Berlin between 1852 and 1856 under, among others, Carl Ritter. After conducting research in Transylvania and Tirol he took part in the Prussian East Asia expedition from 1860 to 1862. Between 1862 and 1868, he examined metal production in the gold mines of California. With the mandate of the

European-American Chamber of Commerce, between the years of 1868 and 1872, he undertook a total of seven journeys through China. The results of these travels were published in five volumes under the title *China: Ergebnisse eigener Reisen und darauf gegründeter Studien* (1877–1912), as well as in the *Atlas von China: Orographische und geologische Karten* (1883). These volumes aimed not only to create a "new and differentiated picture of China" (Zögner 1997, 72; Engelmann 1988, 7–9), but also to connect scientific and practical work. In his cartographic work Richthofen was able to correct a series of mistakes in existing Chinese and Jesuit maps and provide valuable pointers for the development of that faraway land. For instance, he sketched coal deposits potentially accessible for trade and made suggestions for possible railways tracks by incorporating them into his maps (Zögner 1997, 74–75). He hoped, as he wrote in his diary, that one day "the first train in China could be built there" (ibid., 75). He did not randomly choose China as his travel destination: driven by his "part Prussian-German, part . . . total imperialistic sense of mission" he argued that "the last remaining widely untraveled colossal empire outside of Europe" still had the chance to be profiled (Osterhammel 1987, 172).[15]

Richthofen was clear on the meaning of a German, or rather a Prussian, stronghold in China:

> As a free harbor in the hands of a power like Prussia, Tschusan would occupy an imposing position. The Harbor can be fortified with ease and a navy would control the traffic with northern China and Japan. As a place of trade, it would be of great importance. (1907, 4, in Engelmann 1988, 10)

Richthofen did not just allow his work to be classified as colonial romance; rather, as shown above, during his time in China he wrote to the chancellor of the North German Confederation, Otto von Bismarck, to make him aware of the use of such a stronghold. While von Bismarck did not immediately act according to Richthofen's suggestions, August Petermann recognized the relevance of his analyses and, after their publication, was largely responsible for their popularity in Germany (Zögner 1997, 72–75).

A few years later, Richthofen's career began in Germany. In 1875, he became a professor in Bonn and, in 1883, was transferred to Leipzig to succeed Oscar Peschel (Zögner 1997, 11–12). Nonetheless, he was unsatisfied, finding the purely academic aspects of university life hard to bear. As he wrote to the then Prussian cultural minister, he felt himself too "limited by my narrow career and cannot follow my desire to make myself useful for the general interests of the fatherland as much as is possible for the geographers; I cannot be satisfied" (*Letter to Althoff* 1885, in Zögner 1997, 12–13). This situation would change after he was appointed as a professor at the Friedrich Wilhelm University in Berlin. This was a newly instituted chair of Physical Geography and was granted in addition to the original chair of Carl Ritter's held by Heinrich Kiepert (Zögner 1997, 13).

Richthofen's auspices for the realization of a German stronghold in China soon became reality; in 1897, the German Empire took Kiautschou into its possession. Richthofen was thus able to declare: "I have certain pleasure in the result, as it finally happily solved more than 25 years of endeavors" (*Letter to Hettner* 1897, in Engelmann 1988, 10). He enthusiastically celebrated the establishment of a German colony in China with two monographs, *Kiautschou: Seine Weltstellung und voraussichtliche Bedeutung* (1897) and *Schantung und seine Eingangspforte Kiautschou* (1898). Richthofen was also much engaged with lobbying and political advising, getting involved with the practical side of his profession. He was generally less concerned with the founding of a broader German colonial empire and more with the promotion of European influence in Africa; in 1876 he supported King Leopold of Belgium's decision to establish the Association Internationale Africaine, of which Richthofen was one of the founding members. He evidently overlooked the fact that this association was essentially an instrument of propaganda to protect Leopold's plans to build his own private colonial empire in the Congo.[16] A German section of the association was founded in 1876 in partnership with high-ranking aristocrats, militarist, and industrialists. In 1878 it was linked to the *Deutschen Gesellschaft zu Erforschung Äquatorialafrikas*. The German ambassador to Vienna, Prince Heinrich VII Reuß, was nominated its president and Ferdinand von Richthofen asked to act as his deputy (Schulte-Althoff 1971, 64–73). In the periods 1873–78 and 1888–1905, he also officiated as chairman of the *Gesellschaft für Erdkunde* in Berlin (Engelmann 1988, 16). Directly tied to German colonial politics, Richthofen was a member of the Colonial Council. Founded in 1892, it was established to provide advice for colonial administrations concerning concrete questions (Pogge von Strandmann 2002, 32–34).

Although Richthofen was on the cutting edge of geographic research and colonial activities through his engagements with his professional networks, his desire was to strengthen his own position at the Friedrich Wilhelm University and to establish his own institute. Expansion in geographic research had already become necessary for the "strengthening of colonial interests and the unparalleled increase of world traffic." In his opinion, it was,

> therefore, in the interest of the economy and the state, on one hand, to train strengths that are able to act in a strictly scientific way in the research of the individual parts of a large and branched off area and to spread the acquired

> knowledge through teaching. On the other hand, it is also in the state's interest to create a central location where geographical work and trained strengths can be performed under suitable and sufficient conditions, and where the materials can be collected according to different viewpoints so that they can be used at any time in case there is a need for information and processing. (Richthofen [1898] 1983, 157–58)

This "central location" would naturally be under Richthofen's direction and would, along with procuring information from foreign lands, be used to train scientists and travelers. Others would benefit as well, as the "central location" could be used to

> train those officers and state officials in the area of geography who, in their career in the service of the colonies, as consular representatives, in the navy, or in any other way, have problems fall upon them and for which the solution is an extensive knowledge of foreign lands and economic relations. (Richthofen [1898] 1983, 157–58)

Indeed, the training of specialized personnel for the colony had to count as the most important interaction between academic geography and active colonial administration, even though exact statistics on German colonial personnel are not available. Among Richthofen's students who entered into active colonial service, at least one attained prominence as a colonial propagandist, Paul Rohrbach. A natural theologist, he was recruited in Berlin by Richthofen, who prized him because of his practical experience as an explorer. Rohrbach originally planned a large expedition through East Asia and India as preparation for his *Habilitation* in overseas economic studies, a plan Richthofen expressly approved. His plans changed, however, and he went instead to German Southwest Africa as a settlement commissioner. This engagement ended on a negative note for Rohrbach, as he disagreed with the colonial administration; nonetheless, this did not prevent him from becoming one of the principal promoters of colonial politics in the region and an expert on colonial politics. Although he returned from Windhoek with an almost completed thesis, his academic plans were shattered, as Richthofen had died in the meantime and his successor, Albrecht Penck, denied him the *Habilitation*, because he saw the topic as belonging to political economy and not to geography. The deeper reason behind this decision was that Penck did not consider Rohrbach suitable for academic teaching; he was convinced that travel alone did not make a geographer (Mogk 1972, 63–66).[17] Regardless, Rohrbach became one of the most read publicists of the German Empire and was influential in the popularity of some economic arguments used to justify the colonies.[18]

When Richthofen died in 1905, he was replaced by Albrecht Penck. Like

Richthofen, he possessed far-reaching popularity in the field, after having been a professor in Vienna, and having made a name for himself in ice age research and geomorphology. Penck also received a chair at the Friedrich Wilhelm University in 1917–18.[19] What is more, Penck continued the tradition of maintaining a close connection between practical colonial politics and colonial research. In 1904, he collaborated with the Commission for Regional Studies on the Exploration of Protectorates founded by the Colonial Council (Schulte-Althoff 1971, 127). After the German colonial administration in Berlin recognized the use of geographical knowledge for the practice of administration in the colonies, this commission was established to specifically coordinate and professionally support scientific research activities (Schindlbeck 2001, 133–35). After the dissolution of the Colonial Council, the commission was renamed Regional Studies Commission of the Imperial Colonial Office in 1907. The members of the commission were expected to prepare reports addressing colonial questions and lead scientific expeditions. Penck himself was involved in the scientific conception of an expedition to the South Seas, aimed at the exploration of the Bismarck Archipelago to be carried out in 1908 by the Sapper-Friederici Expedition.[20] Even if World War I rendered ongoing and future expeditions problematic, the commission remained in place until 1919 (Schulte-Althoff 1971, 127).

Colonial Revisionism

For Penck and most of his colleagues, the topic of "German colonies" was not over, despite the fact that Germany had lost the war and its colonies under the Treaty of Versailles.[21] This was demonstrated, for example, by the continued existence of the Berlin chair of "Colonial Geography," originally established in 1911 and the first of its kind in Germany. Founded by the explorer, publisher, and patron Hans Meyer, it symbolized the direct link between geography and colonialism. The chair was responsible for managing the section of Colonial Geography at the Geographic Institute (Troll 1969, 10–11).[22] Fritz Jäger, also a former student of Richthofen, was called to cover that chair. He had earlier completed his doctorate under Alfred Hettner in Heidelberg before undertaking countless explorations of East Africa largely financed by the abovementioned colonial commission (for Jäger's biography see Troll 1969). Jäger placed himself very consciously in the service of politics in his inaugural lecture in Berlin:

> In geography and especially in colonial geography, practical use can immediately be seen. . . . It is the mission of geography to sketch the total picture of

> the land. The correct evaluation of the nature of the land and its inhabitants is, however, the most important foundation for correct estimates. The primary purpose of the colonies is to ensure economic utilization and exploitation for the good of the motherland. If no economic use is brought forth, then the colonies are a luxury that cannot be justified economically. Thus colonial geography provides an important foundation for a rational colonial economy. (Jäger 1911, 400–405)

Because the German colonial empire would only last another three years, the actual political influence of the professorship was limited. The chair of Colonial Geography played a role in the colonial revisionism of the Weimar Republic.[23] As a consequence of the economic and financial crisis the Weimer Republic endured, Hans Meyer's assets collapsed and the professorship was then financed by the Prussian state. At the same time, Jäger's tasks concerning the scientific care of ethnic Germans were expanded (Troll 1969, 18). In 1928, Jäger would take the chair of geography at the University of Basel, where he would remain until 1947, when he would be dismissed from his post due to his involvement in the Third Reich and his work for National Socialism in Switzerland (for more detail see Troll 1969, 26–31).[24] After his departure, the chair of Colonial Geography was assigned in 1930 to Carl Troll. Troll, who had completed his doctorate and postdoctoral work in Munich, was an ardent colonial revisionist. Colonial revisionism became rather popular among geographers after Hitler's rise to power in 1933, as shown by the work of Jäger and Penck, as well as that of their colleagues Heinrich Schmitthenner, Erich Obst, Karl H. Dietzel, and Franz Thorbecke.[25]

Certainly, speeches given and literature relevant to the German Geography Day of 1931 had emphasized revisionist interpretations, as had Thorbecke's and Schmitthenner's forceful belief in the importance of the colonies for geographical research. Troll pointedly referred to this relationship also at the 1936 Geography Day:

> How manifold the colonial questions are entwined with geographical science and . . . they will also form a crucial point for future National Socialist colonial politics. . . . What we will strive for with the colonial claim concerning our privileged position is a lawfully acquired place in the sun, a field of activity in the wider overseas world where a genial guide had, in a timely manner, secured our place in foreign politics. (Troll 1937, 122, in Kost 1988, 195–97)

At the same time, Troll saw a fundamental change in the role played by colonial geographic science:

> Unlike the middle of the previous century, you are no longer standing across from the so-called "dark" part of the earth, rather we see instead a rapidly

> transforming African world. . . . It is no longer the duty of European science to send out research expeditions to lesser known areas, rather, it is their responsibility to intervene in colonization and its problems with scientific equipment and scientific views. (Troll 1937, 135–36, in Kost 1988, 195–97)

The influence geographers held in political circles and how those geographers coveted that influence is shown by the research initiatives undertaken by them and particularly by Troll's eleven-month journey to Africa in 1933–34. This expedition is of particular note as the decision to finance it was made at a time when Germany was still suffering from the effects of the world economic crisis of 1929. The purpose of the expedition was to

> gain scientific and colonial knowledge of Africa on the basis of theories imperative for security and legal issues. This is especially important for the scientific study of living conditions and living opportunities in the African mountain lands on a biological and geographical basis. The high regions were chosen because, regarding the question of the development of the tropics, they stood in the foreground of the interests for whites. The mountainous eastern half of Africa, where the highlands laterally cross the tropic zone on both sides of the axis, will undergo particular study. (Troll 1935, 8–10)

The Emergency Organization of German Science, the Cultural Division and the Colonial Division of the Foreign Office, the Prussian Academy of Science, the German Colonial Society, the Charlottenburg Division of the German Colonial Society, and the Bavarian Academy of Science all supported the expedition into the highlands of former German East Africa (Troll 1935, 8).

To express his gratitude for the economic support received and being convinced that this expedition would increase his political influence due to the reputation and position the sponsors held, Troll published his travel report and dedicated it to his funders and "the German compatriots in East Africa." The following introductory lines opened the report:

> Since the unfortunate result of the World War that trimmed our German ground on all sides, and the consequential migration out of the dismembered parts of the German empire, the question of German living space has become the strongest question of our nation. Thus, it is not a coincidence that it stands in line with the programs and aims of our new empire.
>
> A survey ascertaining where an expansion or deepening of the German living space is possible reveals there to be two completely different possible regions: the German east and overseas. Overseas there are again two continents that especially demand our interest: South America and Africa. South America because, in contrast to North America, national traditions are present there that are foreign to Germans. Therefore, Germany is in a position to uphold and keep pure its own ethnic idiosyncrasies and culture as compared

> to North America, as proven by German settlement colonies in southern Chile and southern Brazil. Africa deserves our attention because Germany had acquired a historical right to its own colonial property there and we hope, with German strength and German people, to be able to once again work on this German land; equally, because tropical raw materials were produced in the German colonies that are very much needed by the [current] German market. (Troll 1935, 7)

As it is generally known, Hitler would decide against the reestablishment of German colonies in Africa in favor of expansion in Eastern Europe. Troll, on the other hand, belonged to the circle of "conservative" colonial revisionists that was becoming increasingly influential. As such, he hoped that after World War II, Germany would again acquire colonies in Africa:

> By raising the question of German colonialism, the colonial sciences are pushed into the front row of public interest. Fortunately, through their years of quiet work, they had created perfect conditions for the tasks with which they were being faced. This has nothing to do with the different disciplines that are now seemingly attracted to Africa. The new order of the African colonial world is simply faced with challenges that can only be solved by a deeper understanding of the area and its problems. (Troll 1941, 1)

By this time, Troll had been teaching at the University of Bonn since 1938 and conferred emeritus status. He later became rector in 1960–61.[26] As demonstrated by his African expedition, like Richthofen and Penck before him, Troll did not limit himself to academic research and teaching; he too was happy to act as a political advisor, as an expert, as committee and specialized group chairman for the colonial scientific division of the Reich Research Council, working closely with the colonial political office of the NSDAP (Nationalsozialistische Deutsche Arbeiterpartei, or Nazi Party) which, between 1937 and 1942, made a half a million Reichmarks available for colonial research (Kost 1988, 197n7).

Eastern Expansion

Troll's suggestions for a new spatial organization of entire continents were focused on Africa, something that made them politically weak; concurrently, other geographers worked on planning revisions of the territorial decisions determined in Europe by the Treaty of Versailles, specifically on the changes of the economic and ethnic structure of Eastern Europe. There were two main points around which major geographical analysis pivoted, the first being the question of borders drawn in 1919 and the second the question of

Germans living abroad. As mentioned above, the Berlin chair for Colonial Geography was already concerned with the care of Germans living abroad during the time in which it was funded by the Prussian state. Geographers and policy makers alike were particularly concerned with Germans living in South America and with those in territories in Eastern Europe who, due to the newly established states created by the Treaty of Versailles, such as Poland, were essentially turned into foreign residents. This was one of the foundational problems with Versailles politics; the angst it gave to German society soon turned into an obsessive fixation on the "problem" of borders. In short, Germany had discovered, to its own disadvantage, what indigenous peoples in the colonies had earlier discovered and what Germany had hitherto taken as a self-evident right: to the victor goes the spoils. In this case, in the perceptions of the German larger public and politicians, the victors "rebordered" Europe without adequately considering the ethnic composition of the "indigenous" population.

This angst resulted in Germany's social and political reprobation against the dictated Versailles peace, against the surrender of their territories, and the loss of their colonies. Like many others, Albrecht Penck, had avidly supported the German policy of aggression during World War I. He viewed himself as a spiritual participant to the war and praised the "curative" effect of the battle on the reformation of society through patriotism. In a dualistic, rather simplistic ideology, German troops were often presented as carriers of civilization and human rights, while the *Entente* was responsible for an aggressive war of conquest (Kost 1988, 342).[27] Supporting this view, Penck felt a great disappointment and bitterness for the defeat; he was particularly shocked by the use of African occupational forces by France, as he understood this as a deliberate and humiliating crime against their "white masters" (ibid., 220).[28]

According to Penck, the colonial question was situated in the frame of a larger question regarding the future of the German people. He wanted to "agitate the world and preach to a global conscience that the German colonial problem was only a part of a larger problem: What is the future of our race, what is the future of Europe?" (Penck 1937, 263, in Kost 1988, 219). For him, the answer to this question had racial and ethnic undertones. The white European "race" needed to be protected and supported. In Africa, where, "hitherto at least, the white had not really worked, had rather ruled, administered, or commanded," it was necessary to establish an uninterrupted "bloodstream of the white race." It did "not have to exactly be sizable . . . but if it was abandoned or made smaller, the rule of whites would cease" (Penck 1937, 43–48, in Kost 1988, 224). In Penck's view, this problem was not limited to Africa; it was expounded in Europe as well. According to his creed, Germans living there had been especially wronged and millions now found themselves living as foreigners outside of the new borders of Germany, their homeland. Indeed, in line with the propagandistic war launched in Germany against the "dictate of Versailles," and by looking for evidence of persecution of Germans abroad and for the legitimization of German territorial claims in Eastern Europe, Penck developed his greatest sense of duty.

Thus was the importance of ethnic division and *Lebensraum* developed. Penck focused his analysis particularly on "physical anthropogeography" (Penck 1925)[29] and on "civilization and cultural research." His primary concern was to determine the "ecological load capacity" of the earth by rating crop yields, something that could justify the demand for renewed colonial possessions (ibid.). First and foremost, he supported a form of Social Darwinism stating that nations (must) fight for land and resources. Based on his interpretation of the current state of the planet, Penck claimed that:

> We [are] aware of the narrowness of our own living space. At home it is tight and narrow, and the earth is not large enough. There are not only boundaries of ethnicities and states, but also boundaries of humankind. We may have learned how to rise into the air and to dive into the world's seas and to penetrate into shafts in the depths of the earth's crust. But no invention can physically remove us from the need of our substance to be bound to this surface of the earth. Even if there are no longer inaccessible regions in this modern world, we can only lightly shift the boundaries of living space. The constantly reproducing human is sitting as if he were in a fortress, out of which he cannot come. (Penck 1934, 23–24)

The job of the geographer had shifted, and now included an imperative to educate others about the limited space of the world and the needs of humankind:

> Every national geographer cannot stop that which concerns his own people; rather, he must learn the entirety of both earth and humanity, both effectual powers and dormant powers, and about the war between humans on earth for both food and space. (Penck 1934, 27)

As far as Germany was concerned, this resulted in a renewed determination to quickly reclaim the territories in Eastern and Southern Europe lost after World War I. As a professor, the instrument that stood at Penck's disposal was science, from which he envisaged a new task in the name of the German nation:

> We regard science as the common property of humanity; in a system of supranational sciences, national sciences have no space. But they play a large role

> in the life of the people. National geography is precisely what a nation needs for itself. (Penck 1934, 3)

For Penck, academic geography was not about the free exchange of different opinions; rather, it was about influencing political opinions and providing legitimacy for political actions. Penck believed that the pressure for objectivity in scientific research should never be misunderstood in such a way that "the windows of the own fatherland" are broken, quoting a slogan coined by another geographer, Wilhelm Volz (in Haar 2000, 37; see also Wolf, this volume, chapter 4).

In Leipzig, in 1923, Penck founded the Central Office for Intereuropean Affairs that soon developed into a think tank for revisionist ethnic and cultural research. Its staff acted as intermediaries between academics and politics for members of the Foreign Office and the Ministry of the Interior. The overall objective was to create a central institution where theories of "ethnic and cultural space" could be "established as a paradigm of ethnic science in the time between the wars" (Haar 2000, 31). Consequently, in 1925, this think tank was renamed the German Central Office for Volk and Cultural Research and in 1926 became a foundation (ibid., 26–37) with Penck as its first president.

Despite being problematic if expressed in international circles, the conceptual basis these researchers followed was determined by the assumption that Germany was experiencing extreme "population pressure" since its population was constantly on the rise. In contrast to nineteenth-century imperialism, policy makers, under strong influence from academics like Penck, now targeted prospective German colonial settlements in Eastern Europe. Penck's influence here cannot be overestimated. Aside from his contribution regarding the ecological and ethnic load-bearing capacity of the earth, he also provided a definition for the term "ethnic and cultural space" in a collection of essays that served as the central manifest of the early ethnic *Volkstumsforschung* under the umbrella rule of the "Deutschen Schutzbundes" (Penck 1926).

Penck understood the concept of "ethnic space" to be the areas settled by Germans, where German was spoken, and where the results of German industry played a key role in society. Ethnic space was connected to cultural space because "wherever Germans live socially and use the earth's surface culture appears, whether it results in the development of an ethnic space or not" (Penck 1926, 69, in Haar 2000, 46). Using this terminology, only some of Germany's "ethnic space" was found within Germany's post-Versailles borders. As culture bearers, Germans could thus claim far-reaching boundaries for their empire (for a discussion of this concept see Haar 2000, 46–47).

Penck clearly expressed this point in 1934, after his retirement, when the Leipzig Foundation was disbanded and replaced by other institutions:

> Germany can clearly be identified by its extraordinary cultural landscape. The schooled eye can easily recognize the *German cultural space* . . . by its affectionate cultivation, by the way field, forest, and meadow are nestled in its topography. . . . Like jewels, German villages are disseminated like seeds on an even terrain; they are nestled against the valleys of the mountains and grow in the middle of the forests. Clear paths go throughout the land. On German ground, raw nature is tamed by humanity; mankind is the true master of the land. . . . We recognize three things: Germany with its German cultural space in German Central Europe, German ethnic space, and the German Empire whose boundaries can only be breached at two places. Within this landscape we can see where we are in Germany; with our language we can hear where we are on German land; legislation and regulations instruct us as to where we are within the empire. Germany is the firm pole in which appearances escape. (Penck 1934, 5–8)

This language, where political legitimation is sought through the domination of the forces of nature and geographical cultivation, belongs to the most classical colonial discourse. For example, this is one of the key themes of one of Germany's most important colonial novels, Gustav Frenssen's *Peter Moors Fahrt nach Südwest*, where one of the protagonists justified the 1904 genocide against the Herero in German Southwest Africa:

> These blacks have earned death from God and people, not because they murdered two hundred farmers and rebelled against us, but rather because they did not build any houses or dig any wells. . . . God has let us prevail here, because we are more precious and more ambitious. This does not say much about this black race; we must take care to remain the best and most aware of all the world's races. The world belongs to the most able and the freshest. That is God's justice. (Frenssen 1906, 200)[30]

When viewed with hindsight, one can clearly recognize the effects of this rhetoric on Nazi politics of conquest that would follow, particularly when looking at eyewitness accounts from World War II. Heinrich Himmler's secretary, Hanns Johst, traveled through recently conquered Poland with his boss. While there, he described his own impressions and, very likely, those of Himmler himself, as follows:

> The Poles are not a state building people. They are missing the simplest requirements. I traveled across the land at the side of the Reichführer's SS. A land with so little sense to found settlements that they do not even have the beginnings of villages does not have a claim to any kind of independent

> position of power in the European space. It is a colonial land. (Johst 1940, 94, in Burleigh 2000, 515)

Adolf Hitler also spoke of the "primitive world" that one "merely had to see," in order to know, "that nothing happens there if one does not measure out work for people. The Slavic race is a slave-mass crying out for a master."[31]

It is uncertain exactly how much influence academic geography and its political connections influenced Nazi policy;[32] nonetheless, in terms of legitimizing revisionist thoughts and territorial expansion projections, they certainly played a significant role in preparing the way. This was exactly what successful academics like Penck perceived as their primary task, to gather support for policies that would lead Germany out of the depths of the 1918 defeat and would resolve the seemingly irrefutable demographic dilemma of the "*Volk* without space."

Conclusion

Penck was not new to this kind of instrumentalism of scientific work and the pandering to politics; his training as a geographer and his work under colonialism had given him precisely the tools necessary for this task, also thanks to the established reputation of the discipline during the decades preceding Hitler's revolution. Two particularly important turning points for the direct political involvement of the discipline may be identified: 1884, when a German colonial empire was internationally recognized and established; 1939, when, with the invasion of Poland and the wartime revision of the German borders eastward, new *Lebensraum* for the German people was conquered. In both cases, geographers acted in conjunction with the politicians and influenced their work, including policies of exploitation and plundering.

A glance at the relationship between geography and German colonialism shows the close interlocking relationship existing between the two. Geography, represented by experts like Ferdinand von Richthofen, only too willingly accomplished their contribution to the development of the German Empire and to the exploitation of the conquered territories. This points to rational aspects of colonialism that often threaten to vanish behind pictures of colonial romance and adventure. The application-oriented form of political advising was not limited to the time of colonial empire, from 1884 to 1918; rather, it was responsible for a considerable proportion of the spread of colonial support and remained a prominent voice in colonial revisionism after World War I. Academics such as Carl Troll furthered fantasies of German expansionism and colonization by providing them with a scientific rationale and a popular accessibility. Albrecht Penck, particularly in his work railing against the Treaty of Versailles, excelled in both of these roles in his central position at Friedrich Wilhelm University in Berlin; he pleaded for the reestablishment of the colonies and rendered a weighty contribution to ethnic and cultural research, one of the points of departure for later ethnic policies. Thus he was an important link, among others, between colonial revisionism and ethnic expansion to the East during World War II.

Acknowledgments

I am grateful to Elisabeth Hope Murray, Claudio Minca, and Paolo Giaccaria for translating the original paper from German and editing this version.

Notes

1. For a critical biography of Richthofen see Osterhammel (1987); Van Laak (2001).

2. The Department of Geography was established with the appointment of Richthofen; however, there was an even older Geography Unit harking back to Carl Ritter. In 1922 it was renamed Department for Colonial States and Historical Geography. See Balk (1926, 125–31); Zimmerer (2002b, 125–30). NB: "Department for Colonial States" could also read "Department for Sub-States."

3. As Susanne Zantop (1999) and others (Berman 1998; Friedrichsmeyer, Lennox, and Zantop 1998) have argued, colonial fantasies, interest in travelogues, and in the "discovery" of foreign lands and cultures have a much longer tradition in Germany, confirming that colonialism was not just limited to formal colonial rule. For a theoretical discussion of postcolonial theories on the German example, see Conrad (2002); Eckert and Wirz (2002).

4. For a more detailed discussion of the origin and the motives of the German colonial movement, see Gründer (1995, 25–62). On earlier demands for a German colonial empire, see Fenske (1991).

5. For a discussion of colonial expansion, empire, and such, see Osterhammel (2002).

6. Fabian (2000) clearly shows that this emphasis on rationality was a construction of the travelers, born out of the experience of their own corporeality and emotionality.

7. For why this utopia failed in colonial practice, see Zimmerer (2002a).

8. An exception was Jürgen Osterhammel (1994), who for years pleaded for a stronger consideration of geography.

9. There is comparatively wide research about this, see Burleigh (1994); Weingart, Kroll, and Bayertz (1988). For the international context, see Kühl (1997).

10. To date, only two studies are extensively concerned with the continuity between colonialism and National Socialism: Eckart (1996) and, for the institutional connections, see Wulf (1994).

11. For an introduction to geopolitics, see Diekmann, Krüger, and Schoeps (2000).

12. In 1936, an "Institute for the Teaching of General Defense" was established at the Friedrich Wilhelm University and led by geographer Oskar Ritter von Niedermayer who conducted lectures regarding independent geographic military research. See Jahr (2001).

13. I use this term in the style of Jürgen Osterhammel and Susanne Zantop, who speak of

"armchair geography," or rather "armchair conquest." Osterhammel means the compilation of geographical knowledge without direct fieldwork as experienced at its high point with the work of Carl Ritter.

14. For early controversies within geography, see Honold (2003). For an example of influence of geographic ideas on colonial ideology, see Noyes (2002). For more on imagined spaces and their political implications, see Schultz (2002).

15. Osterhammel clearly describes Richthofen's conviction of the superiority of his own civilization, which was evidenced during his journey by the fact that he carried European food and dinnerware with him to preserve European standards.

16. On the atrocities in the Congo see Nelson (1994). An overview of Belgian engagements in the Congo and its public legitimization through an unparalleled publicity campaign is given by Hochschild (2000).

17. Rohrbach also held significant geopolitical ambitions for the Near East, see Schaller (2003).

18. Out of the multitude of his published works, only those most important for the colonial question are mentioned here: Rohrbach 1907a, 1907b, 1909, 1934.

19. For the biography of Penck see the—for the time after 1918 only very cursory—overview by Engelmann (1988, 23–37).

20. The goal of the new investigation was, among other things, to prove if the soil was suitable for a plantation culture by the examination of growths delivered to Kautschuk (Schindlbeck 2001, 150–51).

21. For the regulation of colonial questions in the Treaty of Versailles see Zimmerer (2001).

22. A further professorship was established in Leipzig in 1915 that took notice of Hans Meyers.

23. Preoccupation due to the loss of colonial empire was not limited to academic geography. Of particular import at the time was the education of youth; German colonial society, supported by geography school teachers, held great influence over what was included in German textbooks and demanded that special emphasis be placed on former German colonies in text and maps (see Brogiato 1998, 405).

24. In a lengthy lawsuit he was able to nullify all relevant reproaches for which he could have been penalized and to secure half of his pension. He stood behind his national political attitude. In Berlin in 1963, he was awarded the golden Gustav-Nachtigal Medal, geography's highest distinction, in recognition of his research in Africa.

25. For a reference to relevant publications of those named, see Kost (1988, 213).

26. After World War II, Troll became one of the most internationally well-known geographers. Between 1960 and 1964, he was, among other things, the president of the International Geographical Union (IGU) as well as the representative for geography at the UNESCO Advisory Committee on Natural Resources Research between 1964 and 1969 and in the International Council of Scientific Unions, and held office as rector of the University of Bonn. He paid special interest to research on geographical problems in developing countries (see Lauer 1970, 12). Connections and continuities between colonial geography and this research in developing countries would be worthwhile investigating. With his retrospect on the history of the Third Reich, which already appeared in 1947, Troll became the principal witness for the salvation of their honor (Troll 1947).

27. On "spiritual mobilization" see Mommsen (1996); Verhey (2000).

28. For the problem of African troops see Koller (2001).

29. This is a reprint of an article appeared the year before. Wide reception evidently was only put into effect with the publication in a journal of geopolitics (see Kost 1988, 95).

30. Also see Brehl (2000). Regarding the connection between the genocide against the Herero and Nazi politics of conquest and extermination see Zimmerer (2003).

31. Hitler, September 17, 1941, "Adolf Hitler, Monologue," 63 [in *Hitler's Table Talk*, compiled by Martin Bormann]. The similarity between the word "*Urwelt*" (primitive world) and "*Urwald*" (primeval forest) also points to an imagined connection to the European penetration of the tropics, a point that could certainly sustain further research. Wirz focused on the field of meaning for the term "*Urwald*" as antithesis to the orderly, disciplined, and civilized European homeland of those dispatched and how it is a central component of colonial justification ideology (Wirz 2000).

32. Fahlbusch sees a direct "line from the beginning of ethnic research to politics of extermination"; Oberkrome is more skeptical (Fahlbusch 1999, 796; Oberkrome 1999). Especially to the person of Penck, further research on his work during the war is necessary. In the postwar literature on Penck there is in fact little reference to his work during the war.

References

Arendt, Hannah. 1951. *The Origins of Totalitarianism*. New York: Harcourt, Brace.

Balk, Norman. 1926. *Die Friedrich-Wilhelms-Universität zu Berlin*. Berlin: Speyer and Peters.

Berman, Russell A. 1998. *Enlightenment or Empire: Colonial Discourse in German Culture*. Lincoln: University of Nebraska Press.

Brehl, Medardus. 2000. "Vernichtung als Arbeit an der Kultur: Kolonialdiskurs, kulturelles Wissen und der Völkermord an den Herero." *Zeitschrift für Genozidforschung* 2: 8–28.

Brogiato, Heinz Peter. 1998. *Wissen ist Macht—Geographisches Wissen ist Weltmacht: Die schulgeographischen Zeitschriften im deutschsprachigen Raum (1880–1945) unter besonderer Berücksichtigung des geographischen Anzeigers*, part 1. Trier: Geographische Gesellschaft.

Burleigh, Michael. 1994. *Death and Deliverance: Euthanasia in Germany c. 1900–1945*. Cambridge: Cambridge University Press.

———. 2000. *Die Zeit des Nationalsozialismus*. Frankfurt am Main: Fischer.

Conrad, Sebastian. 2002. "Doppelte Marginalisierung: Plädoyer für eine transnationale Perspektive auf die deutsche Geschichte." *Geschichte und Gesellschaft* 28: 145–69.

Diekmann, Irene, Peter Krüger, and Julius H. Schoeps, eds. 2000. *Geopolitik: Grenzgänge im Zeitgeist*. Vol. 2. Potsdam: Verlag fur Berlin-Brandenburg.

Eckart, Wolfgang Uwe. 1997. *Medizin und Kolonialimperialismus: Deutschland 1884–1945*. Paderborn: Ferdinand Schöningh.

Eckert, Andreas, and Albert Wirz. 2002. "'Wir nicht, die Anderen auch': Deutschland und der Kolonialismus." In *Jenseits des Eurozentrismus: Postkoloniale Perspektiven in den Geschichts- und Kulturwissenschaften*, edited by Sebastian Conrad and Shalini Randeria, 372–92. Frankfurt am Main: Campus.

Engelmann, Gerhard. 1988. *Ferdinand von Richthofen (1833–1905) und Alfred Penck (1858–1945): Zwei markante Geographen Berlins*. Stuttgart: Franz Steiner.

Fabian, Johannes. 2000. *Out of Our Minds: Reason and Madness in the Exploration of Central Africa*. Berkeley: University of California Press.

Fahlbusch, Michael. 1999. *Wissenschaft im Dienst der nationalsozialistischen Politik?: Die "Volksdeutschen Forschungsgemeinschaften" von 1931–1945*. Baden-Baden: Nomos Verlag.

Fenske, Hans. 1991. "Ungeduldige Zuschauer: Die Deutschen und die europäische Expansion 1815–1880." In *Imperialistische Kontinuität und nationale Ungeduld im 19. Jahrhundert*, edited by Wolfgang Reinhard, 87–140. Frankfurt am Main: Fischer.

Frenssen, Gustav. 1906. *Peter Moors Fahrt nach Südwest: Ein Feldzugsbericht*. Berlin: G. Grote'sche verlagsbuchhandlung.

Friedrichsmeyer, Sara, Sara Lennox, and Susanne Zantop, eds. 1998. *The Imperialist Imagination: German Colonialism and Its Legacy*. Ann Arbor: University of Michigan Press.

Grosse, Pascal. 2000. *Kolonialismus, Eugenik und bürgerliche Gesellschaft in Deutschland, 1850–1918*. Frankfurt: Campus.

Gründer, Horst. 1995. *Geschichte der Deutschen Kolonien*. Paderborn: Ferdinand Schöningh.

———, ed. 1999. *—da und dort ein junges Deutschland gründen: Rassismus, Kolonien und kolonialer Gedanke vom 16. bis zum 20. Jahrhundert*. Munich: Deutscher Taschenbuch.

Haar, Ingo. 2000. *Historiker im Nationalsozialismus: Deutsche Geschichtswissenschaft und der "Volkstumskampf" im Osten*. Göttingen: Vandenhoeck and Ruprecht.

Hildebrand, Klaus. 1995. *Das vergangene Reich: Deutsche Außenpolitik von Bismarck bis Hitler*. Stuttgart: Deutsche Verlags-Anstalt.

Hochschild, Adam. 2000. *Schatten über dem Kongo*. Stuttgart: Klett-Cotta.

Honold, Alexander. 2003. "Flüsse, Berge, Eisenbahnen: Szenarien geographischer Bemächtigung." In *Das Fremde: Reiseerfahrungen, Schreibformen und kulturelles Wissen*, edited by Alexander Honold and Klaus R. Scherbe, 137–61. Frankfurt: Peter Lang.

Jäger, Fritz. 1911. "Wesen und Aufgaben der kolonialen Geographie." In *Zeitschrift der Gesellschaft für Erdkunde zu Berlin*. Berlin: D. Reimer.

Jahr, Christoph. 2001. "Die geistige Verbindung von Wehrmacht, Wissenschaft und Politik: Wehrlehre und Heimatforschung an der Friedrich-Wilhelms-Universität zu Berlin, 1933–1945." In *Jahrbuch für Universitätsgeschichte* 4: 161–76.

Johst, Hanns. 1940. *Ruf des Reiches—Echo des Volkes!* Munich: Eher Nachf.

Koller, Christian. 2001. *Von Wilden aller Rassen niedergemetzelt*. Stuttgart: Franz Steiner.

Kost, Klaus. 1988. *Die Einflüsse der Geopolitik auf Forschung und Theorie der Politischen Geographie von ihren Anfängen bis 1945*. Bonn: In Kommission bei F. Dümmler.

Kühl, Stefan. 1997. *Die Internationale der Rassisten. Aufstieg und Niedergang der internationalen Bewegung für Eugenik und Rassenhygiene im 20. Jahrhundert*. Frankfurt: Campus.

Lauer, Wilhelm. 1970. "Carl Troll zum 70: Geburtstag." In *Argumenta Geographica: Festschrift für Carl Troll zum 70. Geburtstag*, edited by Wilhelm Lauer, 11–17. Bonn: In Kommission bei F. D. Fests.

Mogk, Walter. 1972. *Paul Rohrbach und das "Größere Deutschland": Ethischer Imperialismus im Wilhelminischen Zeitalter: Ein Beitrag zur Geschichte des Kulturprotestantismus*. Munich: Goldmann.

Mommsen, Wolfgang J., ed. 1996. *Kultur und Krieg: Die Rolle der Intellektuellen, Künstler und Schriftsteller im Ersten Weltkrieg*. Munich: Oldenbourg.

Nelson, Samuel Henry. 1994. *Colonialism in the Congo Basin, 1880–1940*. Athens: Ohio University Press.

Noyes, John K. 2002. "Landschaftsschilderung, Kultur und Geographie: Von den Aporien der poetischen Sprache im Zeitalter der politischen Geographie." In *Kolonialismus als Kultur: Literatur, Medien, Wissenschaft in der deutschen Gründerzeit des Fremden*, edited by Alexander Honold and Oliver Simons, 127–42. Tübingen; Basel: Francke.

Oberkrome, Willi. 1999. "Geschichte, Volk und Theorie: Das Handwörterbuch des Grenz- und

Auslanddeutschtums." In *Geschichtsschreibung als Legitimationswissenschaft, 1918–1945*, edited by Peter Schöttler, 104–27. Frankfurt am Main: Suhrkamp.

Osterhammel, Jürgen. 1987. "Forschungsreise und Kolonialprogramm: Ferdinand von Richthofen und die Erschließung Chinas im 19. Jahrhundert." *Archiv für Kulturgeschichte* 69: 150–95.

———. 1994. "Raumerfassung und Universalgeschichte im 20. Jahrhundert." In *Universalgeschichte und Nationalgeschichten: Ernst Schulin zum 65. Geburtstag*, edited by Gangolf Hübinger, Jürgen Osterhammel, and Erich Pelzer, 51–72. Freiburg: Romburg.

———. 2002. "Expansion und Imperium." In *Historische Anstöße*, edited by Peter Burschel, Mark Häberlein, Volker Reinhard, Wolfgang J. Weber, and Reinhard Wendt, 371–92. Berlin: Akademie.

Penck, Albrecht. 1925. "Das Hauptproblem der physischen Anthropogeographie." *Zeitschrift für Geopolitik* 2: 330–48.

———. 1926. "Deutscher Volks- und Kulturboden." In *Volk unter Völkern: Bücher des Deutschtums*, edited by Karl C. von Loesch, 1: 62–73. Breslau: Ferdinand Hirt.

———. 1934. *Nationale Erdkunde*. Berlin: Buchholz and Weisswange.

Pogge von Strandmann, Hartmut. 2002. "Der Kolonialrat." In *Kolonialmetropole Berlin*, edited by Joachim Zeller and Ulrich von der Heyden, 32–34. Berlin: Berlin Edition.

Richthofen, Ferdinand, Freiherr von. [1869] 1999. "Denkschrift an den Kanzler des Norddeutschen Bundes, Bismarck, über ein preußisch-deutsches Engagement in Asien." In *—da und dort ein junges Deutschland gründen: Rassismus, Kolonien und kolonialer Gedanke vom 16. bis zum 20. Jahrhundert*, edited by Horst Gründer, 59–62. Munich: Deutscher Taschenbuch Verlag.

———. 1877–1912. *China: Ergebnisse eigener Reisen und darauf gegründeter Studien*, vols. 1–5. Berlin: Reimer.

———. 1883. *Atlas von China: Orographische und geologische Karten*. Vol. 1, *Das nördliche China*. Berlin: Reimer.

———. 1897. *Kiautschou: Seine Weltstellung und voraussichtliche Bedeutung*. Berlin: Stilke.

———. 1898. *Schantung und seine Eingangspforte Kiautschou*. Berlin: Reimer.

———. [1898] 1983. "Denkschrift betreffend die Erweiterung des Geographischen Instituts der Universität Berlin zu einer Lehr- und Arbeitsstätte der Geographie in ihren wirtschaftlichen Grundlagen und ihren praktischen Beziehungen." In *Die Hochschulgeographie in Preussen, 1810–1914*, edited by Gerhard Engelmann, 157–65. Wiesbaden: Franz Steiner.

———. 1907. *Tagebücher aus China*. Edited by Ernst Tiessen. Berlin: Reimer.

———. 1912. *Atlas von China: Orographische und geologische Karten*. Vol. 2, *Das südliche China*. Edition Max Groll. Berlin: Reimer.

Rohrbach, Paul. 1907a. *Deutsche Kolonialwirtschaft*. Vol. 1, *Südwest-Afrika*. Berlin: Buchverlag der "Hilfe."

———. 1907b. *Wie machen wir unsere Kolonien rentabel? Grundzüge eines Wirtschaftsprogramms für Deutschlands afrikanischen Kolonialbesitz*. Halle: Gebauer-Schwetschke.

———. 1909. *Deutsche Kolonialwirtschaft: Kulturpolitische Grundsätze für die Rassen- und Missionsfragen*. Berlin: Buchverlag der "Hilfe."

———. 1934. *Koloniale Siedlung und Wirtschaft der führenden Kolonialvölker*. Cologne: Schaffstein.

Schaller, Dominik J. 2003. "Die Rezeption des Völkermordes an den Armeniern in Deutschland, 1915–1945." In *Der Völkermord an den Armeniern und die Shoah*, edited by Hans-Lukas Kieser and Dominik J. Schaller, 517–55. Zürich: Chronos.

Schindlbeck, Markus. 2001. "Deutsche wissenschaftliche Expeditionen und Forschungen in der Südsee bis 1914." In *Die deutsche Südsee 1884–1914: Ein Handbuch*, edited by Hermann Joseph Hiery, 132–55. Paderborn: Ferdinand Schöningh.

Schulte-Althoff, Franz-Josef. 1971. *Studien zur politischen Wissenschaftsgeschichte der deutschen Geographie im Zeitalter des Imperialismus*. Paderborn: Ferdinand Schöningh.

Schultz, Hans-Dietrich. 2002. "Raumkonstrukte der klassischen deutschsprachigen Geographie des 19./20. Jahrhunderts im Kontext ihrer Zeit." *Geschichte und Gesellschaft* 28: 343–77.

Troll, Carl. 1935. *Das deutsche Kolonialproblem auf Grund einer ostafrikanischen Forschungsreise 1933/34*. Berlin: Reimer.

———. 1941. "Koloniale Raumplanung in Afrika." In *Das afrikanische Kolonialproblem: Vorträge, gehalten in der Gesellschaft für Erdkunde zu Berlin im Januar–Februar 1941*. Berlin: Zeitschrift der Gesellschaft fem. Vorträge, gehalten.

———. 1947. "Die geographische Wissenschaft in Deutschland in den Jahren 1933 bis 1945: Eine Kritik und eine Rechtfertigung." *Erdkunde* 1: 3–47.

———. 1969. *Fritz Jaeger: Ein Forscherleben*. Erlangen: Fränkische Geographische Gesellschaft.

Van Laak, Dirk. 2001. "Imperiale Infrastruktur: Deutsche Planungen für eine Erschließung Afrikas, 1880 bis 1960." Habilitation thesis, University of Jena.

Verhey, Jeffrey. 2000. *Der "Geist von 1914" und die Erfindung der Volksgemeinschaft*. Hamburg: Hamburger Edition.

Weingart, Peter, Jürgen Kroll, and Kurt Bayertz. 1988. *Rasse, Blut und Gene: Geschichte der Eugenik und Rassenhygiene in Deutschland*. Frankfurt: Suhrkamp.

Wirz, Albert. 2000. "Innerer und äußerer Wald: Zur moralischen Ökologie der Kolonisierenden." In *Der deutsche Tropenwald: Bilder, Mythen, Politik*, edited by Michael Flitner, 23–48. Frankfurt am Main: Campus.

Wulf, Stefan. 1994. *Das Hamburger Tropeninstitut 1919 bis 1945: Auswärtige Kulturpolitik und Kolonialrevisionismus nach Versailles*. Berlin: Reimer.

Zantop, Susanne. 1999. *Kolonialphantasien im vorkolonialen Deutschland, 1770–1870*. Berlin: Erich Schmidt.

Zimmerer, Jürgen. 2001. "Von der Bevormundung zur Selbstbestimmung: Die Pariser Friedenskonferenz und ihre Auswirkungen auf die britische Kolonialherrschaft im Südlichen Afrika." In *Versailles 1919: Ziele-Wirkung-Wahrnehmung*, edited by Gerd Krumeich, 145–58. Essen: Klartext.

———. 2002a. *Deutsche Herrschaft über Afrikaner: Kolonialer Machtanspruch und Wirklichkeit im kolonialen Namibia*. Hamburg; Muenster: LIT.

———. 2002b. "Wissenschaft und Kolonialismus: Das Geographische Institut der Friedrich-Wilhelms-Universität vom Kaiserreich zum Dritten Reich." In *Kolonialmetropole Berlin: Eine Spurensuche*, edited by Ulrich van der Heyden and Joachim Zeller, 125–30. Berlin: Berlin Edition.

———. 2003. "Krieg, KZ und Völkermord: Der erste deutsche Genozid." In *Völkermord in Deutsch-Südwestafrik der Kolonialkrieg (1904–1908) in Namibia und seine Folgen*, edited by Jürgen Zimmerer and Joachim Zeller, 45–63. Berlin: Christoph Links.

———. 2008. "Colonialism and the Holocaust—Towards an Archeology of Genocide." *Development Dialogue: Revisiting the Heart of Darkness—Explorations into Genocide and Other Forms of Mass Violence* 50: 95–124.

Zögner, Lothar. 1997. "Ferdinand von Richthofen—Neue Sicht auf ein altes Land." In *Tsingtau: Ein Kapitel deutscher Kolonialgeschichte in China, 1897–1914*, edited by Hans-Martin Hinz and Christoph Lind, 72–76. Berlin: Deutsches Historisches Museum.

4

The East as Historical Imagination and the Germanization Policies of the Third Reich

GERHARD WOLF

Introduction

It was no secret that, as Goebbels put it, Albert Forster "hated Himmler from the depth of his soul"—a feeling more than replicated by Himmler toward Forster by the way (Fröhlich 1987, 165).[1] The reason for this bitter feud was well known. Looking through their correspondence reveals just how deep this animosity ran while pointing to the main point of contention: the Germanization of the annexed lands, or—to be more precise—how to turn Eastern Europe into the Nazi dystopia of German living space. There was little Forster and Himmler agreed upon and there were arguments aplenty. For example, when they engaged in a quarrel about the prehistory of Forster's fiefdom. As Himmler wrote to Forster on November 20, 1941:

> You write . . . historians assume that the number of aliens intruding into Westprussia after large Germanic parts had left the region during the migration period was not very large. As far as I know this assumption is purely speculative. . . . I cannot comment on your opinion that apart from a few native old families the region has been German until the defeat of the Teutonic Knights in the 16th century. . . . During all of my studies I haven't read anything about such a comprehensive German colonization.[2]

Why did the head of the SS, chief of the German police, Reich Commissariat for the Strengthening of Ethnic Germandom, and the Reichsstatthalter and Gauleiter of Danzig-West Prussia, two of the most powerful men in occupied Poland, get bogged down in an argument about history?[3] The answer, I think it is safe to assume, did not lie just in the past but also in the problematic present, not just with the defeat of the Teutonic Knights in medieval times but also with the difficulties those modern-day *Weltanschauungskrieger* encountered in their attempt to take possession of this land and its population. Still, it is hard to envision another Eastern European landscape quite as steeped in national folklore and myths and so central to the German imaginary than the borderlands to the East. At the fault lines between powerful empires, regions like Western Prussia or Upper Silesia had changed hands many times between Prussia/Germany and its Austrian, Polish, and Russian challengers to the east. A focal point of conflict over centuries, the situation escalated with the emergence of nationalism during the nineteenth century. This is not the place to write about the bitter Germanization campaigns Berlin imposed upon the Polish-speaking minority in eastern Prussia or the humiliation and loss many Germans felt after having to cede those eastern borderlands to a reemerging Poland. However, recalling these developments helps to understand why the struggle for the revision of Versailles was first and foremost a struggle for reclaiming the "lost eastern territories." And it might also help to understand why a radically new interpretation of how this struggle for reconquering and "keeping these lands German" could gain in popularity. If one key appeal of racist ideology is its promise to overcome political strife through science, replacing the struggle for a better future with its technocratic calculation, it is not hard to see why Hitler, among others, found more solace in a decidedly racist worldview. Vehemently criticizing Prussian Germanization policies, Hitler argued that these lands would not be won and secured simply by aiming at the "enforced outward acceptance of the German language" among the native population since Germanization could "only be carried out with the *soil* and never with *men*" (Hitler 1941, 588).[4] Instead, Hitler argued in the (unpublished) sequel to *Mein Kampf*:

> The *völkisch* state, conversely, must under no conditions annex Poles with the intention of wanting to make Germans out of them some day. On the contrary it must muster the determination either to seal off these alien racial elements, so that the blood of its own people will not be corrupted again, or it must without further ado remove them and hand over the vacated territory to its own national comrades. (Taylor 1962, 47–48)[5]

Himmler took his cues from here. Rejecting Forster's view of a German population surviving the end of "German" rule after the defeat of the Teutonic Knights, Himmler held that simply teaching the population to speak the language would not turn these territories into German lands. To do this, to bring these territories "back" into the fold of the German Reich, the population had to be racially sifted and all found unsuitable removed and replaced with racially pure Germans. As he was to find out, however, his call for a racially informed Germanization policy quickly ran into opposition from a broad power alliance that, on the one hand, feared the destabilizing consequences this radicalization would have on German rule, while, on the other hand, were able to counter Himmler in an ideological language that was no less compatible with the discourse at the time. Appropriating a more traditional *völkisch* strand of German right-wing thinking, Forster and others attacked Himmler's notion that a territory was defined by its population, claiming that certainly when it came to western Poland the opposite was the case. Contested borderlands for centuries, German cultural influence and not least German rule had transformed these lands. This civilizing mission was not only carved in stone and could be witnessed in regulated waterways or in a tamed and reshaped landscape, but had also left its imprints on the population living in this environment. In this chapter, I will explore the conflict between racial and *völkisch* ideologues and how they understood the relationship between territory and its population. I will investigate whether the National Socialist Germanization policy owed more to its Prussian predecessor than the statements of some leading actors want us to believe and, more specifically, to which extent this policy was shaped by a spatial vision in which Eastern Europe did not just appear as a uniformly alien land. The ideological topography of Eastern Europe was more complex and multilayered, infused with history that made some regions, like, for example, Galicia, appear very distant, while moving others very close—and none so close, not just in geographical terms, as western Poland, comprising territories that until the end of the World War I had been German, its population German citizens.[6]

Justifying Aggression: Inventing German *Volks- und Kulturboden*

Contrary to widely held beliefs, the Paris peace treaties arguably solidified this focus on territory. Lenin's and Wilson's radical interventions notwithstanding, the (re)emerging states in Eastern Europe owed their existence not primarily to a newfound respect for the national identity of their peoples. As Ulrike Jureit has shown recently, national self-determination might have been more important for the public discussions at the time than for the negotiating tables (Jureit 2012, 197–219). Given the situation in the German-Polish borderlands, for example, it might come as little surprise that neither the German nor the Polish delegations were favoring plebiscites, thus handing over the decision over the future of their territories to the native population. Seen on both sides as so-called in-between people, the loyalty of this "*Zwischenschicht*" or "*ludność labilna*" was questioned in Berlin just as much as in Warsaw. No wonder, then, that both delegations were more concerned with stressing the detrimental effects the secession of, for example, Upper Silesia and Western Prussia would have on either country's economy allegedly—in the German case—scuttling all attempts to generate the reparations imposed by the Allies, or—in the Polish case—defying the country the necessary economic basis to exist.

This is not to say, however, that Germany and Poland shied back from claiming either the population as their own or stressing the deep historical bonds linking the territory in question to either nation whenever it suited their argument. Poland, for example, had spent considerable energy on preparing nationality maps of the eastern Prussian provinces that made the desired headlines in the French press. In this respect, the German delegation clearly had some catching up to do, with the German foreign minister urgently calling on Berlin to furnish him with cartographic material to rebuff Polish claims (Jureit 2012, 195–200). Not that German officials had not been cautioned against ignoring these issues. Albrecht Penck, the director of the Geographical Institute at Berlin University, had contacted the Prussian government already before the end of the war, stressing the importance of academic expertise to substantiate German claims. Germany was not to be caught unprepared again, Wilhelm Volz, geography professor at the University of Breslau, implored the Foreign Office in 1921 after the plebiscite in Upper Silesia, "so that in the hour of justice, which also must come for Germany, we . . . do not have to compile the material again in a frenzy and thereby lose valuable time" (Herb 1997, 66). It was high time to start to collect, validate, and distribute German research on the "bleeding border" and combat studies seen to weaken German claims on the lost territories.[7] Founded in 1922 and run by Volz and Penck, the Mittelstelle für zwischeneuropäische Fragen was envisaged as a clearing house for the new "discipline" of *Ostforschung*—a discipline not to be confused with Slavic studies as it was not interested in generally exploring Eastern Europe in its own right but merely as a playing field for Germany's civilizing mission, "barren wastelands" as dependent on German creativity and ingenuity in the past as they were now (Penck 1925, 64).[8] The challenge these *Ostforscher* faced was the same utterly failed by the German government at Versailles: to convince the world that the majority of the population in the contested territories was German when, not only a rising number of Germans were in fact leaving for the Reich, but also establishing national identity in these border regions had proved difficult in itself because language was not necessarily corresponding with political allegiance, as not least the various plebiscites had shown. While politicians and the public alike could well point to the established notion of *deutscher Volksboden* (a *völkisch* version of land rights based on German settlement) to criticize the exclusion of borderlands inhabited by a German majority like Danzig or the Sudeten territories, it helped little in the fight for territories with a mixed population like Upper Silesia or Western Prussia. German academics proved helpful here, too. Where "the character of the population proved insufficient to reclaim lost territories," they redirected their "attention on the German character of the land itself" (Herb 1997, 50–51). It was again Penck who proved particularly adept in helping to clothe revisionist ambitions in academic parlance by diversifying *Volksboden* into *Volks- und Kulturboden*. As he explained in an article in 1925, focusing only on German *Volksboden* and German language areas would mean repeating the mistakes of the past, ignoring "the German peoples' greatest achievement": German *Kulturboden* (Penck 1925, 69). While German *Kulturboden* might by and large shadow German *Volksboden* in the West, "in the East it reaches beyond the German *Volksboden* almost everywhere forming a belt of land beyond in which the German population while receding behind the peoples speaking other languages was nonetheless imprinting its culture on the land or having done so in the past" (ibid., 68). It speaks to the "competence" of these German settlers, that "a small number . . . can suffice to transform large lands into German *Kulturboden*" (ibid., 64, 69–70). Seen from this perspective, the contested German-Polish borderland was in fact "ancient Germanic land" (ibid., 68). True, Slavic tribes moved into these territories in the early Middle Ages—but into territory previously cultivated by Germanic tribes who had settled there before. They were able to benefit from Germanic cultural achievements but proved unable to develop or sustain them. No wonder, then, that these lands were only "fully developed, when German settlers returned from the West, cultivated it again and taught the Slavic immigrants to do the same" (ibid., 69). The results, Penck believed, could still be seen in the eastern parts of imperial Germany. While it was easy to distinguish "the tidy German from the often fairly poor Polish villages; the intensive German soil cultivation and the accompanying good alleys and roads extended to the Russian border" thus establishing a "cultural border (*Kulturgrenze*) of incisive severity as one will not find again in Central Europe" (ibid., 65, 69). Poles, for Penck, were squatters on foreign land. Germans might no longer form the majority of the population in, for example, Western Prussia or Upper Silesia, but they certainly held the more valid claims to the land that had been allegedly brought to life by and bore the marks of German cultural superiority.

While there can be no doubt that the resounding success of this notion of German *Kulturboden* in interwar Germany was its function in attempting to delegitimize the Treaty of Versailles, it seems equally clear that it also truly shaped the thinking of many Germans at the time. Volz and Penck, in any case, deemed it important enough to rename the Mittelstelle into Stiftung für

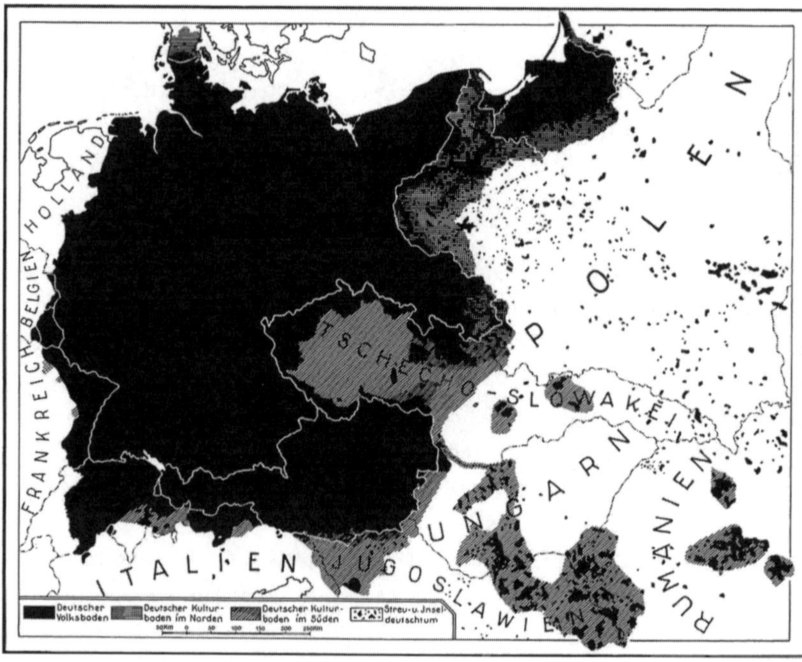

FIGURE 4.1. Albrech Penck, *Karte des deutschen Volks- und Kulturbodens*, 1925. Source: Collectie Steegh-Teunissen, Leiden.

Volks- und Kulturbodenforschung shortly after Penck's publication, firmly establishing this double term of *Volks-* and *Kulturboden* at the center of German revisionist thinking about Eastern Europe.

The Nazi rise to power in January 1933 came as another boost to German *Ostforschung*. In lockstep with Nazi expansionist aims, they were rewarded with lavish support by the government, leaving behind even the most basic academic standard in the process. Professor Erick Keyser from the Technical University in Danzig, another *Ostforscher* and prolific writer, openly stated that his academic interest had always been dictated by the "political necessities and the political life of the immediate present" (Keyser 1942, 93). When Germany invaded Poland in September 1939, Keyser called on his comrades in arms to exchange their desk for fieldwork in the occupied territories, a call heeded in Danzig by, among others, Dr. Detlef Krannhals. When the annexed Polish territories were subdivided into the three provinces of—from north to south—Danzig-West Prussia, Wartheland, and Upper Silesia, Keyser and Krannhals became the principal academic advisers to Forster.

Keyser was arguably most important in providing the German occupation authorities with a narrative of why the incorporation of Reichsgau Danzig-

Westpreußen into the Reich was not an act of aggression but of historical justice. Unsurprisingly, the German occupiers chose not to rely too much on claiming that the majority of the population was German. Instead, Keyser opted for the line pioneered by Penck and Volz and focused on the territory and its history. Like Penck, Keyser, too, maintained that this region was first settled by Germanic tribes who had completely Germanized the land and its population by the second century CE. Slavs only "seeped into [German] *Heimatboden*" after the majority of the Germanic tribes had left the area (Keyser 1942, 94). Keyser and Krannhals might have disagreed whether this territory then became "barren and desolate" again due to Slavic incompetence, since "the cultural distance . . . most likely prevented a blood amalgamation" with the remaining Germanic population (ibid., 94), or whether they did assimilate, thus acquiring "the soil and the means to cultivate it from Germanic hands" (Krannhals 1940, 20). They certainly did not disagree, however, that the onset of the medieval German colonization started to "slowly but surely infuse the . . . Slavic population with German character," turning this land finally into German land (Keyser 1940, 73). Seen from this perspective, the collapse of the State of the Teutonic Order in the mid-fifteenth century and the incorporation of much of this territory into the Polish kingdom appears only as a fleeting interlude. When Frederick II annexed this territory after the first partition of Poland in 1772, it still was German land with "no building of historic importance and no artwork, no canal and no dike, no castle and no church, no town or village plan that was not created by Germans" (ibid., 132).

This self-serving ideological narrative by *Ostforscher* like Keyser and Krannhals became the official government line. Forster himself, in training material for the National Socialist Party in Danzig-West Prussia, impressed upon his rank and file that they all found themselves in a land that Germans had over thousands of years transformed into "a decidedly German *Volks- und Kulturboden*" (Forster 1942, 3–4). Wilhelm Löbsack, Forster's *Volkstumsreferent*, repeated this claim in the lavishly illustrated and semiofficial book *Re-Gained German Land*, adding that the National Socialists had taken over the baton passed down to party and state officials representing "Teutonic Knights of the twentieth century" (Löbsack 1943, 12).

The Power of Ideology: Germanizing Poles

It is testament to the social efficacy of this ideological discourse that it did not simply lose its hold on people's minds even after it had passed its sell-by date and had become redundant. It is ironic that Penck's notion of the "cultural border of incisive severity," allegedly dividing German and Russian *Volksboden*, after all floated and fueled by the desire to garner support for the revision of Versailles, seemed to have started to trouble German policy makers when they could just as well have ignored foreign opinion for the first time. With the war against Poland in its final stages, it was the commission tasked with redrawing Germany's new eastern borders that advocated a rather cautious approach. In his memorandum on the future province of Danzig-West Prussia, its chairman Ernst Vollert left no doubt that, of course, "this old German land had to be turned again into German land as swiftly as possible." It was only German land, however, that according to Vollert should be incorporated into the Reich—and not, as he pointed out, "former Russian territories."[9] Much to the dismay of the Nazi elite, Vollert's commission was not alone, with notable *Ostforscher* like Dr. Theodor Schieder also opposing the expansion of Silesia beyond the pre-1919 borders. In Silesia, this debate continued even beyond the formal annexation of western Poland on October 8, 1939, with the local *Gauleiter* of Silesia demanding a return to the old borders as late as February 1940.[10]

Obviously, after Hitler and Göring had opted for a large-scale annexation that particularly in the new province of the Wartheland incorporated large swathes of land that had never been part of the German Empire, the doubts of both bureaucrats with the Reich Ministry for the Interior and the civilian administration were doomed to failure. More, it seems, would have been necessary to dispel the notion of the alleged fundamental difference dividing former German from Russian territory. Arguably, Penck's cultural border reemerged in the guise of the new police border that divided the annexed territories. Closely following the pre-1919 borders, it was set up to control the influx of goods and people from the eastern into the western part.[11] More significantly, however, it survived in the Germanization process, the most important task Hitler had assigned to the *Gauleiter* in the annexed provinces.

For the German leadership, the army and SS units on the ground had already started the Germanization of these territories. While Wehrmacht and SS commanders were often at odds about how best to secure the conquered territory, there is little doubt that both were engaged in a war that went far beyond the limits set by international law and were systematically mistreating Polish prisoners of war and targeting the civilian population.[12] SS Einsatzgruppen were particularly murderous, soon hunting down and killing many tens of thousands of Poland's political and social elite, with the territory of what later was to become the province of Danzig-West Prussia as a main focal point.[13] The native population quickly realized that the only way to escape this explosion of violence German military units brought about was to turn to German civilian authorities installed by the Wehrmacht who were up and running even before the last Polish units had capitulated. The acting head of the county of Dirschau close to Danzig—apparently on his own initiative and without central instructions—quickly presented a selection procedure that divided the population into three groups: ethnic Germans, Jews, and Poles. German local authorities were advised that only those could be recognized as "*Volksdeutsche*" who had "openly expressed their German identity and supported Germandom" in the past. They were to be issued with "identity cards to protect them from measures taken against Jews and Poles." Jews, however, were not mentioned anywhere else in these guidelines, most likely because most of them had already been either killed or expelled from Dirschau. Where they were mentioned one could also find what is so conspicuously missing here: *any reference to racial criteria*. In fact, racial terminology was not extended to identify Poles. This is all the more remarkable as the guidelines are most comprehensive when dealing with the Polish population, which instead was to be subdivided into two groups according to political and territorial criteria, separating those who had "moved here from Congress Poland and Galicia" or had "actively fought Germandom" from those who "were and are residing in Western Prussia and had not taken part in fighting Germandom."[14] These distinctions eventually also made it into the census conducted by the German occupation authorities in December 1939.[15]

This process emerging in Danzig-West Prussia—initially more due to the quickly evolving situation on the ground than to any preconceived plans of how to deal with the native population—was mirrored by more or less similar screening procedures in the other two provinces to the south, in the Wartheland and Silesia, and, importantly, was also supported from Berlin. In an attempt to coordinate and not least to regain the initiative in a selection process that de facto prejudged any decision about the individual's citizenship, the Interior Ministry of the Reich intervened on November 25. In a first step, German authorities in the annexed territories were issued instructions on how to identify ethnic Germans who were to be given Reich citizenship. In line with the practice already established on the ground, this was again made dependent primarily on the individual's social and political conduct. As the Reich Ministry of the Interior stipulated with reference to an earlier regulation passed after the invasion of Czechoslovakia:

> Of German ethnicity [*deutscher Volkszugehörigkeit*] is who declares himself to be part of the German *Volk* if only this declaration is backed up by certain facts like language, education, culture etc. Under the given circumstances a clearer definition of this term German ethnicity is not possible. However, it

will generally pose no difficulty to decide accordingly whether somebody is of German ethnicity or not.[16]

The importance of individual behavior was stressed even more by the ministry's reminder that

> German ethnicity . . . does not necessarily presuppose that someone is fully or partially of German stock. Since the declaration to belong to the German *Volk* is crucial it is in fact possible to accept as German who is partially or fully of alien stock. Vice versa it is possible in particular circumstances that due to his statement somebody has to be seen as belonging to an alien people even he is of partial or full German stock.[17]

But what about the large majority of the population and particularly those who while not meeting the strict criteria for ethnic Germans might still be considered—hence the term soon to be employed: "re-Germanizable?" How long could the Germans afford to wait until they, too, should be identified and incorporated into the German *Volksgemeinschaft*, thus bolstering the numbers on the ground? No longer, argued Arthur Greiser, *Reichsstatthalter* and *Gauleiter* in the Wartheland, defying the Reich Ministry for the Interior and demanding that they should be screened right away instead of waiting until the identification of the ethnic Germans was completed.[18] Greiser did agree, however, that the criteria should again focus on social and political conduct. Himmler's intervention finally escalated this discussion, especially as he supported Greiser's demand to speed up the process but then introduced a new criterion explicitly ignored so far: *race*. As Himmler protested with the Reich Ministry of the Interior, it was not acceptable to decide according to "an only outward profession to belong to the German *Volk* (language, upbringing, culture, etc.)." Instead, the "prime criteria" has to be the "positive establishment of racial belonging (*rassischen Zugehörigkeit*)."[19]

With no clear idea what to do with the native population when invading Poland, German authorities seemed to have been even more divided a few months later. The question dominating population policies in the annexed land, that is, who was or could become German, increasingly elicited a variety of answers, delaying negotiations in Berlin for more than another year. Forster, it seems, used this paralysis at the center for implementing his own vision of how to Germanize his province, and on December 14, 1940, announced the so-called "re-Germanization program" (*Wiedereindeutschungsaktion*). In an attempt to deflect recent criticism claiming that his administration was too lenient with handing out German citizenship, Forster insisted that this new initiative was aiming solely at "unearthing buried and recovering lost Germandom" and, therefore, not so much about Germanization but "in essence about re-Germanization." The selection criteria of Forster's decree tell a different story, however, with little attention paid on how to prevent the inclusion of unsuitable persons. Instead, the emphasis was clearly on whom to include: thirty thousand families with a list of desirable qualities. Political reliability was one of them. Members of ethnic German organizations were explicitly targeted, as were parents who had sent their children to a German-speaking school. After having killed the political elite, Forster even felt confident enough to open the door for former rank and file members of Polish political parties. Most importantly, however, was something else: the thirst of the local economy for skilled labor. While increasingly troubling the entire German economy, expulsions, deportations, and not least mass murder had left the labor market in the annexed territories particularly volatile. Local authorities were thus informed that "for the time being this Germanization program is limited to farm hands, workers, craftsmen, and peasants." If some of them fell short meeting other requirements like, for example, providing a gapless genealogy, the guidelines directed the view on the applicant's "professional quality and performance":

> It is possible to infer from typical German skills and talents . . . that there is an infusion of German blood (for example technical skill, a sense of properly maintaining and organizing tools in the house and on the farm). In this context the necessity to consider the personal and homely cleanliness has to be pointed out.[20]

Privileging those who promised to be politically reliable and economically indispensable certainly proved congenial to the demands of an occupying power. It would be too simple, however, to dismiss Forster's excursions into history as merely an ideological smokescreen put up to deceive the regime's ideological hardliners. Opting for an inclusive Germanization policy was unquestionably in Forster's political interest—but it was also unquestionably in line with earlier Prussian policies and not least with two decades worth of revisionist *Ostforschung*, insisting on the German character of these territories and, by extension, its population. Germanization theoretically presupposed German descent, but local authorities were offered little help in making this judgment. German-sounding names were important, but then the rejection of persons with Polish names was explicitly ruled out since, allegedly, many current "bearers of Polish family names many years ago had German family names." In practice, it seems, German descent was translated into a simple question of geography, and it was only here where the guidelines became more specific:

> In principle, the Germanization should only extend to the native population of Western Prussia. This means that parents or grandparents of the indigenous family have to come from Western Prussia.[21]

Forster went even further by announcing that each county would soon receive a detailed list of "administrative districts, villages, and neighborhoods, which were identified as regions of lost Germandom . . . in cooperation with the academics Dr Krannhals and Professor Keyser." The screening should commence there with the lists continuously updated by the "political science in our province." Although I was unable to locate these lists, it is clear that spatial considerations were pivotal in considering people for Germanization. This did not change even after a settlement had been reached in Berlin leading to the establishment of the German People's Register (*Deutsche Volksliste*) in March 1941, the comprehensive framework regulating the transfer of German citizenship to the native population in annexed western Poland.[22] It is testament to the power of the regional *Gauleiter* that Forster introduced the *Deutsche Volksliste* in Danzig-West Prussia only after completing the *Wiedereindeutschungsaktion*, and even then he changed very little about the actual screening process. The Reich Ministry for the Interior gave up very soon in trying to bring the practice in Danzig-West Prussia in line with the selection in the Wartheland and Upper Silesia. Even Himmler, while more persistent and urging his local representatives to exert their influence, eventually resigned too.

At least on paper, the new regulations had given Himmler's SS apparatus a bigger say. As the SS was to find out, however, the right to dispatch at least one representative to every panel makes less of a difference when confronted with loyal party members carefully selected by Forster. During the first session of the *Deutsche Volksliste* in the district of Thorn, the local party leader, Warras, left no doubt that he expected the committee members to simply continue the previous inclusive selection procedure. After all, as Warras explained:

> The native population had been systematically driven into Polish hands by ill-advised German governments, making them Polish on the outside although they are German at the core and it will be up to National Socialist educational work to turn these people, many of which don't even know who they belong to, into valuable German people. . . . It doesn't matter whether little mistakes are made or not, the direction adopted by the Gauleiter is correct.[23]

Competing interpretations were not appreciated. On an inspection mission through his province, Forster also attended a meeting of the *Deutsche Volksliste* commission in Neumark and immediately lost his temper when the SS representative, SS-Rottenführer Wellnitz, objected to the first applicant already vetted by the party. Forster told Wellnitz in no uncertain terms that he was talking "rubbish" and demanded he withdraw his objection. When Wellnitz refused, pointing out that he acted in complete accordance with the guidelines set out by the Reich Ministry of the Interior, Forster shouted at him that he was "too stupid to interpret the decree"—and kicked him out.[24] Himmler fared little better in another attempt to restrict access to the *Deutsche Volksliste* when trying to amend the selection process on September 30, 1941, to exclude those he deemed "racially unsuitable."[25] Although Forster was not able to prevent people in his province to be screened by the SS Race and Settlement Office, he instructed party and state officials that they were "not binding for the decision . . . of the *Deutsche Volksliste*."[26]

Conclusion

By the end of the war the *Deutsche Volksliste* had registered roughly three million people.[27] While it is true that Germanization policies were often neither uniform nor in line with the instructions set out by, for example, the Reich Ministry of the Interior, my analysis of Forster's Germanization policy allows, I think, for some comments on Nazi population policies in general.

The practice of the *Deutsche Volksliste*, and this is true for Danzig-West Prussia as well as for the two other provinces, did not conform to Hitler's demand that Germany "must under no condition annex Poles with the intention of wanting to make Germans out of them" (Taylor 1962, 47–48). Instead, the *Deutsche Volksliste* arguably stands for the single most expansive assimilation project in German history. Nor does it fit neatly into much of the historiography on National Socialist population policy that highlights exclusionary rather than inclusionary practices. One principal reason for the latter is what I would call the "racial turn" in the study of National Socialism. Sped up by the collapse of the Eastern Bloc that helped to de-legitimize Marxist and structural approaches in general, this new approach seemed to offer a way forward; it allowed for analyzing Nazi mass crimes like the systematic extermination of European Jewry and Soviet prisoners of war, of inmates of mental asylums and so-called asocials—the various but interconnected aspects of a policy aimed at the creation of a "racial state" (the emblematic title of a study published by Michael Burleigh and Wolfgang Wipperman in 1991). It offers little to understand the *Deutsche Volksliste* and its selection process, however, which was clearly not based on racial criteria. In order to do so, I would argue, we must embed the National Socialist living space dystopia more broadly in German imaginations of Eastern Europe and, more specifically, Germany's eastern borderlands. Instead of race the occupiers were emphasizing social and political behavior, or more precisely, the willingness to accept German rule and integrate into the German *Volksgemeinschaft* to be established in these lands—a logic, I would claim, resembling the Germanization policies in the eastern provinces of imperial Germany. What underpinned much of National Socialist occupation policy in annexed Poland was a deeply held belief about Germany's right to this land. For the occupiers, laborious and persistent German colonists acquired this land, allegedly transforming *terra nullius* into German *Volks- und Kulturboden* and thereby pushing the cultural frontier further eastward. No wonder, then, that this came to shape National Socialist Germanization policy. In Danzig-West Prussia this is clear to see with, for example, Keyser and Krannhals instructing each county where to locate "lost Germandom." While these documents might not have survived the war, the message they conveyed has—enshrined in the results of the *Deutsche Volksliste*'s registration practice. In Danzig-West Prussia, for example, more than two-thirds of the population was registered across the province, but only 20 percent in the two counties of Leipe and Rippin, which until 1919 had belonged to Russia. In Upper Silesia the difference was even greater, with 89 percent of the population registered in territories formerly belonging to Prussia or Austria and only 2 percent in Russian Galicia.[28] The notion of a *Kulturgrenze*, delineating the reach of the German *mission civilisatrice*, loomed large in National Socialists Germanization policies, deeply inscribing space into definitions of Germanness.

Notes

1. For an early and still valid assessment of Forster's relationship to Himmler, see Levine (1969).

2. Himmler to Forster, November 20, 1941, German Federal Archive Berlin (GFA-B) R 49/36a, 56–59.

3. On Himmler's admiration for Henry I and the Teutonic Knights, see Kroll (1998, 237–40, 242–44); Longerich (2008, 279–89); or his talk delivered at the occasion of the SS moving into the former headquarters of the Order in Vienna in January 1939, in Smith and Peterson (1974, 50–51).

4. Italics in the original. The potential for violence encapsulated in this notion was already noticed by Raphael Lemkin when developing the concept of genocide in his *Axis Rule in Occupied Europe: Laws of Occupation—Analysis of Government—Proposals for Redress* (1944).

5. See also Wirsching (2001).

6. More succinctly on German conceptions of Eastern Europe, see Dickinson (2008).

7. On the Stiftung für Volks- und Kulturbodenforschung, see Herb (1997) and Haar (2008).

8. On *Ostforschung* see Zernack (1980, 542–59); Mühle (2006, 110–37); and Krzoska (2008, 452–63).

9. Vollert's memorandum *Suggestion on the Territorial Demarcation of Western Prussia*, undated, probably from October 6, 1939, GFA-B R 1501/5401, 31–40.

10. See Schieder's memorandum, undated, most likely from October 7, 1939, GFA-B R 153/291; on the timing, see Ebbinghaus and Roth (1992). On the discussion between Breslau and Berlin see, for example, Gauleiter Josef Wagner asking the Reich Interior Ministry to move the borders back, February 2, 1940. See also Steinbacher (2000, 111–13).

11. The police border was only removed on May 12, 1942; Heydrich to heads of government districts, 12/05/1942, State Archive Kattowice (SAK) 117/32, 50.

12. On German warfare, see Böhler (2006) and Rossino (2003).

13. On SS Einsatzgruppen, see Krausnick (1985); Mallmann and Musial (2004); and Mallmann, Böhler, and Matthäus (2008).

14. Head of the county Dirschau to state officials, October 14, 1939, Instytut Pamięci Narodowej Warsaw (IPN) NTN/200, 111–13.

15. The final results under IPN NTN/190, 23.

16. Decree Reich Ministry of the Interior, 29/03/1939, RMBliV (1939), 783 (emphasis in the original).

17. Ibid.

18. The so-called *Deutsche Volksliste*, or German People's Register, introduced in the Wartheland on October 28, 1939; see Greiser's decree from the same day, State Archive in Poznan (SAP) 406/1105, 1.

19. Note by Dr. Walter on Himmler's protest to the Reich Ministry of the Interior on January 13, 1940, 20/05/1940, GFA-B, R 49/61, 47.

20. Forster's decree on the *Wiedereindeutschungsaktion*, 14/12/1940, GFA-B, R 49/76, 2–29. On the importance of cleanliness, see Reagin (2007, 181–217); on the use of distinguishing between Germans and Poles, see Orłowski (1996, 319–46).

21. Forster's decree on the *Wiedereindeutschungsaktion*, 14/12/1940, GFA-B, R 49/76, 2–29.

22. Decree on the *Deutsche Volksliste* and the German citizenship in the annexed eastern territories, 4/03/1941, RGBl, part 1, 1941, 118–20. See also the implementing regulations signed by the Reich Minister of the Interior, Wilhelm Frick, 13/03/1941, SAP 406/1105, 9–28.

23. Minutes of the meeting, 24/09/1941, IPN NTN/201, 50.

24. Higher SS and Police Leader Hildebrandt to Himmler, 23/10/1941, GFA-B, R 49/36a, 52f. More on the ensuing clash in Schenk (2000, 205–12).

25. Himmler's decree from September 30, 1941, SAP 406/1114, 5–6.

26. Forster to the offices of the *Deutsche Volksliste*, 9/2/1943, State Archive Bydgoszcz (SAB) 9/380, 243.

27. Overview by the SS *Stabshauptamt* over how many people were registered by the *Deutsche Volksliste* on 1/4/1944, 27/9/1944, GFA-B R 49/467.

28. Overview over the *Deutsche Volksliste* in the annexed territories on 31/3/1943, GFA-B, R 186/32.

References

Böhler, Jochen. 2006. *Auftakt zum Vernichtungskrieg: Die Wehrmacht in Polen 1939*. Frankfurt: Fischer.

Burleigh, Michael, and Wolfgang Wippermann. 1991. *The Racial State: Germany 1933–1945*. Cambridge: Cambridge University Press.

Dickinson, Edward Ross. 2008. "The German Empire: An Empire?" *History Workshop Journal* 66: 129–62.

Ebbinghaus, Angelika, and Karl Heinz Roth. 1992. "Vorläufer des 'Generalplans Ost': Eine Dokumentation über Theodor Schieders Polendenkschrift vom 7. Oktober 1939." *1999: Zeitschrift für Sozialgeschichte des 20. und 21. Jahrhunderts* 7: 62–92.

Forster, Albert. 1942. "Die Volkstumsfrage im Reichsgau Danzig-Westpreußen." In *Die Volks-*

tumsfrage im Reichsgau Danzig-Westpreuße, edited by Albert Forster and Wilhelm Löbsack, 3–5. Danzig: Gauschulungsamt der NSDAP Danzig-Westpreußen.

Fröhlich, Elke, ed. 1987. *Die Tagebücher von Joseph Goebbels: Sämtliche Fragmente*, series 3, vol. 6. Munich: K. G. Saur.

Haar, Ingo. 2008. "Leipziger Stiftung für deutsche Volks- und Kulturbodenforschung." In *Handbuch der völkischen Wissenschaft*, edited by Ingo Haar and Michael Fahlbusch, 374–82. Munich: K. G. Saur.

Herb, Guntram Henrik. 1997. *Under the Map of Germany: Nationalism and Propaganda, 1918–1945*. London: Routledge.

Hitler, Adolf. 1941. *Mein Kampf*. New York: Reynal and Hitchcock.

Jureit, Ulrike. 2012. *Das Ordnen von Räumen: Territorium und Lebensraum im 19. und 20. Jahrhundert*. Hamburg: Hamburger Edition.

Keyser, Erich. 1940. *Geschichte des deutschen Weichsellandes*. Leipzig: Hirzel.

———. 1942. "Die Erforschung der Bevölkerungsgeschichte des deutschen Ostens." In *Deutsche Ostforschung: Ergebnisse und Aufgaben seit dem ersten Weltkrieg*, vol. 1, edited by Hermann Aubin et al., 90–104. Leipzig: Hirzel.

Krannhals, Detlef. 1940. "Reichsgau Danzig-Westpreußen: Die Entstehung einer deutschen Landschaft." *Schriften der Adolf Hitler-Schule: Schulungsburg Danzig-Jenkau der NSDAP* 26. Danzig: Kasemann.

Krausnick, Helmut. 1985. *Hitlers Einsatzgruppen: Die Truppen des Weltanschauungskrieges, 1938–1942*. Frankfurt: Fischer.

Kroll, Frank-Lothar. 1998. *Utopie als Ideologie: Geschichtsdenken und politisches Handeln im Dritten Reich*. Paderborn: Schöningh.

Krzoska, Markus. 2008. "Ostforschung." In *Handbuch der völkischen Wissenschaft*, edited by Ingo Haar and Michael Fahlbusch, 452–63. Munich: K. G. Saur.

Lemkin, Raphael. 1944. *Axis Rule in Occupied Europe: Laws of Occupation—Analysis of Government—Proposals for Redress*. Washington, DC: Carnegie Endowment for International Peace.

Levine, Herbert S. 1969. "Local Authority and the SS State: The Conflict over Population Policy in Danzig-West Prussia, 1939–1945." *Journal of Central European Affairs* 11: 331–55.

Löbsack, Wilhelm. 1943. "Reichsgau Danzig-Westpreußen—der Weichsel und Ordensgau." In *Wiedergewonnenes deutsches Land*, edited by Otto H. Spatz, 9–48. Munich: J. F. Lehmanns.

Longerich, Peter. 2008. *Heinrich Himmler*. Berlin: Siedler.

Mallmann, Klaus-Michael, Jochen Böhler, and Jürgen Matthäus. 2008. *Einsatzgruppen in Polen: Darstellung und Dokumentation*. Darmstadt: Wissenschaftliche Buchgesellschaft.

Mallmann, Klaus-Michael, and Bogdan Musial. 2004. *Genesis des Genozids: Polen 1939–1941*. Darmstadt: Wissenschaftliche Buchgesellschaft.

Mühle, Eduard. 2006. "Der europäische Osten in der Wahrnehmung deutscher Historiker: Das Beispiel Hermann Aubin." In *Traumland Osten: Deutsche Bilder vom östlichen Europa im 20. Jahrhundert*, edited by Gregor Thum, 110–37. Göttingen: Vandenhoeck and Ruprecht.

Orłowski, Hubert. 1996. *Polnische Wirtschaft: Zum deutschen Polendiskurs der Neuzeit*. Wiesbaden: Harrassowitz.

Penck, Albert. 1925. "Deutscher Volks- und Kulturboden." In *Volk unter Völkern: Bücher des Deutschtums*, edited by Karl C. von Loesch, 62–73. Breslau: Ferdinand Hirt.

Reagin, Nancy. 2007. *Sweeping the German Nation: Domesticity and National Identity in Germany, 1870–1945*. Cambridge: Cambridge University Press.

Rossino, Alexander B. 2003. *Hitler Strikes Poland: Blitzkrieg, Ideology, and Atrocity*. Lawrence: University Press of Kansas.

Schenk, Dieter. 2000. *Hitlers Mann in Danzig: Gauleiter Forster und die NS-Verbrechen in Danzig-Westpreußen*. Bonn: Dietz.

Smith, Bradley F., and Agnes F. Peterson, eds. 1974. *Heinrich Himmler: Geheimreden 1933 bis 1945*. Berlin: Propyläen.

Steinbacher, Sybille. 2000. *"Musterstadt" Auschwitz: Germanisierungspolitik und Judenmord in Ostoberschlesien*. Munich: K. G. Saur.

Taylor, Telford, ed. 1962. *Hitler's Secret Book*. New York: Grove Press.

Wirsching, Andreas. 2001. "'Man kann nur Boden germanisieren': Eine neue Quelle zu Hitlers Rede vor den Spitzen der Reichswehr am 3. Februar 1933." *Vierteljahrshefte für Zeitgeschichte* 49: 517–50.

Zernack, Klaus. 1980. "Bemerkungen zur Geschichte und gegenwärtigen Lage der Osteuropahistorie in Deutschland." In *Europa Slavica—Europa Orientalis*, edited by Klaus-Detlev Grothusen and Klaus Zernack, 542–59. Berlin: Duncker and Humblot.

5

Race contra Space: The Conflict between German *Geopolitik* and National Socialism

MARK BASSIN

Introduction

In the Anglo-American world at least, German *Geopolitik* has since World War II carried with it a certain mystique. Emanating in the popular imagination from a mysterious, Munich-based "Institute for Geopolitics," it is commonly assumed that its practitioners not only stood in immediate proximity to the National Socialist leaders in Germany under the Third Reich, but moreover that they in fact supplied key ideas and theories, on the basis of which the Nazis formulated their overall foreign policy goals. The role the geopoliticians were thought to have played in the formation of the Nazi *Weltanschauung* could indeed hardly be overestimated: an article in *Life* magazine by the American who interrogated Haushofer after the war—someone who obviously ought to know—purported among other things to detail "how Haushofer gave Hitler a philosophy" (Walsh 1946, 107). Certain factors suggest that this view is plausible: in the person of Karl Haushofer, at least one geopolitician unquestionably did stand close to the leaders of the National Socialist movement and the movement itself. Moreover, the fact that a vitally important aspect of geopolitical doctrine—the theory of *Lebensraum* and all that it implied—clearly was taken up by the Nazis and indeed "put into practice," as it were, after 1939 suggests not only contacts but positive influences. As Diner (1984) quite correctly emphasizes, geopolitics shared numerous affinities with National Socialism, as indeed did most of German conservative sentiment at this time. At the same time, however, there always remained fundamental differences between the two doctrines. Recent research—most notably H.-A. Jacobsen's (1979) two-volume collection of Haushofer's writings, correspondence, and diary excerpts—suggests that these differences in fact increased in intensity after the Nazis came to power in 1933, and that subsequent relations with the geopoliticians were affected by them. The depiction of Haushofer as the man "who gave Hitler his ideas" is at this point clearly untenable. Indeed, rather than complicity, a picture is now emerging of an uneasy and tense relationship between the geopoliticians and the Nazi state, a relationship characterized by ever-increasing suspicion and intolerance on the part of the latter toward the former. The geopolitical "mystique" thus betrays more and more a mythical basis, and not only because, as it turns out, the infamous "Institute for Geopolitics" never in fact existed. More importantly, it is increasingly obvious that geopolitics failed of its own or was not permitted to play the key role ascribed to it.[1]

This chapter addresses the question as to why this should have been the case. The argument is made that elements in the philosophical or ideological orientations of the various movements in question were at least one significant factor. Geopolitics, deriving from F. Ratzel's political geography of the 1890s, was imbued with the latter's scientific materialism and environmental determinism. It was this fact that laid the groundwork for a conflict with National Socialism, which took much fundamental inspiration from a very different tendency: the antirationalist, antimaterialist *völkisch* movement of the nineteenth century. For the Nazis, the individual—imbued with a specific set of racial characteristics—was supreme, while Haushofer and his colleagues stressed environmental influences. For them, considerations of inherited genetic qualities were, if not entirely irrelevant, at least clearly not of primary importance. Fatefully for the geopoliticians, the Nazi state, true to its character as a totalitarian system, was particularly sensitive to and intolerant of precisely these sorts of ideological divergences.

Ratzel, Environment, and Race

Although the relationship of Friedrich Ratzel to the geography of the Nazi period has traditionally been a matter of some controversy, recent research leaves little question that the "father" of modern human geography was at the same time the grandfather of German *Geopolitik* (e.g., Jacobsen 1979, 1: 241–46ff.; Faber 1982, passim). Many of the fundamental insights of the geopoliticians, while not necessarily Ratzel's original creations, were nevertheless taken over directly from his unique, geographically oriented articulation of them. Chief among these were the organismic view of the state, and the notion of *Lebensraum*, or living space. On a somewhat more abstract but no less important level, German geopolitics after 1918 persisted in the Ratzelian tradition of environmental determinism. It is to this latter fact, I argue, that the subsequent tensions with National Socialist ideology may ultimately be traced.

Determinism in Ratzel's thinking implied two things. On the one hand was the view, inspired by nineteenth-century post-Darwinian scientific materialism, that the operation of the universe was "determined" by immanent natural laws. These laws were unconditional and applied equally to all organic phenomena on earth: that is, to the human realm as well as to all other forms of plant and animal life. It was from this perspective that the study of the natural sciences gained such overwhelming importance at the time, for through the study of the natural non-human world it was possible to ascertain the laws which then would provide the key to understanding the workings of human society. In this way, the metaphoric view of human society as an organism acquired its meaning: an organism subject to the same natural laws of growth, development, and decline as everything else. Significantly, Ratzel's professional training, up to and including his doctoral degree, was entirely in the natural sciences of zoology and geology. His general human geography was then built on this foundation, and his "scientific" political geography was specifically concerned to identify and analyze "the laws of the territorial growth of states" (Ratzel 1896, 97ff.). The second element in Ratzel's determinism was the more familiar notion, taken from Carl Ritter, of the physical environment as a molding influence on the character and development of human society. This perspective was fully apparent already in his earliest work in human geography—his "cultural geography" of the United States (Ratzel 1878–80, 2: 45–47; Bassin 1984, 13–16)—and he developed it yet more fully as the basis of his *Anthropogeographie* (e.g. Ratzel 1899–1912, 1: 48, 198–99). Indeed, Ratzel's self-acknowledged crowning achievement—the application of the laws of anthropogeography to the political life of mankind—had as a principal goal the demonstration and analysis of environmental influence on the historical evolution of political life. By geographical environment Ratzel meant not only the tangible properties of the local milieu: its flora, fauna, topography, climate, soil, and resource endowment, and so on. Of equal or greater importance were more abstract considerations of relative location, distance, and the simple factor of the availability of *Raum*, or space.

Each of these two elements in Ratzel's determinism suggests a similar philosophical perspective that is critical for my argument. In both cases the explanatory emphasis is on factors *external* to human society—either on the working of incontrovertible natural laws or the effects of the physical environment. Influence and control comes in either case from without, and the implication is that humans have little ability to contravene either. The significance of Ratzel's view on this matter is that it went very much against at least one intellectual tendency in European and American thinking, important already during Ratzel's time and fated to become much more so in subsequent decades: that of modern racism.

Concern with race and heredity in the nineteenth century was widespread and was expressed in a multitude of differing theories and outlooks. One of the key problems facing those trying to understand the evolution of races and the differences between them was the so-called dilemma of "nature or nurture:" the relative significance, that is, of factors of inborn, inherited genetic makeup as opposed to external or environmental influence. In some theories of race both of these factors were considered in some sort of combination, but by the end of the century a strong current in racial thinking had emerged clearly favoring the former and discounting the latter. The qualities attributed to different races, it was argued, derived wholly from their differing genetic constitutions. The character of the individual and the racial group to which he or she belonged was therefore already predetermined in the biotic plasma out of which it came, and nothing could be done to change or overcome this inherited genetic complex. Moreover—and quite importantly—the various racial types were seen as essentially fixed. Alteration or mutation within these types could occur in only one way, that is, by affecting the genetic constitution itself. This could be done deliberately and selectively in a constructive manner (toward which end the science of eugenics was developed), or could result from what became the quintessential racial nightmare: careless and uncontrolled interbreeding among the races.

By the turn of the century, this particular form of racial thinking had become quite widespread. It was propagated, among others, in the popular works of Arthur de Gobineau and Houston Stewart Chamberlain. On a scholarly level, racism was honored as a science and elevated to the status of a university discipline in, for example, the fields of phrenology, craniology, and eugenics (Gould 1981, 30–145). The important point for our purposes is that this particular brand of racial thinking ran directly and explicitly counter to environmentalism. Through stressing the primary importance of inherited genetic qualities and the immutability of races, environmental conditions became logically irrelevant. The character of society and the individual was not determined, indeed could not even be influenced, by the physical environment, but was rather entirely predetermined through biological inheritance.[2] At the very most, environmental factors were allowed a distant primeval role in human prehistory, when the various races were originally cast (e.g., Schrepfer 1936, 967), but even this was by no means universally accepted and in any event did not challenge the primacy of genetic factors in the present.

Ratzel was deeply aware of the extent to which contemporary racial theories contravened the spirit of his own views, and it is much to his credit that he criticized and rejected them as explicitly as he did. Humankind was "despite all differences of race" nevertheless a single family, the members of which shared the fundamental attributes of reason, language, religion, and a certain cultural fund. The ubiquitous ability to mate fruitfully indicated that all possessed a similar biological constitution as well (Ratzel 1899–1912, 1: 204; 1901–2, 2: 617). As his environmentalist bias might suggest, he specifically attacked the notion of the immutability of races (*Unveränderlichkeit der Stammformen*); like all other organisms, humans mutated and evolved under environmental influence, and continue to do so (Ratzel 1906, 487). He went on to attack the notion of racial purity, so important at the time, as mere wishful thinking, for all modern peoples, through migration, growth, and physical expansion, have absorbed innumerable foreign elements that have fundamentally altered their original stock (Ratzel 1906, 465–67). The Germans themselves were not racially pure, Ratzel maintained, and consisted of at least two distinct racial types, one related to the Mongolian (Ratzel 1906, 474). Moreover, this mixing was not at all detrimental, but rather positive, for the resulting hybrid is in fact stronger and more able to succeed. Gobineau and Chamberlain, along with the "race fanatics" in general, he attacked specifically for their "backwardness" and "erroneous conception of reality" (Ratzel 1906, 485–87).

Ratzel's objections to racial thinking were not limited to scientific biological disputes. With even greater passion he attacked the application of these ideas to state life—that is, the conviction that ethnic or racial affinity should somehow serve as the basis for national unity. This so-called *Nationalitätenpolitik* had by the late nineteenth century come to inspire broad layers of the population in Central and Eastern Europe, and within this context Ratzel's vociferous rejection of it was nothing short of heretical. Rather than strive to include all members of a given ethnic group in one state, he argued, the state should rather seek its basis and boundaries in its *geographische Grundlage*, or the configuration of the physical environment (Ratzel 1901–2, 2: 676; 1923, 25).

For Ratzel, national unity and coherence was a product, not of ethnic or racial homogeneity, but rather of geographical conditions which create similar conditions of life. The German peasant grows grain and potatoes "from the Alps to the North Sea: his house, his barn, his views on life, even his stove . . . are the same throughout all of Germany" (Ratzel 1906, 476). It was this similitude in environmentally conditioned economic activity that was the source of German unity, and not ethnic affinity at all.[3] Switzerland provided an even clearer example of this principle, where no fewer than four distinctly different races were firmly united on the basis of geographical conditions into a strong and enduring political unit (Ratzel 1878, 190; 1923, 141). Ratzel's rejection of race as a significant factor in the constitution of states led him to oppose for the most part the national chauvinism and aggression which accompanied the *Nationalitätenpolitik* of the time. (This was despite his membership of ultrachauvinistic organizations such as the Pan-German League.) Rather, his was a plea for toleration of neighboring peoples and national minorities, and, indeed, he greeted the prospect that with ever-increasing international intercourse, the differences between nations and races would grow less and less (Ratzel 1878, 192–93; 1901–2, 2: 676; Faber 1982, 395).

German *Geopolitik* and the Ratzelian Tradition

The term *Geopolitik*, or geopolitics, was coined around the turn of the century by the Swedish political scientist Rudolf Kjellén. It was only during World War I, however, that his work and ideas became broadly popular in Germany.[4] A more or less distinct though hardly intellectually homogeneous "school" of geopolitics then emerged in the post-1918 period, largely through the organizational efforts of the retired Bavarian general Karl Haushofer. *Geopolitik*, as it took shape in the writings of Haushofer and others, was very much a part of the general conservative mood prevalent in Germany in the 1920s, characterized by an uncompromising opposition to the Versailles treaty and the Weimar government, which were seen as inextricably linked. As a quasi-intellectual, quasi-political movement, *Geopolitik* quickly became widespread, with numerous adherents among academics and political activists alike. Prominent in the latter group was Haushofer's war comrade and student, Rudolf Hess, who at the same time was a founding member of Hitler's National Socialist Party and eventually became the Führer's second-in-command.

The full array of ideas and beliefs that characterized geopolitics as an ideology is a topic in itself and cannot be treated within the confines of the present chapter.[5] Discussion here is limited to a few key considerations. While the years since Ratzel's death in 1904 had witnessed fundamental changes in Germany's position in Europe and the world, still the legacy of his political geography remained in important ways decisive for *Geopolitik* in the 1920s and 1930s. This legacy was transmitted both in the work of Kjellén, who although never having studied with Ratzel was nonetheless a true disciple, as well as in that of Haushofer himself. Despite some striking differences in doctrine, Ratzel was consistently honored as the brilliant mason who laid the foundation for what was to become the science of *Geopolitik*. Even after the outbreak of World War II, Haushofer paid homage to Ratzel by editing, as one of his last published works, a collection from the grand master's writings (Ratzel 1940).

As with Ratzel, the *Weltanschauung* of the geopoliticians after 1918 was founded in large measure on notions of scientific materialism and determinism. In a summary statement written in 1925 on the nature of his geopolitics, Haushofer maintained that the course of political events was underlain by fixed laws. The general nature of these laws could be discovered through rational inquiry, and with this knowledge prognoses and an overall plan of action for a state could be formulated (Haushofer 1979a, 511). This point is important, for from it Haushofer derived the ultimate task and at the same time justification for geopolitics. The discipline set out precisely to determine these laws and chart a national plan of action in accordance with them (Haushofer 1979b, 547). Like Ratzel, Haushofer argued that such a science had to begin by incorporating, indeed basing itself on, biology in a sort of marriage of the natural and political sciences, much in the spirit of the late nineteenth century (Haushofer 1979a, 512–14; Diner 1984, 3–5). As a logical corollary to his view of a world determined through the operation of independently functioning natural laws, Haushofer asserted that geopolitical knowledge is objective and therefore equally true and valid for all: "Genuine, scientific geopolitical knowledge must necessarily remain independent of the convictions of all political parties, and must be equally true for the most radical left as well as the right" (Haushofer 1979b, 545). Haushofer took this seriously, and was outspoken in his admiration not only for the geopolitical insights of conservative spokesmen for the rival British Empire, such as Halford Mackinder (upon whose "geopolitical genius" he unceasingly lavished praise), but for those of leading Soviet Marxist ideologists as well (e.g., Haushofer 1979b, 546). Indeed, as is seen later, the archconservative Haushofer was capable of remarkably high praise even for the geopolitical analyses presented in the works of contemporary German communists.

The second element that was observed in Ratzel's determinism—an emphasis on the primacy of environmental influences—was characteristic of Haushofer as well. The study of *Geopolitik* demonstrated "the dependence of all political events on the enduring conditions of the physical environment" (Haushofer 1979a, 512), and it is clear that the laws governing these events relate for the most part to environmental influences. Haushofer was at pains to demonstrate the ultimate inability both of human society and the individual to resist the course determined by nature. While conceding that great and strong leaders can on occasion affect the course of history—by arousing the passions of the masses they can temporarily "raise them above the conditions of the natural world" (ibid., 511)—in the end the environment will win out, and the society will settle back into the proper course of historical development dictated by it. At this point, he explained in a radio address delivered in 1931:

> Geopolitics takes the place of political passion, and development dictated by natural law [*naturgesetzliche*] reshapes the work of the arbitrary transgression of human will. The natural world, beaten back in vain with sword or pitchfork, irrepressibly reasserts itself in the face of the earth. This is geopolitics! (Haushofer 1979b, 547)

A determinism essentially identical to Haushofer's was expressed in Richard Hennig's *Geopolitik: The Theory of the State as a Living Being*, which appeared in 1928. Hennig shared his colleague Haushofer's perspective in regard to the natural laws which determine the functioning and development of human society, and perhaps even outdid the latter in his zeal to identify these in terms of environmental determinism:

> This is precisely the strength of the new approach [i.e., geopolitics]: *historical and political events in the fate of the peoples of the earth are to a considerable extent subject to unalterable natural laws* . . . . In the same way that the science of economics speaks of an "iron law" of wages, so is it also possible to speak of an iron law of geopolitical [i.e., environmental] influences. (Hennig 1928, 7; emphasis in original)

Indeed, the influence of the natural environment on the conduct of human political affairs was nearly absolute. The young Soviet republic, for example, possessing the same physical-environmental conditions as had the Russian empire, consequently pursued the identical state policies. Like Haushofer, Hennig explicitly minimized the significance of the individual in the face of such dominating natural laws. While stopping short of denying entirely the effect an individual or government might actually exercise, he circumscribed this considerably. Bismarck was certainly a great leader, but the German states would have been united even without his leadership. He, like Luther or Columbus before him, had simply been the "executor of natural laws," and his greatness lay in the ability to recognize and follow them (Hennig 1928, 7–8).

Within this framework, the position of the progenitors of German geopolitics on the significance of race was essentially similar to that of their mentor. To be sure, Ratzel's outspoken panhumanism did not survive World War I, and Haushofer—no doubt with a careful eye on the young National Socialist movement which he hoped to influence—was at pains to always include some sort of *Rassenlehre* or theory of race as a category when he outlined his new science (e.g., Haushofer 1979a, 513). Nevertheless, this appeared always as a lesser element, and one directly dependent on the all-important factor of *Raum*, or environment. In an early work of his on the geopolitics of Southeast Asia, the development of races is treated as a consequence of ongoing conditioning by the physical environment. Races are not biologically fixed or locked, but rather are constantly evolving, mixing, and mutating under this external influence (Haushofer 1923, 92–93). In the manifesto cited above he explicitly identified the environment as a race-forming (*Rassenbildende*) factor. In a most telling passage, he characterized theories of racial superiority, along with imperialism, as merely another ideology—that is, opinion and not factual knowledge—subject to the "objective" scrutiny of *Geopolitik* (Haushofer 1979a, 517–18).

In the work of Hennig, always less equivocal than Haushofer, the question of race simply had no place. He did not consider it as an issue relevant to his geopolitics, and indeed largely avoided the term. He explained important factors such as "national character" (*Volkscharakter*), popularly interpreted in terms of racial qualities, without any reference to genetic inheritance, but rather as a product of the physical environment. Hennig approvingly cited attempts by Herder and others to link this character to factors such as climate (Hennig 1928, 2, 62). Yet the key to the historical development of a people was not to be found in the national character, even if this were environmentally derived: it lay rather in the environment itself and the possibilities and restrictions it presented. In an example that was subsequently to cause him no little difficulty, Hennig explained the emergence of the northern Germans as the dominant power in the Baltic region, and the associated ascendance of the Hansa, not in terms of innate national qualities but rather out of the economic imperative to conquer and control the richest fishing regions and the most lucrative markets (ibid., 53). This final point is indicative of Hennig's tendency, traces of which were apparent in Ratzel as well, to interpret environmental influence as not always direct, but rather as mediated through economic patterns and activities. In this spirit, Hennig further argued that it had been the exigencies of Germany's economic life—the need to create a unified market—that lay behind its eventual unification in the nineteenth century (ibid., 6).

Thus, German *Geopolitik* remained in important ways true to the tradition it claimed, that of nineteenth-century determinism and scientific materialism. This is not to suggest that Haushofer and other geopoliticians were unambiguously rational thinkers and materialists, and this was not in fact the case. Haushofer's evocation of the "forces of blood and soil" in his professional writing and his poetry, or his fascination with astrology, suggest fundamentally irrational, indeed mystical, elements (e.g., Jacobsen 1979, 1: xi). He explicitly rejected an excessive materialism in the form of "geographical sociology" (Haushofer 1928a, 23–24). Hennig also, as will be seen, disclaimed the accusation of total materialism (Hennig 1936, 973). In a somewhat different way, this had been true of Ratzel himself, especially in his later years.

The important point is that, like the latter's political geography, the enterprise of the geopoliticians was founded in the spirit of a materialist, law-seeking science.

The Ideology of National Socialism

In order to understand the conflicts that eventually emerged between the ideas just discussed and those of the National Socialist state, some attention must be given to the intellectual or ideological underpinnings of the latter. Even more than geopolitics itself, National Socialism was a loose and highly opportunistic amalgam of often unrelated and even contradictory ideas, prejudices, and beliefs, which even the most comprehensive and competent analyses have been unable to unravel entirely (e.g., Neumann 1944; Arendt 1966; Bracher, Sauer, and Schulz 1969; Fest 1970). Nevertheless, certain general characterizations of direct relevance for the theme of this chapter can be made.

The corpus of ideas and beliefs taken over by National Socialism derived to a significant degree, as George Mosse (1964) and others have endeavored to show, out of the so-called *völkisch* movement in Germany. This movement originated in the mid-nineteenth century as a reaction to the social dislocation and overall sense of alienation that accompanied the processes of industrialization and urbanization at this time. Taking much inspiration from the Romantic movement, *völkisch* thinkers rejected rationalism and rigorous logical thinking (then still associated with the Enlightenment of the eighteenth century), and maintained that genuine truth and knowledge must be sought rather in human emotions and subjective feeling (Mosse 1964, 13ff.). From the 1860s on, the original rejection of Enlightenment rationalism was redirected against the scientific materialism which came to dominate intellectual activity in the wake of the Darwinian revolution. Modern science, it was felt, was blinded and deceived by its empiricism, inductive method, and excessive attention to individual facts. Real insight into the nature of the world could never come from the cold mechanistic laws which were the goals of science, for the universe operated not on the basis of such laws but rather in harmony with primeval, subjective values.

Völkisch thinkers longed to see the world as an integrated, well-functioning totality, founded on genuine spiritual harmony. A view of such a world, they felt, was obscured by the petty pursuits of academic science. Darwin had ultimately delivered "bricks, not buildings," as the *völkisch* writer Julius Langbehn put it, with which the "spirit of the whole" could never be understood. Science per se was not disclaimed, but it was felt that real science should be based not on empiricism but on intuition, sentiment, and an "unbounded subjectivity" (cited in Stern 1961, 125).[6] It followed that science should become art, and indeed, for the *völkisch* movement, the artist was seen as the principal articulator of the quest for inner spiritual values. The title of Langbehn's major work, *Rembrandt as an Educator*—which became something of a *völkisch* manifesto—betrayed this preoccupation with art and the artist (Mosse 1964, 39–44).

A central focus of concern for the *völkisch* movement was the corporate body from which it derived its name: the *Volk*, or people. The idea itself of the *Volk* was of course a Romantic notion par excellence, having been developed in the writings of Herder, Rousseau, and others. The overall significance of this concept was immense, and it suffices to indicate one important element. By the late nineteenth century, the *völkisch* movement had crossed lines with the proponents of racial ideas discussed above. The theories of Gobineau and Chamberlain were taken over as their own, with the result that the *Volk* came to be seen in racial terms (Mosse 1964, 87–107). With this, not only was the panhumanism of the Enlightenment rejected, so too was the "benign plurality" with which the early Romantics had treated ethnographic and cultural diversity among peoples. Fully in the spirit of contemporary racial ideas, all this was replaced by a vision of humankind divided into mutually exclusive and irreconcilable categories. In addition to the standard criteria of cultural patterns and language which had been used (and indeed still are) in order to identify different national groups, the ties which bonded a people or *Volk* together were now understood in genetic or biological terms as well. In this way each group was seen as necessarily and fundamentally different from all other peoples.

Above and beyond this categorization, a specific hierarchy was now established, based on the conviction that certain races were "higher" or more worthy than others. The qualities which denoted the relative worthiness of one group over another were conferred by birthright, or inborn, and thus were at once both absolutely guaranteed and inalienable. Only one thing could threaten the otherwise unerodable qualities of the superior race: the danger of genetic adulteration through interbreeding with different races. Consequently, the notion of racial purity became at this time not only important, but a literal obsession, and led eventually to laws specifically prohibiting procreative activity with races perceived to be less worthy.

Another concept of importance for *völkisch* thinkers was that of *Verwurzelung*, or rootedness: the rootedness of the individual and the *Volk* in their native soil and landscape. This was a reaction to the greatly increased mobility, physical as well as social, that accompanied the industrialization and urbanization of the day and seemed to bring with it ever greater alienation and dislocation. Rootedness, in contrast, suggested enduring contact and harmony with the vital forces of nature which comprised the totality so anxiously sought. Of special significance was the native *Heimat* or home region—the specific location where the *Volk* was rooted and where it maintained its elemental ties with the natural world. This *Heimat* was visualized above all in terms of the *Kulturlandschaft*, or cultural landscape: a primeval environment molded and given character through prolonged harmonious habitation by the *Volk*. The cultural landscape was the medium through which the *Volk* was linked to the natural world and from which it derived its virtues and energies. Indeed, one could hardly be visualized without the other: "The [cultural] landscape . . . became a vital part of the definition of the *Volk* through which it retained continuous contact with the life spirit of the transcendent cosmos" (Mosse 1964, 15ff.).[7]

Those aspects of the ideology or *Weltanschauung* of National Socialism that are important to this study can be understood for the most part within this context of *völkisch* convictions and beliefs. First and foremost was the overriding obsession with race that lay at the bottom of all National Socialist thought and action. This involved on one hand the arrogant conviction in the absolute superiority of their own, the so-called Aryan race, but on the other a panic-stricken hysteria at the prospect of its genetic dilution and degeneration through contact with other, lesser races. It was therefore entirely logical that the archenemy designated by the Nazis—the Jewish people—was defined strictly in racial terms. The continuity of racial thinking from the nineteenth century to the Nazi leaders in the 1930s was unbroken, and moreover quite conscious, as indicated by the homage Hitler paid by kissing the hand of a dying Houston Stewart Chamberlain (Mosse 1964, 93). The point can hardly be overemphasized: racism formed the essence and rationale for the National Socialist movement.

As well, National Socialism persisted in its rejection of rational thought and of all science that operated in accordance with it. The Romantic and *völkisch* notion that humankind had become "enslaved by reason" was promoted energetically, and Hitler declared that a materialist science had a devastating effect because it "leads away from the instincts" (Fest 1970, 250). A common alternative or antidote to this remained the recourse to art, which could supply genuine insights and truths about the real nature of the world. It is not an insignificant observation that Hitler's first and passionately pursued career goal was to be a painter (Toland 1976, 37–69), and that he was never entirely to renounce this early ambition.

Geopolitics and National Socialism

From the very beginning of its rapid development in the early 1920s, geopolitics stood in a close relationship to the Nazi movement. Haushofer, who resigned his commission and turned to academics only after Germany's defeat in 1918, was an outspoken conservative and German nationalist. This was hardly uncommon at the time, however, and by itself did not necessarily indicate an overall sympathy with the platform of the young Nazi Party. Nevertheless, many elements were shared: rejection of the terms for the German surrender, a determination to reestablish German territorial integrity after its truncation at Versailles, and an implacable hostility to the Weimar Republic and refusal to accept it as the genuine representative of the German people. Less immediate, but no less significant, was the chauvinistic dream of the creation of a greater pan-Germany, which would gather together all the German peoples of Europe for the first time into a unified ethnic state (Jacobsen 1979, 1: 178ff.). Finally, the conviction was common that Germany faced a crisis of overpopulation, the sole solution to which was the acquisition of yet more *Lebensraum* for national development. With his *Geopolitik* Haushofer believed he possessed the science which could help Germany along its difficult course of regaining lost stature and realizing its dreams. In a much-quoted passage, he asserted that geopolitics ought to become the "geographical conscience of the state," and he preached it eagerly to all who would listen.

Among those who listened were some of the leading figures of the National Socialist movement. Haushofer's closest contact was with Rudolf Hess, who had fought under him during the war and then studied political geography and geopolitics with him at the university in Munich. The two developed an extremely close personal relationship, Haushofer even offering refuge from the police when his young friend was being sought after the failed "Beer Hall Putsch" in 1923 (Toland 1976, 237). While notions of Germany's vital need for an expanded *Lebensraum* had been common for decades in the *völkisch* movement and elsewhere, and it is quite incorrect to see them as actually originating with Haushofer, still his "scholarly" articulation of this issue without question had its significance. His influence was mediated for the most part through Hess (Fest 1974, 217; Jacobsen 1979, 1: 224–57), although Haushofer did on a number of occasions meet briefly with Hitler. In the early years of World War II he proudly recounted how he had left a "well-read" copy of Ratzel's *Political Geography* behind him after a visit in 1924 to Landsberg prison, where from his cell Hitler was busy dictating the first draft of *Mein Kampf* to his assistant, Hess (Ratzel 1940, xxvi).

Yet despite these close contacts and apparent affinities, neither Haushofer nor his geopolitics was ever an integral part of the National Socialist movement. At heart a careful conservative, he could not but be thoroughly uncomfortable with Hitler's nihilistic radicalism, and found numerous elements in the latter's program entirely unacceptable. Important among these was the obsession described above with the question of race. Although unquestionably bigoted, and often contemptuous of national and ethnographic differences, Haushofer's interpretation of these differences was largely cultural, and he shared with the other geopoliticians an unwillingness to adopt the biological racism of the National Socialists. Thus, he did not follow their lead in making this the basis and central concern of his *Weltanschauung*. His reserve arose in part because the Nazi views on race were, from an intellectual standpoint, so clearly at odds with his determinism, in part also perhaps from the fact that his wife Martha was herself half-Jewish. This meant that his family was not racially pure and under the Third Reich could hold the dubious status of "honorary" Aryans (*Ehrearier*) thanks only to the special intervention of their patron Hess (Jacobsen 1979, 1: 170, 451).[8]

Yet the problems behind Haushofer's failure to enter the National Socialist mainstream did not come entirely, or even primarily, from his side. For however fundamental the ideological differences separating them may have been, Haushofer was more than anxious to be of use for this up-and-coming political movement, and unceasingly demonstrated his willingness to put his geopolitics at its disposal. Even their crude racism did not deter him in this endeavor. In contrast, National Socialism, despite its scattered, inconsistent, and eclectic nature, was quite emphatic in its insistence on the strictest of ideological conformity from its adherents and those it trusted. Haushofer's thoroughly genuine sympathies with the Nazis in regard to a need for a fundamental revision of the contemporary German status quo, the creation of a pan-German ethnic state, and even the overtly expansionist drive to increase the country's *Lebensraum* were not enough; more important was the increasingly obvious fact that he did not completely share the *völkish* values they esteemed so highly. A good example of this is offered by the principal organ of German geopolitics—the *Journal of Geopolitics*, founded by Haushofer in 1924. Although conservative and anti-Weimar in its general tone, it was nevertheless an often eccentric collection. Into the early 1930s it contained, along with tracts sympathetic to the Nazis, the writings of Karl Wittfogel, then still an important intellectual in the German Communist Party, or the liberal Jewish historian and specialist on Near-Eastern affairs, Hans Kohn (Kohn 1930; Wittfogel 1932).[9] Intellectual pluralism of any sort was of course anathema to the Nazis, most especially that involving Communists and Jews. Thus Haushofer, despite his own unconcealed desire to place his *Geopolitik* in the service of the Nazi state, was viewed all along with suspicion. This was a suspicion which could not but have been enhanced by his wife's heritage, which Hitler himself never overlooked (Jacobsen 1979, 1: 451).

After the Nazi ascendance to power in 1933, *Geopolitik* was subject to a heightened array of tensions and pressures. On the one hand, the state officially endorsed the existence of this field, and in official proclamations stressed the need for geopolitical education both for party cadres and the German people in general (e.g., *Verordnungsblatt der Reichsleitung der NSDAP*, January 15, 1934, cited in Zeck 1936, 975). The lack of teachers with a proper education in geopolitics was repeatedly pointed out, and the need for creating additional chairs of geopolitics affirmed as one remedy ("Aussprache über Geopolitik" 1935, 715–18). This might have augured well for Haushofer, were it not for the problem that what the National Socialists understood under the term *geopolitics* did not necessarily coincide with the discipline as he and others had conceptualized and developed it in the 1920s. For along with the apparent endorsement came a much intensified ideological scrutiny, and increasingly obvious efforts were made to steer it in a direction more acceptable to the new regime.

As in all other areas of German life, this ideological *Gleichschaltung* of geopolitics was at first probing and tentative. A number of brief statements had already appeared in the *Journal of Geopolitics* in 1933, stressing the importance of obviously *völkisch* elements such as rootedness and *Heimat*: geopolitics was the study of the "unity of the *Volk* with its motherland," its task being "to demonstrate the rootedness [of the *Volk*] in the earth's soil, to observe its power in political life, and to make the *Volk* conscious of it" ("Denkschrift" 1933, 301–3). These initial prescriptions were general and vague, and Haushofer, always anxious to comply, gave them his hearty endorsement. Of greater significance were the activities of the so-called *Arbeitsgemeinschaft für Geopolitik*, or Working Group for Geopolitics, founded in 1931 and actively supported by, among others, Haushofer's publisher Kurt Vowinckel. This group, made up primarily of party members, acted as an outpost of "genuine" National Socialist doctrine within the field of geopolitics, and after 1933 came increasingly to fill the role of a sort of ideological watchdog. Significantly, Haushofer did not belong to this group, but cooperated by giving it regular space in his journal (Troll 1947, 19–20; Schöller 1957, 3).

As the Nazis consolidated their power, the *Arbeitsgemeinschaft* grew more outspoken, and eventually a meeting between them and selected geopoliticians was held in 1935 in Bad Saarow for the purposes of ideological orientation. The high-level sponsorship of this meeting, shared by the Central Institute for Education (*Zentralinstitut für Erziehung und Unterricht*) and the party itself, was indicative of just how important a geopolitics "properly pursued" was for the Nazi state. No question was left as to the direction the movement was expected to take; as one speaker succinctly declared, "geopolitics is well-suited to deepen *völkisch* attitudes." Under Haushofer's conciliatory chairmanship, agreement on these points appeared to be general, but the principal concern of the party representatives was explicitly stated: geopolitical education had to incorporate "racial thinking" (*Rassedenken*) as well as "spatial thinking" (*Raumdenken*)—space in this case referring primarily to the geographical environment ("Aussprache über Geopolitik" 1935, 713–15). The deficiencies on the part of the geopoliticians in regard to racial considerations had clearly not escaped notice.

Indeed, these deficiencies—entirely understandable in terms of the intellectual roots of geopolitics outlined above—were in the long run to prove the overriding source of friction between the field of geopolitics and the state. Contrary to appearances, this issue had by no means been resolved at the Bad Saarow meeting, and matters finally came to a head in the following months. From a number of different sides, geopolitics came under scathing attack for the errors and "false ways" (*Irrwege*) in its ideological underpinnings. Although implicated, Haushofer himself was not the main focus, and the brunt of the stinging criticism was directed rather against his colleague Richard Hennig, whose coauthored *Introduction to Geopolitics* appeared in a third edition in 1934, and a fourth in 1935. The attacks often began by implicitly associating Hennig with liberalism, which the new regime in principle categorically rejected (e.g., Gehl 1934–35, 589; Zeck 1935, 14). Given Hennig's attachment to an organismic theory of the state, such an association was spurious, but was significant in that along with it came a much more legitimate charge, that of an antiquated "mechanistic way of thinking" and moreover of a far-reaching materialism "that indulged in a theory of environmental determinism (*Milieuthorie*)" (Schrepfer 1936, 966). The chief objection here was to the "inflexible rules" (*feste Regeln*) on which the geopoliticians professed to base their science,[10] and the strictly causal relationship that this suggested between man and the environment.

> [Materialism] equates in principle human connection with the environment (*Raumgebundenheit*) with the [assumed] determining influence of this environment (*Raumbedingtheit*). On this basis it attempts to explain the activities

> of the state and the course of political events as the result of factors in the natural environment and the necessary outcome of the operation of laws, and disregards the vital forces which reside in the *Volk* and the race. (Schrepfer 1936, 966)

As is apparent, the National Socialists were quite willing to acknowledge the connection of man with the environment, for this was entirely in line with the *völkisch* emphasis on the rootedness of the *Volk* in the natural landscape. However, the suggestion that this relationship might be subject to inflexible laws, and involve human subjection through dependency on the environment, violated notions of the primacy of racial strength and initiative.

Hennig's failure to recognize the quintessential importance of race provided the focus for the sharpest attacks. His detractors pointed out that he did not draw the necessary distinctions between races and seemed not to appreciate the cultural hierarchy into which all groups had to fit. One commentator, writing in the journal of the National Socialist teachers' organization, noted with horror that in Hennig's work the Germanic tribes (*Stämme*) were "without any trace of racial thinking equated with Apaches, Mohicans, and the wild peoples of Africa, who have not yet come under the domination and educational influence of the white man" (Gehl 1934–35, 589). The essential independence of racial qualities from environmental influence was repeatedly affirmed: "the [same] geographical environment is not [necessarily] able to produce the same races" (Bahr 1934–35, 110). Predictably, Hennig's explanation of the rise of German power on the North Sea and the Baltic seemed a major slight of Germanic racial virtue, and brought the full wrath of his attackers down upon him. In Hennig's view, it was maintained, the Hansa was built not on the "daring spirit" of traders from Lower Saxony, but rather on the herring catch (e.g., Gehl 1934–35, 590). One writer explained the "correct" position on this problem in detail, at the same time providing a concise statement of the fundamental point of friction between National Socialism and geopolitics:

> Men do not become seafarers because the natural environment is conducive for this, rather: men who sense strength and energy in themselves go to sea. The inhabitants of the Alps and the Himalayas, the Caucasus and the Hindu-Kush, are by character defiant of life (*lebenstrotzig*) not because they must constantly struggle with the natural environment of their high mountain homes: rather, they climbed into the mountains [in the first place] because they sensed in themselves the strength to defy the natural elements. It is first of all due to inner qualities, in other words to the quality of the genetic inheritance (*Bluterbe*), that some people become seafarers, others mountain inhabitants. . . . *The environment influences the manner in which the life-energies*

> (Lebensenergien) *are expressed, but it does not create them. These energies are rather innate* (ursprünglich) *values, and come from the racial inheritance* (rassische Erbe). (Zeck 1936, 976; emphasis added)

As noted, the brunt of the attacks on geopolitics were not directed against Haushofer himself, and indeed criticism of Hennig by the *Arbeitsgemeinschaft* appeared even on the pages of the *Journal of Geopolitics* (e.g., "Geopolitik und Kochkunst" 1936). Nevertheless, the grand master did not emerge entirely unscathed. A minor party functionary, W. Staudinger, writing under the pseudonym Wilhelm Seddin (Jacobsen 1979, 2: 254), published an article in 1936 entitled "The Mistaken Ways of 'Geopolitics' as a World View." This piece was directed not to a scholarly audience but rather to the general public, and appeared in the popular Nazi publication *Das Volk*. That Haushofer, although unnamed, was in fact the clearly intended subject of the broadside was apparent not only from the cartoon caricatures illustrating the article, but from its content as well. The rejection of "geographical materialism" in the works discussed above was repeated, as was the castigation of geopolitics for its failure to appreciate the primacy of factors of race (Seddin 1936, 60–62).

Certain points not made in the other attacks, however, served to implicate Haushofer more directly. Seddin had harsh words for what he called the "fondness" (*Liebe*) of geopoliticians for the Soviets. This was a reference to Haushofer's persistent conviction, inspired, among other sources, very much by his reading of Mackinder, that Germany would benefit most of all from an alliance with the "Heartland," embodied by the Soviet Union (Jacobsen 1979, 1: 277; Diner 1984, 11–16). Finally, and most importantly, the author attacked the attempts to present geopolitics as something useful, even necessary for the National Socialist state: "That after 1933 geopolitics immediately made broadly conceived efforts to do business with the new state is known in every alley." Indeed, Haushofer went so far as to pass his new science off as the "National Socialist theory of the state." This, Seddin concluded, was nothing short of impudent. He was unconvinced by the language of "blood and soil" which geopolitics—i.e., Haushofer—adopted after the Nazi takeover, implying that it was used insincerely and opportunistically. In any event, the familiar and unbridgeable opposition between race and environment remained dominant. In the interpretation of Haushofer and others, the German "geopolitical environment" included all who inhabit it, and the vital distinctions as to "whether or not [these inhabitants] have a profession, or whether they are German or Jewish" were ignored (Seddin 1936, 63).

The geopoliticians themselves differed in their reactions to all this negative attention. Haushofer, increasingly fearful of losing further stature, and persisting moreover in his original desire to indeed "do business" with the Nazi government, did not publicly respond. His correspondence with Hennig, however, indicates that he was both aware of and quite upset by the attacks (e.g., Jacobsen 1979, 1: 255–56). Hennig, on the other hand, confronted his detractors directly. In a brief but concise statement published some months after the spate of articles just discussed, he took up the criticisms leveled against him and geopolitics. In view of the time and the general climate in the country, his response was rather admirable, for it was a full and vigorous reaffirmation of geopolitics as he had portrayed it in 1928. On the question of materialism, he stated outright that one of the tasks of geopolitics was to demonstrate "that in very numerous cases material factors, the qualities of the soil, mineral, botanical, and animal resources have influenced the policies of states, have brought about their development and occasioned even wars" (Hennig 1936, 970). He disclaimed the accusation of "geographical materialism" by explaining that he was not discounting spiritual values (ibid., 973); but this was clearly a question of semantics, insofar as the statement just quoted would more than suffice to win him that title from the Nazis.

On the question of race, his statements were equally explicit. He maintained that racial characteristics were first and foremost a "product of cultivation and development by the environment." These genetic characteristics were subject to change due to environmental influence, and one of the primary tasks of geopolitics was precisely the study of this process (Hennig 1936, 972). The study of race in and of itself (*Rassenkunde*) has no particular place in geopolitics (ibid., 974), and in a letter to Haushofer he exclaimed, "a boundary between geopolitics and a science of race now has to be very clearly drawn" (Jacobsen 1979, 2: 254). Using the same term as Ratzel, he denigrated his detractors by characterizing them as "preachers of race" (*Rasse-Prediger*), and paraphrased their sermonizing sarcastically in the following manner:

> Truly heroic *Völker* never think about money or merchandise when they go to war. Rather they seek out the battle for the sake of the battle, in order to win fame and because [their] blood and race drive them to do it. (Hennig 1936, 970)

This, of course, was precisely what the ideological spokesmen for the state would have maintained, and one may well wonder if Hennig realized that with his sarcasm he was mocking the holiest of *völkisch* values.

In conclusion, a point made earlier may be stressed. The attacks on geopolitics were directed against specific treatments and interpretations. That there should, in principle, be such a field was never questioned, and, indeed, most of the attacks cited above contained indications of just what a properly constituted National Socialist geopolitics ought to look like. Predictably, the model for this was thoroughly *völkisch*. Geopolitics was no longer to be a political science, but rather a "basic principle" (*Grundprinzip*) and an "inner attitude" (Schrepfer 1936, 924), which ought not to speak in terms of laws at all, much less those of the natural environment (Gehl 1934–35, 590). An important task was to seek out "timeless forces and timeless values" such as the "spatial or environmental destiny" (*Raumschicksal*) of a people—in short, geopolitics should be reconstituted metaphysically as an art (Zeck 1936, 975). Needless to say, it was commonly held to be of the utmost importance to found the field on the basis of the science of race.

In another regard, it was felt that the most fruitful field of application for geopolitical study was not the traditional concern with the state as a single unified entity, but rather various aspects of civic life on a more local level: the study of German villages, cities, economic regions, and so on (Schrepfer 1936, 924). This point is of considerable interest. It has been noted that the rejection by the Nazis of environmental determinism did not amount to the rejection of a man-land connection per se. Insofar as the new view of man focused above all on inherent racial qualities and virtues, the reaction away from environmentalism often went far in the opposite direction, toward a focus on conscious human impact in reshaping this environment. Thus, in a truly National Socialist geopolitics the object of study would no longer be the unaltered physical environment and its influences on human society, rather the anthropogenically altered environment, in other words the *Kulturlandschaft* or cultural landscape (e.g., Gehl 1934–35, 590). The idea of the cultural landscape could not have fitted more perfectly into the ideological framework we have been discussing. In the first place, it was evidence par excellence of the duration and intensity of the rootedness of human communities in the land. Moreover, because it was the product of deliberate human activity, the elements characterizing it could be derived from the racial qualities of its creators: "Like the pages of an open book, the [cultural] landscape informs us as to the essential characteristics of the race which had formed it" (Bahr 1934–35, 111; Sauer 1934–35, 580–81). The cultural landscape was understood in its classic sense in terms of house types, field forms, altered vegetation patterns, and so on. Quite apart from the ideological considerations of race contra environment, this new conception had nothing in common even with the formal scope of traditional geopolitics, and in fact amounted to a particular variety of what today is studied as "cultural geography." From the mid-1930s on, whatever stature and authority geopolitics still retained was progressively eroded. Hennig came out with yet another edition of his work in 1938, in which the pressure that had been brought to bear was apparent. While still insisting on the primary goal of the work as the study of environmental influences, he was now quick to explain that "spiritual forces of various types" were fully active as well in the determination of political and historical processes. He had high words of praise for the activities of his former detractors, the *Arbeitsgemeinschaft für Geopolitik*, in fostering geopolitical study in Germany (Hennig and Körholz 1938, ix). His solicitude went even further: in addition to epigraphs from the pen of Adolf Hitler, he now included entire sections devoted to previously unthinkable topics, such as "The Role of Great Individuals in History" (ibid., 10, 164–68). Haushofer himself continued to lose favor, and by the end of the decade some of his work was even banned. With the ill-conceived flight of Rudolf Hess to Scotland in 1941, he lost his all-important patronage and came under direct suspicion. He was interrogated by the Gestapo and finally, after the complicity of his son in a conspiracy against Hitler in 1944 was uncovered, actually interned for some weeks in Dachau. Interrogated after the war's end by the occupying American forces (Walsh 1946) and under obvious suspicion as to his role in the Nazi state, he and his wife took their lives together in 1946 (Jacobsen 1979, 1: 443–47).

Conclusion

Thus it is clear that, despite all intentions, geopolitics as conceived and developed by Haushofer and others in the 1920s did not play the role of a state science under the Nazis, and could not have done so. Geopolitics, deriving essentially from the scientific materialism of the nineteenth century, was conceptualized as a law-seeking discipline. Its fundamental goal was to study and reveal the determining elements in the natural environment that served to influence and mold human society, in particular the evolution of political states and the course of relations between them. The human element, whether individual or as a group, was seen as subject to this external influence. The ideological orientation of National Socialism, on the other hand, came largely from the *völkisch* movement, and differed in fundamental ways from geopolitics. Materialist, law-seeking science was in principle rejected in favor of a Romantic recourse to the emotions and sentiments. Moreover, the essentially subjective human element in the form of the *Volk* held the dominant position. By infusing the *Volk* with racial qualities, emphasis was moved from external influence to innate, inherited biological qualities which were supreme and could not be affected through environmental factors. After the Nazis came to power in 1933, these divergences in perspective became increasingly problematic, and resulted in official attacks upon geopolitics and the loss of whatever significance it may have had. To be sure, the term *Geopolitik* continued to be quite popular, but as practiced by those closer in spirit to the Nazis it had little in common with the actual tradition of the discipline, developed from Ratzel through Kjellén to Haushofer and his colleagues. It was only by altering its essence that it could continue to appear useful.[11]

In attempting to correct exaggerated notions of the influence the geopoliticians exercised under the Third Reich, however, it was by no means the intent of this chapter to exonerate Haushofer or his colleagues. However great or limited their tangible effect, their enterprise represented a cynical and opportunistic use of a scholarly mantle to justify and indeed purify their underlying political objectives. While not a racial fanatic, Haushofer's passionately held vision of a pan-Germany was unmistakably expansionist. Moreover, with his "scientific" teachings of Germany's need for an expanded *Lebensraum* he without question contributed to the chauvinist atmosphere in which the Nazis formulated their foreign policy, and ultimately put it into effect after 1939.

A number of questions remain unresolved and present interesting challenges for further research. I have maintained that Haushofer was not the originator of the concept of *Lebensraum* in National Socialist thinking. Indeed, writers far more popular than Haushofer propagandized this idea entirely independently—writers such as Hans Grimm (1928) in his *A People without Space* (Smith 1983). The question remains: where did the notion come from? The answer lies in an examination of the nineteenth-century roots of German national-political and colonial thinking. Ideas about the need for Germany to expand its territorial base to accommodate a growing population are apparent in writings as early as those of Friedrich List in the 1840s, if not earlier (Smith 1985). Rather than the contribution of German geography—whether from Ratzel (who did in fact coin the term), Haushofer, or others—to German public opinion, notions of *Lebensraum*, on the contrary, found their way into geography *from* the larger political-social context. The "scientific" cloak which then was draped on the idea by geographers and other scholars was fascinating, and may well in its turn have had an influence of some sort; more significant, however, seems the insight that "science" was in this case but a reflection of the larger national context.

Acknowledgments

This chapter is based on ideas first developed in a talk given in 1984 at the Institut für europäische Geschichte (Abteilung Universalgeschichte), in Maim, West Germany. I would like to thank the Institut for its generous support of my research. In addition, I am most grateful to David Livingstone, Queens University, Belfast, and Yi-Fu Tuan, University of Wisconsin–Madison, for their perceptive comments on an earlier draft.

Notes

1. In retrospect it is apparent that this mystique was the result of studies that were obviously—if unavoidably—biased; carried out during World War II, not only was available information limited, but the subject of study was a national enemy as well (Dorpalen 1942; Strausz-Hupé 1942; Fifield and Pearcy 1944). Until recently, these remained the most important English-language sources on the subject.

2. This is not to suggest that a combined concern for both the geographical environment *and* race as determining factors could not coexist within a single framework (see note 11). However, as George Mosse has demonstrated in his *Toward the Final Solution: A History of European Racism* (1985), proponents of late nineteenth-century racial science, at least in Germany, were commonly vociferous in their insistence on the primacy of racial factors to the explicit exclusion of external or environmental influences. In a most interesting way, Mosse traces this opposition to its roots in the Enlightenment (1985, l–34, and passim).

3. As I have demonstrated in detail in another article (Bassin 1987), Ratzel's consistent adherence to this view led him to explicitly reject what later became known as the principle of "*ein Volk, ein Land*," or the conviction that all members of an ethnic group rightfully belong within one single state structure. This is a most significant point, and one that the geopoliticians after World War I, with their cherished dreams of a greater pan-Germany, were to betray. Indeed, to many convinced German nationalists of the 1890s, who had made their peace with Bismarck's *kleindeutsch* solution only briefly, if at all, Ratzel's reasoning must have seemed very much a sort of national treason.

4. Kjellén's *The Great Powers of the Present Day* (1915), for example, first appeared in German translation one month after the outbreak of the war, and within the year had already gone through nine printings.

5. Of the numerous English-language works that have appeared in the last few years dealing with the history of political geography and geopolitics, a few (e.g., Brunn and Mingst 1985; Parker 1985) give satisfactory if often highly generalized overviews. The most recent German and European work on this subject (e.g., Bakker 1967; Gollwitzer 1972–82; Matern 1978; Jacobsen 1979; Wolff-Powęska 1979; Schöller 1982; Diner 1984; Fox 1985; Klein 1985) is on the whole more comprehensive and nuanced.

6. This is not to suggest that scientists were not also part of the *völkisch* movement. Among others, the ultra-Darwinist and founder of the modern study of ecology, Ernst Haeckel, along with his Monist league, shared basic *völkisch* values, particularly a rabid racism (Gasman 1971). Indeed, the holistic inspiration underlying the concept of ecology fitted quite well with the yearning for totality just discussed. In contrast, Ratzel held firmly and consistently to his scientific principles. Even though he was an early admirer and a student of Haeckel, Ratzel later

denounced him. Quite in line with his orientation as presented above, Ratzel remained in important ways aloof from the *völkisch* movement, and attempts to identify him within it, as, for example, in Smith's excellent study (1980, 55ff.), are not convincing.

7. As suggested here, the concept of the cultural landscape originated in the early 19th century as part of this *völkisch* movement. Its progenitors in large measure were not academics, but German nationalist writers of various backgrounds. It was only later in the century that it was taken over as a scholarly pursuit in the form of *Landschaftskunde* (Overbeck 1965, passim). Thus, the origins of modern cultural geography have to be sought, in part at least, far outside the strict borders of the discipline itself, in the growth of European nationalist thinking.

8. After the war Haushofer insisted he had always been critical of Hitler and indeed claimed to have refused to review *Mein Kampf* in his journal when it first appeared (Walsh 1946, 117). Jacobsen (1979, 1: 451) characterized Haushofer's view of Hitler somewhat less equivocally as "ambivalent." On the question of race, Diner (1984, 6–10) is able to demonstrate a definite anti-Semitism (directed, as was common, primarily against the so-called *Ost-Juden*, or Jews from Poland and other territories to the east, and not against assimilated German Jews such as his wife's family). It remains nevertheless clear that the exaggerated racial notions of the Nazis were foreign to him, and the concessions that he was to make to them in his writings apparently caused him no little grief. Confronted by his American interrogator after the war with his own statements sympathizing with the racial ideas of the Nazis, the elderly geopolitician and ex-general broke down into tears (Walsh 1946, 110).

9. Haushofer's attitude toward Wittfogel was remarkable. He was attracted to the latter's work out of his interest in Asian society, which Wittfogel interpreted from the standpoint of historical materialism. Despite Wittfogel's trenchant Marxist perspective, Haushofer found his writings fascinating and compelling, and consistently gave them highly enthusiastic reviews in his *Journal of Geopolitics* (Haushofer 1927; 1928b; 1931). The culmination came in 1932, when Haushofer reprinted excerpts from Wittfogel's article "Geopolitics, Geographical Materialism, and Marxism" (Wittfogel 1929; Ulmen 1978, 540), which had appeared in a scholarly organ of the German Communist party, *Under the Banner of Marxism* (published simultaneously in Russian in the Soviet Union), and was in fact an attack on geopolitics overall, including that of Haushofer! (When Wittfogel was interned with other Communists in 1933, his wife sought assistance, among others, from Haushofer, who apparently raised his case in vain with Hess [Ulmen 1978, 162].) Not only is this a rather enigmatic illustration of Haushofer's belief in "objective" knowledge; more importantly, his attraction to Wittfogel came from real affinities in their perspectives, based on the materialism and determinism to which both of them, albeit with differing nuances, subscribed. This unexpected affinity with Marxist determinism is a point of potentially great significance which deserves further research.

10. Indeed, this objection led certain National Socialist commentators to reject even the organismic view of the state, which for the geopoliticians, as for Ratzel before them, represented one of the foundations of their entire system (even in the title of his 1928 work, Hennig referred to the state "as a living being"). The problem with the organismic view was that it locked its subject into a predetermined, unalterable pattern determined by laws of organic growth: youth, maturity, and decline. An article in the leading National Socialist youth magazine, *Will and Power*, pointed out that this amounted to heresy, for it would mean that "the attempt of the National Socialist revolution to create a new and better state would be meaningless and pointless" (Zeck 1935, 14).

11. The history of environmentalism in geography in fact often followed a rather different pattern from the one indicated in this essay, in that it combined elements of geographical de-

terminism *and* some sort of racial theory more or less harmoniously within one unified perspective. This was especially true of the Anglo-American tradition, which was at the same time the most heavily influenced by explicitly environmentalist ideas. The works, for example, of Ellsworth Huntington (1945, 61–76), Thomas Griffith Taylor (1927, passim), and Nathaniel Shaler (Campbell and Livingstone 1983, 272–77) all demonstrate a dual concern: to show environmental and genetic influences in some sort of combination. Indeed, Huntington—the quintessential determinist—was one of the founding fathers of eugenics in the United States. In view of this, the German example might have been anomalous, but more research is clearly needed on this question. What is certain is that Ratzel was outspokenly opposed to racial thinking, and whatever willingness the later geopoliticians may have shown toward adopting and integrating these ideas in some way was meaningless in the face of the absolute ideological demands imposed by the state.

References

Arendt, Hannah. 1966. *The Origins of Totalitarianism*. London: Allen and Unwin.

"Aussprache über Geopolitik in Bad Saarow." 1935. *Geographische Wochenschrift* 3: 713–17.

Bahr, K. 1934–35. "Die Rassenfrage im Erdkundeunterricht." *Die Deutsche Höhere Schule* 2 (4): 109–15.

Bakker, Geert. 1967. *Duitse Geopolitiek, 1919–1945: Een Imperialistische Ideologie*. Assen: Van Gorcum.

Bassin, Mark. 1984. "Friedrich Ratzel's Travels in the United States: A Study in the Genesis of His Anthropogeography." *History of Geography Newsletter* 4: 11–22.

———. 1987. "Imperialism and the Nation State in Friedrich Ratzel's Political Geography." *Progress in Human Geography* 11 (4): 473–95.

Bracher, Karl D., Wolfgang Sauer, and Gerhard Schulz. 1969. *Die Nationalsozialistische Machtergreifung: Studien zur Errichtung des Totalitären Herrschaftssystems in 1933–34*. Cologne: Westdeutscher Verlag.

Brunn, Stanley D., and Karen A. Mingst. 1985. "Geopolitics." In *Progress in Political Geography*, edited by Michael Pacione, 41–76. London: Croom Helm.

Campbell, J. A., and D. N. Livingstone. 1983. "Neo-Lamarckism and the Development of Geography in the United States and Great Britain." *Transactions of the Institute of British Geographers* 8 (3): 267–94.

"Denkschrift: Geopolitik als nationale Staatswissenschaft." 1933. *Zeitschrift für Geopolitik* 10: 301–05.

Diner, Dan. 1984. "'Grundbuch des Planeten': Zur Geopolitik Karl Haushofers." *Vierteljahrshefte für Zeitgeschichte* 32: 1–28.

Dorpalen, Andreas. 1942. *The World of General Haushofer: Geopolitics in Action*. New York: Farrah and Rinehart.

Faber, Karl-Georg. 1982. "Zur Vorgeschichte der Geopolitik: Staat, Nation, und Lebensraum im Denken deutscher Geographen vor 1914." In *Weltpolitik, Europagedanke, Regionalismus: Festschrift für Heinz Gollwitzer zum 65. Geburtstag*, edited by Heinz Dollinger, Horst Gründer, and Alwin Hanschmidt, 389–406. Münster: Aschendorff.

Fest, Joachim E. 1970. *The Face of the Third Reich*. Translated by Michael Bullock. New York: Pantheon.

———. 1974. *Hitler*. New York: Harcourt, Brace and Jovanovich.

Fifield, Russell H., and George E. Pearcy. 1944. *Geopolitics in Principle and Practice*. Boston: Ginn.

Fox, William. 1985. "Geopolitics and International Relations." In *On Geopolitics: Classical and Nuclear*, edited by Ciro E. Zoppo and Charles Zorgbibe, 15–44. Nato ASI Series D, No. 20. Dordrecht: Martinus Nijhoff.

Gasman, Daniel. 1971. *The Scientific Origins of National Socialism*. London: Macdonald.

Gehl, W. 1934–35. Review of *Einführung in die Geopolitik*, by Richard Hennig and Leo Körholz. *Die Deutsche Höhere Schule* 2: 589–90.

"Geopolitik und Kochkunst." 1936. *Zeitschrift für Geopolitik* 12: 651.

Gollwitzer, Heinz. 1972–82. *Geschichte des weltpolitischen Denkens*. Vols. 1 and 2. Göttingen: Vandenhoeck and Ruprecht.

Gould, Stephen J. 1981. *The Mismeasure of Man*. New York: Norton.

Grimm, Hans. 1928. *Volk ohne Raum*. Munich: A. Langen.

Haushofer, Karl. 1923. Zur Geopolitik der Selbstbestimmung: Südostasiens Wiederaufstieg zur Selbstbestimmung. Munich; Leipzig: Rösl.

———. 1927. Review of *Das erwachende China*, by Karl A. Wittfogel. *Zeitschrift für Geopolitik* 4: 187–90.

———. 1928a. "Grundlage, Wesen, und Ziele der Geopolitik." In *Bausteine der Geopolitik*, edited by Karl Haushofer, Erich Obst, Hermann Lautensach, and Otto Maull, 2–48. Berlin: K. Vowinckel.

———. 1928b. Review of *Aufzeichnungen eines chinesischen Revolutionärs* by Sun-yat Sen. Edited and with introduction by Karl A. Wittfogel. *Zeitschrift für Geopolitik* 5: 608–9.

———. 1931. Review of *Wirtschaft und Gesellschaft Chinas*, by Karl A. Wittfogel. *Zeitschrift für Geopolitik* 8: 170.

———. 1979a. "Politische Erdkunde und Geopolitik" (originally published 1925). In *Karl Haushofer: Leben und Werk*, edited by Hans A. Jacobsen, 1: 508–24. Boppard-am-Rhein: Harald Boldt.

———. 1979b. "Zur Geopolitik" (originally published 1931). In *Karl Haushofer: Leben und Werk*, edited by Hans A. Jacobsen, 1: 542–52. Boppard-am-Rhein: Harald Boldt.

Hennig, Richard. 1928. *Geopolitik: Die Lehre vom Staat aus Lebensform*. Leipzig: B. G. Teubner.

———. 1936. "Geographischer Materialismus?" *Zeitschrift für Erdkunde* 4: 969–74.

Hennig, Richard, and Leo Körholz. 1938. *Einführung in die Geopolitik*. 5th ed. Leipzig: B. G. Teubner.

Huntington, Ellsworth. 1945. *Mainsprings of Civilization*. New York: Mentor Books.

Jacobsen, Hans A., ed. 1979. *Karl Haushofer: Leben und Werk*. Vols 1 and 2. Schriften des Bundesarchives 24. Boppard-am-Rhein: Harald Boldt.

Kjellén, Rudolf. 1915. *Die Grossmächte der Gegenwart*. Translated by Carl Koch. Leipzig; Berlin: B. G. Teubner.

Klein, Jean. 1985. "Reflections on Geopolitics: From Pangermanism to the Doctrines of Living Space and Moving Frontiers." In *On Geopolitics: Classical and Nuclear*, edited by Ciro E. Zoppo and Charles Zorgbibe, 45–76. Nato AS1 Series D, No. 20. Dordrecht: Martinus Nijhoff.

Kohn, H. 1930. "Übersicht über den Vorderen Orient." *Zeitschrift für Geopolitik* 7: 791–94.

Matern, Rainer. 1978. "Karl Haushofer und seine Geopolitik in den Jahren der Weimarer Republik und des Dritten Reiches." PhD diss., Karlsruhe University.

Mosse, George L. 1964. *The Crisis of German Ideology: Intellectual Origins of the Third Reich*. New York: Grosset and Dunlap.

———. 1985. *Toward the Final Solution: A History of European Racism*. Madison: University of Wisconsin Press.

Neumann, Franz. 1944. *Behemoth: The Structure and Practice of National Socialism*. New York: Oxford University Press.

Overbeck, Hermann. 1965. "Ritter-Riehl-Ratzel: Die grossen Anreger zu einer historischen Landschafts- und Länderkunde Deutschlands im 19. Jahrhundert." *Kulturlandschaftsforschung und Landeskunde*: 88–103. Heidelberg: Heidelberger Geographische Arbeiten 14.

Parker, Geoffrey. 1985. *Western Geopolitical Thought in the Twentieth Century*. London: Groom Helm.

Ratzel, Friedrich. 1878. "Die Beurteilung der Völker." *Nord und Süd: Eine deutsche Monatsschrift* 6: 177–200.

———. 1878–80. *Die Vereinigten Staaten van Nord-Amerika*. Vols. 1 and 2. Munich: R. Oldenbourg.

———. 1896. "Die Gesetze des räumlichen Wachstums der Staaten: Ein Beitrag zur wissenschaftlichen politischen Geographie." *Petermanns Mitteilungen* 42: 97–107.

———. 1899–1912. *Anthropogeographie*. Vols. 1 and 2. Stuttgart: J. Engelhorn.

———. 1901–2. *Die Erde und das Leben: Eine vergleichende Erdkunde*. Vols 1 and 2. Leipzig; Vienna: Bibliographisches Institut.

———. 1906. "Nationalitäten und Rassen." *Kleine Schriften*, 2: 462–87. Munich: R. Oldenbourg.

———. 1923. *Politische Geographie, oder die Geographie der Staaten, des Verkehrs, und des Krieges*. 3rd ed. Munich: R. Oldenbourg.

———. 1940. *Erdenmacht und Völkerschicksal*. Introduction by Karl Haushofer. Stuttgart: A. Kroner.

Sauer, O. 1934–35. "Rasse, Volk, und Lebensraum." *Die Deutsche Höhere Schule* 2: 580–81.

Schöller, Peter. 1957. "Wege und Irrwege der politischen Geographie und Geopolitik." *Erdkunde* 11: 1–20.

———. 1982. "Die Rolle Karl Haushofers für die Entwicklung und Ideologie nationalsozialistischer Geopolitik." *Erdkunde* 36: 160–67.

Schrepfer, Hans. 1936. "Geopolitik und Erdkunde." *Zeitschrift für Erdkunde* 4: 919–25, 966–69.

Seddin, W. 1936. "Der Irrweg einer 'Geopolitik' als Weltanschauung." *Das Volk: Kampfblatt für völkische Kultur und Politik*, May, 59–64.

Smith, Woodruff D. 1980. "Friedrich Ratzel and the Origins of Lebensraum." *German Studies Review* 3: 51–68.

———. 1983. "The Colonial Novel as Political Propaganda: Hans Grimm's *Volk ohne Raum*." *German Studies Review* 6: 215–36.

———. 1985. *The Ideological Origins of German Imperialism*. Oxford: Oxford University Press.

Stern, Fritz. 1961. *The Politics of Cultural Despair: A Study in the Rise of the Germanic Ideology*. Berkeley: University of California Press.

Strausz-Hupé, Robert. 1942. *Geopolitics: The Struggle for Space and Power*. New York: Putnam's.

Taylor, Thomas G. 1927. *Environment and Race*. London: H. Milford.

Toland, John. 1976. *Adolf Hitler*. New York: Ballantine Books.

Troll, C. 1947. "Die geographische Wissenschaft in Deutschland in den Jahren 1933 bis 1945: Eine Kritik und Rechtfertigung." *Erdkunde* 1: 3–47.

Ulmen, G. L. 1978. *The Science of Society*. The Hague: Mouton.

Walsh, Edmund A. 1946. "The Mystery of Haushofer." *Life Magazine*, September 16, 106–20.

Wittfogel, Karl A. 1929. "Geopolitik, geographischer Materialismus, und Marxismus." *Unter dem Banner des Marxismus* 1: 17–51; 4: 484–522; 5: 698–735.

———. 1932. "Geopolitik, geographischer Materialismus, und Marxismus." Excerpts from Wittfogel (1929). *Zeitschift für Geopolitik* 9: 581–91.

Wolff-Powęska, Anna. 1979. *Doktryna geopolityki w Niemczech*. Studium niemcoznawcze Instytutu Zachodniego no. 34. Poznań.

Zeck, H. F. 1935. "Geopolitik?" *Wille und Macht: Führerorgan der nationalsozialistischen Jugend* 3 (8): 14–17.

———. 1936. "Was ist Geopolitik?" *Zeitschrift für Erdkunde* 4: 974–77.

6

Back Breeding the Aurochs: The Heck Brothers, National Socialism, and Imagined Geographies for Non-Human *Lebensraum*

CLEMENS DRIESSEN AND JAMIE LORIMER

> *Volk* without space and space without wildlife! These are both bad things.
>
> ULRICH SCHERPING (1958, 135)

> The extinct aurochs has arisen again as a wild German species in the Third Reich.
>
> LUTZ HECK (1939, 537)

Introduction

The aurochs is the wild ancestor of domestic cattle. It went extinct in 1627. In the 1920s two German zoologists—Lutz Heck and Heinz Heck—embarked on separate programs to bring it back to life. The brothers believed that the *Erbmasse* (hereditary material) of the aurochs could be recombined through "back breeding," thereby reversing the process of domestication. To forge their "reconstituted aurochs" they selected cattle breeds for their desired wild characteristics: one with the right horn shape, another with the correct coloration, a third for its anatomy and behavior, and so on. Lutz and Heinz were the sons of Ludwig Heck, the famous director of the Berlin Zoo, and became high-profile figures themselves. In the 1930s Lutz's project gained patronage from the Nazi elite, in the person of Hermann Göring. His wild cattle became entangled with wider Nazi efforts to conserve natural landscapes and to reintroduce indigenous species to the occupied lands of Eastern Europe.

This chapter reflects on the symbolic and material roles the brothers' scientific programs played in the Nazi geographies of occupied Eastern Europe. Lutz managed to align back breeding the aurochs with a broader legitimation for the Nazi plans for Eastern expansion. His interest in back breeding conjoined mythological, geographical, and ecological desires for hunting large animals, restoring natural landscapes, and ensuring industrial and agricultural autarky. In conversation with a wider network of Nazi spatial planners, Lutz and other nature conservation officials argued for expanding "non-human *Lebensraum*." This would involve comprehensive *Landschafts-*

pflege (landscape care) beyond the mere protection of animals and plants. In publications he called for the "complete reshaping" (for example, L. Heck 1942) of the newly conquered Eastern land in accordance with the mythical Germanic landscapes described in the ancient *Nibelungenlied*. This required reintroducing charismatic animals and addressing the "*Versteppung*" (becoming-Steppe) of the landscape that he believed to be caused by the wrong type of racialized ("Slavic") land management. His project to recreate extinct primordial wildlife was firmly integrated within wider discourses and practices of planning and conservation that emphasized the Germanic character of the European landscape. In contrast, our archival research suggests that Heinz differed markedly from his brother in his back-breeding practices and discourses. He was much more ambivalent about National Socialism, the Germanic provenance of the aurochs, or the need for German expansion to benefit landscapes and wildlife in the East.

The novel ecologies and future geographies imagined and created as part of the Heck brothers' projects touch upon a series of themes, practices, and disciplines that have come to concern historians of German science under National Socialism. These include zoology, animal breeding, and eugenics, alongside nature conservation, spatial planning practice, and geopolitical thought. We draw on and develop these literatures in our analysis of the brothers' projects. We flag the character and significance of a particular mode of "reactionary modernism" (Herf 1984) in nature conservation during this period that entangled mythologized and nationalistic hunting practices with the science of animal breeding and land management. However, as we will demonstrate, the differences between the brothers' endeavors complicate an overly schematic understanding of the ways in which natural science became entangled in National Socialism. The histories of back breeding draw attention to a series of paradoxes, tensions, and inconsistencies in Nazi ideology and conservation policies. In conclusion we bring this discussion to the present to explore whether the contemporary "Heck cattle" that descended from those animals bred during this period should be considered tainted by their Nazi genealogy. This is a timely question given their involvement in a revived interest in bovine back breeding and the ecological restoration or "rewilding" of Eastern Europe (Lorimer and Driessen 2013).

Lutz, Göring, and the Hunt

Lutz and Heinz Heck were born in 1892 and 1894 respectively and grew up in the family home in the Berlin Zoo. They witnessed their father develop a plot of land with a few cages into one of the world's foremost zoos ("Urmacher unerwünscht" 1954) and after World War I, both set out to follow in his footsteps. The eldest brother, Lutz, rose to prominence in Berlin in the 1920s and took over the position of zoo director from his father in 1932. Heinz moved out of Berlin to pursue his career and in 1928 was granted the role of reestablishing the Munich Zoo. Here he created the first "geo-zoo" featuring naturalistic dioramas of regional landscapes with multiple animal species (Hirsch and Wiesner 1986).

To determine their breeding goals the brothers took cues from the Spanish cave paintings known at the time, as well as several "images and artworks" and "precise contemporary descriptions" (see L. Heck 1936, 247; and H. Heck 1936, 9). They added bone and skull finds to help establish size and horn shape (L. Heck 1936, 284). In Berlin the emphasis was on Spanish and French fighting cattle valued for their "ferocious character" (ibid., 255). Lutz traveled extensively through southern France and Spain in the late 1920s and early 1930s in search of specimens. Heinz claimed to have started back breeding as early as 1921 but without a predefined plan. In an account of his efforts he explained that "Hungarian and Podolian steppe cattle and Scottish Highland cattle were mated with grey and brown Alpine breeds . . . and with piebald Friesians. To save time I bought a few crosses. . . . All were thrown into one pot, so to speak" (H. Heck 1951, 120; cf. H. Heck 1934a, 13). Both claimed success in bringing back the aurochs during the mid-1930s.

Back breeding was well underway before the Nazis seized control, but it took the patronage of the Nazi authorities for it to come to fruition. Lutz became a paying member (*Förderndes Mitglied*) of the SS soon after the NSDAP (Nationalsozialistische Deutsche Arbeiterpartei, or Nazi Party) came to power and joined the party on May 1, 1937[1] (Neumärker and Knopf 2007; Gautschi 2010b). This in itself is sadly not very remarkable. Recent historical analyses have shown that large numbers of German geographers, spatial planners, conservationists, and biologists took advantage of the possibilities the Nazi regime offered to further their personal careers and their causes (e.g., Rössler and Schleiermacher 1993; Deichmann 1996; Uekötter 2006; Barnes and Minca 2013). However, Lutz's involvement runs deeper. It didn't take him long to befriend and gain patronage from the Nazi authorities. He dropped off animals at Goebbels's household (Fröhlich 2008, I, 4, 138), presented Göring with a pair of lion cubs, which he personally collected after they had grown too big to handle (Neumärker and Knopf 2007), and was in conversation with Albert Speer about plans for an ideal zoo on the outskirts of Berlin.[2]

Lutz found a kindred spirit and his strongest ally in the figure of Hermann Göring. He shared Göring's passion for big game hunting and his fascination with Germanic myths such as those recounted in the *Nibelungenlied*. This epic tale, in which Teutonic knights dwelling in the primordial forest hunt deer, elk, wisent, and aurochs (Reichert 2005, 149), was prominent in the German nationalist self-understanding cultivated by the Nazis (Stoehr 2000, 167–68).[3] For Göring, the mythology of the lost aristocratic hunter apparently generated the need to restore the heroic ecology of Germanic leadership. Lutz and Göring went on numerous hunting parties together, during which they sought to relive these myths, wearing traditional dress and even carrying spears. Lutz documented their adventures in a series of intimate photographic portraits. On occasion they were also photographed together (see figure 6.1).

In 1938 Lutz released his first new *aurochsen* into the Rominten Heide hunting reserve in the presence of Göring.[4] A year later, Lutz vividly described

FIGURE 6.1. Lutz Heck and Göring carrying spears on a hunting trip, December 1934. Source: Bundesarchiv, Image 102-04224. Photographer: Georg Pahl.

in the hunting journal *Wild und Hund* an encounter between Göring and an aurochs bull. He celebrated Göring reveling in the violent character of the animal, as a "*Sinnbild der Urkraft*" (symbol of primal force) (L. Heck 1939, 537). Lutz defined the real aurochs not solely in genetic or hereditary terms but also by virtue of its aggressive behavior and development in its correct environment. It is only after having revealed its ferocious character that Lutz claims: "The extinct Auerochs has arisen again as German wild species in the Third Reich" (ibid.). The desire for bovine ferocity and combative hunting explains his selection of Spanish fighting cattle in his back breeding.

In addition to being prime minister of Prussia and commander in chief of the Luftwaffe (the German air force), Göring had also taken upon himself the positions of Reichsjägermeister (Reich hunt master) and Reichsforstmeister (Reich forest master) (Gritzbach 1942; Gautschi 2010a). Lutz was keen to both exert and highlight his own influence on his patron's nature conservation efforts. His eulogy for Göring as a conservationist (L. Heck 1943a) was summarized in the Nazi biology journal *Freude am Leben*:

> The *Reichsmarschall*, already soon after the takeover of power in 1933, and after foundational conversations with [Lutz Heck], had far-reaching plans put into effect. With his unique power and warmhearted love of the German forest and wildlife, and in his powerful efforts to retain the German people's age old nature values and strengthen their *Heimat*[5] feeling, *Reichsmarschall* Göring put himself to the task of protecting the endangered, German, large, wild species such as the wisent, aurochs, and *steinbock*. Yes, even for the last indigenous predators: bear, wolf, and lynx. (L. Heck 1943b, 7)

With such sycophancy, Lutz soon became central to Göring's efforts to repopulate the expanding Reich with his favored indigenous wildlife. In 1935, Göring opened a 60-hectare public park near his Carinhall residence in the Schorfheide conservation area north of Berlin. Here Lutz was charged with breeding the wisent (European bison) back from near extinction for use in future reintroductions (Rubner 1997). With Lutz's assistance Göring also sought to reintroduce elk in the Schorfheide (Gritzbach 1942). The aurochs back-breeding project promised to complete the list of animals Siegfried hunted in the *Nibelungenlied* and to restore their imagined lost Teutonic ecosystem.

Nature Conservation, Hunting, and the Nazis

A consensus is emerging from recent debates among historians that the "green" image of the Nazis, popularized in some quarters, was mostly a mat-

FIGURE 6.2. Lutz Heck (left) and Hermann Göring (right) studying a relief map with wildlife figurines, during a visit to the 1937 International Hunting Exhibition in Berlin. The mounted animal in the background is a wisent; on the table is an "aurochs" horn. Source: The official exhibition catalog, *Waidwerk der Welt, Erinnerungswerk an die Internationale Jagdausstellung Berlin 1937 2.–28. November*, Paul Parey, Berlin, 1938.

ter of paying lip service (Lekan 2004; Landry 2010). While prominent Nazis were keen to publicly promote environmental and conservation causes, in practice few projects were realized, and the interests of industry and armament often received priority (Brüggemeier, Cioc, and Zeller 2005). Nevertheless, the inordinate amount of time that Göring spent on his hunting estates and his patronage of expensive conservation efforts indicate that concerns over nature played some role in his decision making.[6] Göring celebrated hunting culture as an issue of national prestige in the International Hunting Exhibition he organized in Berlin in 1937 (see figure 6.2).

Uekötter dismisses accomplishments in Nazi nature protection as "merely the accidental by-product of Göring's penchant for hunting" (Uekötter 2006, 135), but this claim may unduly underestimate the centrality of hunting in the understanding and practice of many conservationists at the time. German hunters in the late 1930s considered hunting essential for maintaining the quality of a natural area. Hunters prevented the unchecked growth of game populations that would otherwise damage the forest, while selecting the right deer at the right time in order to optimize the antlers of the major bucks. Hunting became a tightly regulated and elite practice under the new national hunting law passed in 1934. Decisions on who could shoot which animals and when were from then on controlled from Berlin (Bode and Emmert 2000). Wild animals were thus subject to closely controlled "breeding" (*Hege*) and conservation practices, which integrated the creation of appropriate environmental conditions, the selection of individuals, and the management of population size, structure, and dynamics. Hunting offered a conception of nature in which human (ideally German) management and care was central to the optimal functioning of the "ecosystem" in relation to the environmental preconditions of an area.

While Göring was a keen advocate of this approach to land management, his wider responsibilities generated a mixed and sometimes conflicting set of allegiances—specifically to reconcile the creation of hunting reserves with the demands of the armament industry and the Four Year Plan, including the demand for agricultural autarky and wartime timber production. Moreover, his desire for expansive and uncultivated territory for hunting in the East was in continuous tension with Himmler's geography of *völkisch* resettlement as envisaged in Konrad Meyer's *Generalplan Ost* (General Plan for the East) (Barnes and Minca 2013). This focused on the domestication of the wilds and the creation of agricultural landscapes, for example, through the reclamation of marshland (Blackbourn 2006) and the implementation of Christaller's central place theory (Barnes and Minca 2013). Working in this context, as Göring's henchman, Lutz Heck had to align his interest in animal reintroductions with wider Nazi goals. He sought to present the newly conquered areas as offering renewed opportunities for German people and nature. In the rest of this chapter we will argue that nature conservation, as Göring and Lutz Heck imagined it, should be considered as significant in motivating, legitimating, and to a lesser degree actually shaping the colonization of Eastern Europe.

Nature Conservation as Geopolitics and Landscape Planning for the *Volk*

In 1938 Göring appointed Lutz as the head of the Obersten Naturschutzbehörde im Reichsforstamt (Nature Protection Authority within the Forest Service).[7] This made him a central figure among nature conservationists. In December of that year, Konrad Meyer's journal *Raumforschung und Raumplanung* dedicated a special issue to forestry and nature conservation (issue 11–12). This featured articles by the conservationists with whom Lutz was now formally collaborating, including the landscape architect Heinrich Wiepking-Jürgensmann, and Göring's Oberstjägermeister (chief hunt master) Ulrich Scherping. Wiepking-Jürgensmann emphasized how German culture—from myths and fairy tales to major authors and composers—and thus the German people were the expression of their forest landscape. This Germanic human being was said to organically emerge from the plant and animal world. In a contribution titled "*Wild und Raumordnung*" (wildlife and spatial planning) Scherping described how modern technological developments and the hard necessity for Germany to till every square foot of usable soil "in its struggle for food security" (*im Kampf um seine Ernährungsfreiheit*) had reduced the *Lebensraum* of the animal world (Scherping 1938, 542).

When Germany invaded Eastern Europe in the following year, Lutz was in a powerful position to contribute to shaping plans for the eastern "Germanic" land. Six months after the occupation of Poland, he wrote in an article published by the official Nazi newspaper (the *Völkischer Beobachter*) that "landscape protection is *Volk* protection, since here Nature protection works for the most precious possession we have, our greater German *Heimat*" (L. Heck 1940). Just as the true resurrection of the aurochs for Lutz was not a purely genetic affair, so the true German *Volk* is imagined as both the outcome of, and essential to, being rooted in the right landscape. Here the Nazi eugenicist project gets a spatial dimension, in which the right environment, achieved through landscape conservation and design, is just as important for *Volkspflege* as racial hygiene aimed at purging the hereditary base.

In May 1942 Lutz signed a formal agreement with Meyer on integrating his Forest Protection Agency with the Dienststelle des Reichskommissars für die Festigung des Deutschen Volkstums (the Commissariat for the Strengthening of Ethnic Germandom, RKFDV),[8] under which Meyer's *Generalplan Ost* resided (Oberkrome 2004, 241–42; see also Radkau 2003, 36). Soon after, he published an article in the journal *Neues Bauerntum*, a publication (also edited by Konrad Meyer) aimed at those involved in extending the German population into the East (L. Heck 1942). This particular edition was dedicated to Himmler's *Allgemeine Anordnung Nr. 20/VI/42*—a general standing order laying down the aims and regulations for landscape planning in the "integrated Eastern areas" (cf. Rössler and Schleiermacher 1993). Lutz stresses how the landscape of the "*Ostraum*" ("Eastern space") itself would need to be made Germanic, in addition to the creation of new German villages and towns. He explains how this will be achieved by integrating the regional authorities of *Landschaftspflege* and *Naturschutz* into a single organization. This would explicitly align the Reichsforstmeister's (Göring) nature conservation with the Reichsführer SS (Himmler) spatial planning efforts.

Lutz argues that nature conservation should move beyond the mere care of rare species to landscape-scale conservation, linking zoology and landscape planning, and turning geopolitics into a more-than-human affair. In sketching the task ahead he evokes the monotonous, boundless landscape seen by soldiers from the train to the eastern front and suggests that:

> The vast Eastern space, as ardently desired settling area, must therefore be comprehensively conquered again by us for a second time through complete transformation. The reshaping of this dull and strange landscape into a German one must be our most important goal. For the first time in history the imprinting of a cultural landscape will be consciously taken up by a people (*Volk*). (L. Heck 1942, 214)

A few pages earlier in the same issue of this journal, Konrad Meyer had stressed the need for Germanic "*Gestaltungsdenken*" (formative ideas) and "*Gestaltungskraft*" (formative power) on the "shapeless landscape" to "replenish the racial substance" of the East (Meyer 1942, 205). In this and previous issues of the journal these authors sought to demonstrate how "non-Aryan" peoples had destroyed the Polish landscape (see especially Wiepking-Jürgensmann 1942). Using maps and aerial imagery the articles claimed to prove the degenerative processes of "*Versteppung*" (becoming-steppe) associated with deforestation and the resultant soil erosion (see figure 6.3). Here the need for both human and non-human *Lebensraum* was legitimated by the racial and cultural superiority of the German *Volk* in land management (see Blackbourn 2006). In different ways these authors argued that the German *Volk* should "retake" possession of these lands and extend their beneficial care to the robbed soil. By highlighting the treeless and eroded state of the land under Polish rule and inhabitation, the eastern expansion of *Germanhood* and the ethnic cleansing of these areas were understood and promoted as (what we now would call) an "ecological restoration" project.

Colonizing the East was presented as an issue of *Landschaftsgestaltung* (landscape formation) and *Landschaftspflege* (care for the landscape). This causal link between *Volk* and landscape was thought to operate both ways. For Wiepking-Jürgensmann, who worked for both Lutz's Nature Protection Department and Himmler's RKFDV, *Landschaftspflege* (care for the landscape) equaled *Volkspflege* (care for the *Volk)*, since "humans are the product of two forces, genes and the environment" (Oberkrome 2004, 242). This discourse and its associated practice of "landscape formation as caring for the *Volk*" integrated both the geographical determinism of some geopoliticians (Bassin 2005) and the emphasis on innate racial *völkisch* qualities that was so central to Himmler's and Hitler's ideology. Thus, it promoted a spatialized form of eugenics that mirrored the hunter's management of wildlife through the control of populations and environmental conditions. Here shaping the

Ausgeraubte Landschaften des weiteren Ostens

Ein Bildbericht zum Aufsatz Prof. Wiepking-Jürgensmann: Das Landschaftsgesetz des weiteren Ostens.

Bild links unten: Abschwemmung im Lößgebiet der Karte 1. Oberflächenwasser drang, auf größerer Breite, tief ein und verursachte einen Bergrutsch, der, neben dem Gratenriß, kennzeichnend für unverbaute Lößgebiete ist. Foto: Wiepking.

Bild rechts oben: Lößtrichter im Gebietsraum der Karte 1. So fängt es an! Oberflächenwasser drang an einer Stelle tief ein. Das Wasser brach seitwärts, zur Böschung, durch und bildete einen typischen, ganz flachauslaufenden Lößkegel. Foto: Wiepking.

Bild Mitte: Typische Flachdüne in Polesien. Restlos unverbaute Landschaft.

Bild rechts unten: Podolische Platte, Sseret-Tal vor der Einmündung in den Dnjestr. Die Schwarzerde über dem Löß ist deutlich sichtbar. Der Lößhang zeigt tiefe Regenschluchten. Das Schneewasser wird direkt von der Platte dem Flußtal zugeführt.

FIGURE 6.3. Page from Konrad Meyer's journal *Neues Bauerntum*, showing the "robbed empty landscapes" of the East. The images depict areas in eastern Poland and the Ukraine, the erosion of which is claimed to lead to human induced "*Versteppung*" and climate change. Source: "Bildbericht: Ausgeraubte Landschaften des weiteren Ostens," *Neues Bauerntum* 34 (1): 5.

environment created both German land and a population naturally at home in its *Heimat* landscapes (Scherping 1938). In this way a particular conjunction of nature conservation, economic requirements, a sense of international geopolitical justice, and a *völkisch* racist understanding of what it means to inhabit a landscape were combined into a single murderous logic that promoted ethnic cleansing as a form of landscape restoration.

Lutz Heck took part in this project of restoring Germanic landscapes in the East in his formal role as the head of Nature Protection. Lutz's discourse echoed many of the most poisonous elements of National Socialist ideology. But he did not extend his rhetoric to advocate more explicitly murderous forms of racism like other central figures competing for power in the landscape planning and nature conservation bureaucracies. For example, the *Autobahn* landscape architect Alwin Seifert (Zeller 2005, 161), Wiepking-Jürgensmann who we encountered earlier, and the conservationist Walther Schoenichen[9] added staunch anti-Semitic and anti-Slavic rhetoric to this mix. Similarly, although Lutz was keen to connect his back breeding to the work of contemporary eugenicists, and proudly claimed that both Erwin Baur and Eugen Fischer had responded sympathetically to his project (L. Heck 1939), he did not link his discussions of genetics, breeding, and domestication to the popular discourse of racial hygiene associated with these figures and others such as Konrad Lorenz (Sax 1997).

Białowieża and the War for Non-Human *Lebensraum*

Göring commissioned a film on the European bison (wisent) that was released in 1941. It opens with a close-up of an ancient copy of the *Nibelungenlied* and is set against the backdrop of the ongoing war. The narrator explains that "the swift victory over Poland has brought a welcome return of pure blooded wisents to Germany. . . . Giving justified hope for the conservation of the wisent, the strongest wild animal and the last witness of former Germanic primordial forest."[10] Here the war in the East is celebrated for its liberation of authentic wildlife from incapable management. Hopes are expressed that early successes will lead to a return to the style of big game hunting associated with ancestral Germany. Furthermore, the film offers a rationale for the war as the ("re-")possession of the famous ancient forests of Białowieża in Poland and Askania Nova in the Ukraine.[11] Göring had visited Białowieża on hunting expeditions every year between 1933 and 1938 and was keen to gain control (Neumärker and Knopf 2007, 55). Soon after the forest was conquered he gave orders to reintroduce bears, and some of Lutz's back-bred aurochsen were transferred there in 1942 (L. Heck 1943a).

In the spring of 1941 Göring ordered that the forest be expanded from 100,000 to 260,000 hectares. He turned it into an official hunting ground that was cleared of people to prevent them disturbing the wildlife (Gautschi 1997, 386). Ulrich Scherping, the hunting authority leader, and Walter Frevert, who was formerly Göring's chief forester at his Rominten Heide estate, were made responsible. Under orders from Göring, Frevert's foresters embarked upon a ruthless campaign to burn down villages, evict villagers, and hunt down and execute "bandits"—local partisans and Jewish resistance fighters (Gautschi 2010a, 232; 2010b, 118). Protecting the recently introduced aurochsen provided an extra motive for this "bandit hunting," as Göring had personally ordered that any "poachers" in the area be shot (see Blood 2010). The Führer Directive No. 46 of August 1942, titled "Instructions for Intensified Action against Banditry in the East," categorized "bandits" as legally killable (Bode and Emmert 2000, 154n302). While the Hunting Commando (Jagdkommando) order of the same month called for the application of hunting techniques traditionally used for stalking wild game to people (Blood 2010). In the spring and summer of 1943, Frevert went to Białowieża on a special mission for the "pacification and evacuation" of the Białowieża Forest on behalf of the Reichsmarschall (Gautschi 1997; Blood 2010). Soon after, in the winter of 1943/44 Lutz met with Frevert and Scherping for a wild boar hunting party in the forest (see figure 6.4). Later that year Frevert reported in a letter: "All is fine with me here. Like before I am leading a hunter-battalion and shooting bandits dead as if on a conveyor belt" (Neumärker and Knopf 2007, 132).

Lutz continued transporting animals throughout the Reich right up until the very end of the war. In an official letter to a general on the Eastern Front he implored him to remain on the lookout for rare wild herbivores. He also persevered with his back breeding, repositioning his wild horse back-breeding program as part of the war effort by proposing a plan to create a hardy fighting horse. This resembled the project of the Ahnenerbe explorer and zoologist Ernst Schaefer, who was breeding a hardy "steppe horse" (Hagen 1950, 92–93) for an imagined elite breed of frontier SS men. These so-called *Wehrbauer* were modeled on Teutonic knights. Schaefer hoped they would pioneer landscape development and farming while policing the Reich's extended borders on horseback (Kater 2006, 217). As the German Reich began to implode, his primordial forests and their wildlife became one of Göring's main concerns. He ordered his personal regiment to shoot as many wild animals as possible—including Heck's aurochs—in their retreat from Rominten Heide. He charged his personal Luftwaffe division with protecting his Schorfheide estate from the advancing Soviets. When these forces were summoned by

FIGURE 6.4. A wild boar hunting party in Białowieża forest, presumably winter 1943/44. On the left stands Göring's forester Walter Frevert. Third from the right, with hands in pockets and smoking a pipe, is Lutz Heck. To his right stands Ulrich Scherping, the hunting authority leader. Source: personal archive A. Gautschi.

Hitler to defend Berlin instead, Göring himself shot the wisents before fleeing to Bavaria (Gautschi 1997, 361).

Heinz Heck: Cosmopolitan Back Breeding?

It is clear that Lutz closely aligned his mode of nature conservation and rare species (back) breeding with violent territorial expansionism and the reshaping (*Gestaltung*) of landscapes legitimated by *völkisch* and mythological imaginations. These were practices fully in step with National Socialist ideology. However, if we take a closer look at the ideological commitments and zoological practices of his younger brother Heinz, the back breeding of the aurochs emerges in a different light. For one thing, Nazi authorities seriously doubted his personal allegiance to Nazism. In November 1936 Heinz applied to the *Reichsschrifttumskammer* for formal approval to be allowed to publish and give public lectures. This led to a protracted inquiry into his "moral and political reliability" involving a series of exchanges between the regional Nazi Party headquarters and other Nazi authorities.[12] In January 1937 he was reported to be "a difficult to fathom character."[13]

Serious allegations had generated numerous suspicions. Not only had he briefly been married to a Jewish woman during World War I, but also he was suspected to have been a member of the Communist Party and for this reason to have been interned as a political prisoner in Dachau in 1933. At the time of the inquiry he was said to be separated from his wife and cohabiting with an English woman whose Aryan descent was questioned and who allegedly maintained friendly relations with Jews.[14] Zoo employees who were questioned on his opinions of National Socialism claimed to have heard him say, "I'd prefer one Jew over ten Nazis."[15] And for no apparent reason he had shot a lion he received as a gift from Rudolf Hess, an act interpreted as deeply suspicious and clearly in contrast with his brother's use of animals to deepen relations with Nazi leaders. Heinz eventually received his permission to write in June 1937, as a friendly official vouched that he had not been in Bavaria at the time of his supposed stay in Dachau. Many of the other allegations were unresolved.

Heinz was a prolific writer. Every month he published several articles on the background of the animals in his zoo in his own popular animal magazine. Unlike his brother he did not deploy Nazi discourse to discuss his back-breeding efforts or other interests. In contrast to Lutz's rabid advocacy of *völkisch* nationalism and violent spatial practices, Heinz's work evokes a set of cosmopolitan historical as well as geographical imaginations of primeval wildlife. It provides peaceful rationales for European conservation. For example, when it discusses the *Urwildpferde* (primordial wild horses) and how wild horses figured in the European past he notes that "there was no Bavaria back then" (H. Heck 1934a, 5). It affords no special role for the Germanic nation or *Volk* in landscape conservation, expresses no geopolitical desire for expanding German nature conservation territories, and (after 1939) does not celebrate the incorporation of land in the East. He attributes the extermination of rare animals by local inhabitants to the complex political situation of nomadic, hunt-loving peoples (ibid., 3) rather than any hereditary or cultural inability to appreciate nature and wildness.

The Heck cattle Heinz produced in Munich were also different from those created in Berlin. In his breeding choices, Heinz did not use fighting cattle and seems not to have been planning on hunting his aurochsen. He used more Central and Eastern European cattle, which are less connected to an exclusively Aryan/German imagination of the ur-cow. Heinz's back-bred animals were not used in reintroduction projects in conquered areas, and he claimed to have been solely motivated by educational and scientific reasons, personal curiosity, and some consolation that the violent extinction of animals due to humanity's "mad rage for destruction of himself and all other creatures" might be undone (H. Heck 1951, 122). Furthermore, the lenient sense of genetic purity expressed by Lutz in his rationale for back breeding was even more pronounced with Heinz. He wrote gleefully that he had been "crossing all kinds of cattle races in a way that would have horrified a pedigree breeder" (ibid., 121). He seemed to have reveled in using terms such as *Mischling* (mixed-race) and *Bastard* to describe his aurochs breeding products. These terms were loaded due to their centrality in Hitler's *Mein Kampf*, where he called upon Germans to "first and foremost halt any further bastardisation" (Hitler 1942, 444). Indeed, framed in this way, the uncontrolled mixing and joyful bastardization of what in the public mind were pure breeds could just as well be considered subversive to the strict notions of racial and genetic purity outlined in the Nuremberg laws.

Conclusion

Lutz's back-bred aurochsen in Germany did not survive the bombing of the Berlin Zoo and Göring's retreat. A few were reported to still roam Białowieża in the late 1940s (H. Heck 1951), straying into the Soviet side of the area before dying out (Wang 2012). However, some of the cattle Heinz back bred in Munich did survive the war, languishing in a few zoos and nature reserves. Their descendants have recently risen to prominence as popular tools for a new mode of European nature conservation known as rewilding (Fraser 2009; Marris 2009). These Heck cattle, as they are now known, have proved to be hardy grazers, able to survive with limited human care. They are often released as aurochs-surrogates with the aim of returning bovine grazing and other ecological processes to marginal landscapes throughout Europe, including those once coveted by Göring and the National Socialists (Schwartz 2006; Lorimer and Driessen 2013). The desirability and veracity of rewilding with Heck cattle is contested and, as a consequence, their Nazi history has been subject to popular interest. Publicists and critics invoke Hitler's geographies to parody the cattle's past and contest their authenticity (see figure 6.5). However, present-day Heck cattle are descended from those bred by Heinz rather than Lutz (H. Heck 1951). Their Nazi genealogy is complicated and their possible ideological baggage much more ambivalent.

We believe it to be nonsensical to arbitrate as to the guilt of contemporary cattle—and would flag the ironical (and contradictory) nature of any claim that they should not be allowed to flourish in a wilder Europe by virtue of any purported Nazi history. However, this episode highlights both the continued fascination with the historical geographies of National Socialism and the need to offer nuanced analysis of their character and consequences. The story of the Heck brothers offers a compelling and hitherto underexamined chapter in the history of German natural science and conservation under National

FIGURE 6.5. Drawing accompanying an article with the title "The Herd Reich," discussing the arrival of Heck cattle on a Devon farm, *The Sun* newspaper, April 22, 2009.

Socialism. Our analysis suggests that their entanglements are by no means as clear-cut as the sensational (but lazy) journalist accounts illustrated above would suggest. In various ways and to differing degrees both brothers became caught up in the discourse and practices of National Socialism. Lutz in Berlin was most clearly aligned with Göring's, if not Himmler's, mode of romantic nationalism. In his practice and presentation of back breeding and animal reintroductions he developed a distinct mode of "reactionary modernism." This conjoined a scientific understanding of animal breeding and protoecological land management with a reinvention of mystical modes of hunting.

Although he was much more marginal to and ambivalent about the Nazi project, even Heinz was unable to resist the patronage the Nazi elite offered. Together with his brother he joined Himmler's Ahnenerbe—a shadowy scientific research organization that sought to use science (archaeology, cultural geography, and even musicology) to legitimate mythological understandings of the Aryan people (Kater 2006). Both Lutz and Heinz were part of a research project named "*Wald und Baum in der Arisch-Germanischen Geistes- und Kulturgeschichte*" (Forest and tree in the Aryan-Germanic history of thought and culture), and were commissioned to write popular books on the zoological and cultural history of the aurochs.[16]

In sorting out the character and consequences of the Nazi geographies presented in this story, we would like to conclude by reflecting on this troubled and troublesome place of myth. The wartime propaganda potential that the back-breeding discourse offered the Nazi elites derived not so much from its particular scientific truth or ecological impact, but from the power and performance of the myths it embodied about German Nature and the Nation. Lacoue-Labarthe and Nancy (1990) have drawn attention to the centrality of myth to Nazism and argue that it is not so much the content of the myth that mattered but its performative function, especially the belief in its efficacy in producing a national identity. Perhaps the most dangerous consequence of the back-breeding project was that it allowed Göring to stalk through the forest carrying a spear, legitimizing his policies and the Nazi war efforts by the purported truth of Germanic mythology—taken as serious as, and fully incorporated within, a form of (at the time) widely respected natural science. The tale of the Heck brothers serves as a cautionary tale to contemporary back breeders and "rewilders." Although they are steeped in science, they must acknowledge the power, provenance, and persistence of myths of the wild and offer new, cosmopolitan, and democratic imaginations for European nature.

Notes

1. NSDAP membership nr: 3.934.018.

2. BArch R 4606/493, November 29, 1937.

3. A legend also taken up by Richard Wagner in his *Ring des Nibelungen*. The notion of *Nibelungentreue* (Nibelungen loyalty)—the total dedication to fight until the last man—was used by Göring in addressing the Wehrmacht just before the collapse of Stalingrad.

4. BArch NS 21/G120, 1124.

5. The notion of *Heimat* (which to some extent translates as "homeland") denoted a combination of regional folk culture and personal rootedness through ancestry. The Nazis extended a "nationalized" interpretation of *Heimat* in an attempt to connect ideas of race, *Volk*, landscape, nation, and German superiority. Applegate (1990) has argued that under Nazism "Heimat ceased to mean much of anything." See also Uekötter (2006).

6. Göring spent so much time hunting, even at crucial moments in the war, that according to some of his generals it amounted to "treason" (Gautschi 1997, 55; 2010a, 209). Hitler and other prominent Nazis, however, derided hunting (Rubner 1997, 172).

7. On April 20, 1938, he was made an honorary professor on the occasion of Hitler's forty-ninth birthday. *Der Biologe*, Nr. 8, Berlin 1938, 279.

8. The RKFDV was the authority headed by Himmler dedicated to colonizing the East with people of proper Aryan descent.

9. In a 1942 book Schoenichen emphasized the *völkisch* importance of Lutz's back-breeding project (see Schoenichen 1942).

10. *Wisente* (directed by Schulz 1941; authors' translation).

11. Göring also had his east Prussian hunting estate Rominten Heide expanded into Polish territories during the war (Scherping 1958; Rubner 1997, 130).

12. BArch NS 15/138, 18–31; BArch PK E38, 2845–916.

13. BArch NS 15/138, 31; BArch PK E38, 2862, cfr. 2912.

14. BArch PK E38, 2874.

15. BArch PK E38, 2874, 2884.

16. BArch NS 21/G120, 1071, 1074, 1082, 1111, 1112.

References

Applegate, Celia. 1990. *A Nation of Provincials: The German Idea of Heimat.* Berkeley: University of California Press.

Barnes, Trevor J., and Claudio Minca. 2013. "Nazi Spatial Theory: The Dark Geographies of Carl Schmitt and Walter Christaller." *Annals of the Association of American Geographers* 103 (3): 669–87.

Bassin, Mark. 2005. "Blood or Soil? The Völkisch Movement, the Nazis, and the Legacy of Geopolitik." In *How Green Were the Nazis? Nature, Environment, and Nation in the Third Reich*, edited by Franz-Josef Brüggemeier, Mark Cioc, and Thomas Zeller, 204–42. Athens: Ohio University Press.

Blackbourn, David. 2006. *The Conquest of Nature: Water, Landscape, and the Making of Modern Germany.* London: Jonathan Cape.

Blood, Philip W. 2010. "Securing Hitler's Lebensraum: The Luftwaffe and Białowieża Forest, 1942–1944." *Holocaust and Genocide Studies* 24 (2): 247–72.

Bode, Wilhelm, and Elisabeth Emmert. 2000. *Jagdwende vom Edelhobby zum ökologischen Handwerk.* Munich: C. H. Beck.

Brüggemeier, Franz-Josef, Mark Cioc, and Thomas Zeller, eds. 2005. *How Green Were the Nazis? Nature, Environment, and Nation in the Third Reich.* Athens: Ohio University Press.

Deichmann, Ute. 1996. *Biologists under Hitler.* Cambridge, MA: Harvard University Press.

Fraser, Caroline. 2009. *Rewilding the World: Dispatches from the Conservation Revolution.* New York: Metropolitan Books.

Fröhlich, Elke, ed. 2008. *Die Tagebücher von Joseph Goebbels: Sämtliche Fragmente.* Part I, *Aufzeichnungen, 1923–1941*, Vol. 4. Munich: K. G. Saur; Berlin: De Gruyter.

Gautschi, Andreas. 1997. *Die Wirkung Hermann Goerings auf das deutsche Jagdwesen im Dritten Reich.* Göttingen: Georg August Universität.

———. 2010a. *Der Reichsjägermeister: Fakten und Legenden um Hermann Göring.* Melsungen: Neumann-Neudamm Verlag.

———. 2010b. *Rominten 1500 bis 1945: Ein alphabetisches Merkbuch.* Melsungen: Neumann-Neudamm Verlag.

Gritzbach, Erich. 1942. *Hermann Göring, Werk und Mensch.* Munich: Zentralverlag der NSDAP, Franz Eher Nachf.

Hagen, Walter [Wilhelm Hoettl]. 1950. *Die geheime Front: Organisation, Personen und Aktionen des deutschen Geheimdienstes.* Linz: Nibelungen-Verlag.

Heck, Heinz. 1934a. "Der Ur, ein vergessenes Tier." *Das Tier und wir* 3: 1–16.

———. 1934b. "Urwildpferde." *Das Tier und wir* 7: 1–7.

———. 1936. "Über die Rückzüchtung des Urs." *Das Tier und wir* 10: 8–14.

———. 1951. "The Breeding-Back of the Aurochs." *Oryx* 1 (3): 117–22.

Heck, Lutz. 1936. "Über die Neuzüchtung des Ur oder Auerochs." *Berichte der Internationalen Gesellschaft zur Erhaltung des Wisents* 3 (4): 225–94.

———. 1939. "Die Neuzüchtung des Auerochsen." *Wild und Hund* 37: 537–39.

———. 1940. "Neue Aufgaben des Naturschutzes: Schutz dem Landschaftsbild!" *Voelkischer Beobachter* 82, March 22, 5.

———. 1942. "Behördliche Landschaftsgestaltung im Osten." *Neues Bauerntum mit Landbaumeister* 34 (6): 213–15.

———. 1943a. "Hermann Göring, der Schützer des deutschen Urwildes." *Wild und Hund* (39–40): 154–57.

———. 1943b. "Urwild im Grossdeutschen Reich." *Freude am Leben* 20 (1): 7–9.

Herf, Jeffrey. 1984. *Reactionary Modernism: Technology, Culture, and Politics in Weimar and the Third Reich.* Cambridge: Cambridge University Press.

Hirsch, Fritz, and Henning Wiesner. 1986. *75 Jahre Münchner Tierpark Hellabrunn: Eine Chronik.* Munich: Tierpark Hellabrunn.

Hitler, Adolf. 1942. *Mein Kampf: Zwei Bände in einem Band.* Munich: Zentralverlag der NSDAP, Eher.

Kater, Michael H. 2006. *Das "Ahnenerbe" der SS, 1935–1945: Ein Beitrag zur Kulturpolitik des Dritten Reiches.* Munich: Oldenbourg Verlag.

Lacoue-Labarthe, Philippe, and Jean-Luc Nancy. 1990. "The Nazi Myth." *Critical Inquiry* 16 (2): 291–312.

Landry, Marc. 2010. "How Brown Were the Conservationists? Naturism, Conservation, and National Socialism, 1900–1945." *Contemporary European History* 19 (1): 83–93.

Lekan, Thomas M. 2004. *Imagining the Nation in Nature: Landscape Preservation and German Identity, 1885–1945.* Cambridge, MA: Harvard University Press.

Lorimer, Jamie, and Clemens Driessen. 2013. "Bovine Biopolitics and the Promise of Monsters in the Rewilding of Heck Cattle." *Geoforum* 48: 249–59.

Marris, Emma. 2009. "Conservation Biology: Reflecting the Past." *Nature* 462: 30–32.

Meyer, Konrad. 1942. "Der Osten als Aufgabe und Verpflichtung des Germanentums." *Neues Bauerntum* 34 (6): 205–8.

Neumärker, Uwe, and Volker Knopf. 2007. *Görings Revier: Jagd und Politik in der Rominter Heide.* Berlin: Christoph Links Verlag.

Oberkrome, Willi. 2004. *Deutsche Heimat: Nationale Konzeption und regionale Praxis von Naturschutz, Landschaftsgestaltung und Kulturpolitik in Westfalen-Lippe und Thüringen (1900–1960).* Paderborn: Schöningh.

Radkau, Joachim, and Frank Uekötter. 2003. *Naturschutz und Nationalsozialismus.* Frankfurt: Campus.

Reichert, Hermann, ed. 2005. *Das Nibelungenlied.* Berlin: De Gruyter.

Rössler, Mechtild, and Sabine Schleiermacher, eds. 1993. *Der "Generalplan Ost": Hauptlinien der nationalsozialistischen Planungs- und Vernichtungspolitik.* Berlin: Akademie Verlag.

Rubner, Heinrich. 1997. *Deutsche Forstgeschichte, 1933–1945: Forstwirtschaft, Jagd und Umwelt im NS-Staat.* St. Katharinen: Scripta Mercaturae Verlag.

Sax, Boria. 1997. "What Is a 'Jewish Dog'? Konrad Lorenz and the Cult of Wildness." *Society and Animals* 5: 3–21.

Scherping, Ulrich. 1938. "Wild und Raumordnung." *Raumforschung und Raumordnung* 2 (11–12): 541–42.

———. 1958. *Krone des Waidwerks: Von der jagdlichen Erfüllung durch Hege, Gedanken und Erfahrungen.* Hamburg: P. Parey.

Schoenichen, Walther. 1942. *Naturschutz als völkische und internationale Kulturaufgabe: Eine Übersicht über die allgemeinen, die geologischen, botanischen, zoologischen und anthropologischen Probleme des heimatlichen wie des Weltnaturschutzes.* Jena: Gustave Fischer.

Schwartz, Katrina Z. S. 2006. *Nature and National Identity after Communism: Globalizing the Ethnoscape.* Pittsburgh: University of Pittsburgh Press.

Stoehr, Ingo R. 2000. "(Post)Modern Rewritings of the Nibelungenlied: Der Nibelungen Roman and Armin Ayren as Meister Konrad." In *Medieval German Voices in the 21st Century: The Paradigmatic Function of Medieval German Studies for German Studies,* edited by Albrecht Classen. Amsterdam: Rodopi.

Uekötter, Frank. 2006. *The Green and the Brown: A History of Conservation in Nazi Germany*. Cambridge: Cambridge University Press.

"Urmacher unerwünscht." 1954. *Der Spiegel*, June 23. Accessed September 11, 2013. http://www.spiegel.de/spiegel/print/d-28956824.html.

Wang, Michael. 2012. "Heavy Breeding." *Cabinet* 45. Accessed October 10, 2013. http://www.cabinetmagazine.org/issues/45/wang.php.

Wiepking-Jürgensmann, Heinrich F. 1942. "Das Landschaftsgesetz des weiteren Ostens; Steppe, Wald und Wiese als Lebensgrundlagen (incl. 'Bildbericht: Ausgeraubte Landschaften des weiteren Ostens')." *Neues Bauerntum* 34 (1): 5–18.

Wisente. 1941. Directed by Ulrich K. T. Schulz. Bundesfilmarchiv, Berlin. Berlin, Germany: Universum Film (UFA).

Zeller, Thomas. 2005. "Molding the Landscape of Nazi Environmentalism: Alwin Seifert and the Third Reich." In *How Green Were the Nazis? Nature, Environment, and Nation in the Third Reich*, edited by Franz-Josef Brüggemeier, Mark Cioc, and Thomas Zeller, 147–70. Athens: Ohio University Press.

SECTION 2

Spatial Planning and Geography in the Third Reich

7

National Socialism and the Politics of Calculation

STUART ELDEN

Introduction

In their study of *The Nazi Census*, Aly and Roth suggest that,

> It was neither through the ideology of blood and soil nor through the principle of guns and butter, upheld until the end of 1944, that the National Socialists secured their might or carried out their destructive activities. It was the use of raw numbers, punch cards, statistical expertise, and identification cards that made all that possible. Every military and labor column existed first as a column of numbers. Every act of extermination was preceded by an act of registration; selection on paper ended with selection on the ramps. (2004, 1; see Burleigh 2002 [1988], 8)

The study of the technical, rational, scientific techniques utilized by National Socialism has been made in numerous studies (see also, e.g., Bauman 1989; Black 2002; Aly and Heim 2003), and though these need to be balanced with the work that shows that the claims to efficiency are overstated (e.g., Broszat 1981; Kershaw 2000), the use to which technology was put is important to a full understanding of this complex phenomena (see also Arendt 1963, 1973; Adorno and Horkheimer 1973). Indeed, despite the undoubted irrationalism of much of the foundations of Nazi thought, what is continually found is research that sought a scientific ground for, in particular, its racist and expansionist claims (see, e.g., Gasman 2004).

However, the question goes beyond merely an understanding of the techniques that made particular actions possible. Rather the question is one of the conditions of possibility for the techniques themselves, a deeper problematic that seeks to account for the technical, rational, scientific mindset itself. The question of conditions of possibility is a Kantian term, where in the *Critique of Pure Reason* he seeks to uncover the grounds for synthetic a priori knowledge.

His question is, "how are synthetic a priori judgements possible?" (Kant 1988, B19). This is not merely a question of epistemology, but, precisely because it enquires into *how* "what is" *is*, a question of ontology, as Heidegger has argued (1997).

Ontology as the investigation of how "what is" is, the question of conditions of possibility, can also be found in a number of thinkers following in Kant's wake, including Nietzsche, Heidegger, and Foucault. But in their work it is radically transformed, as the question is historicized. As Nietzsche suggests, we should replace the Kantian question, "how are synthetic a priori judgements possible?" by the question, "why is belief in such judgements necessary?" (1966, §11). In other words, how, in certain times and places, are certain beliefs and orderings necessary in order to effect particular outcomes. This is one way to read both Heidegger's later work on the history of being and to understand Foucault's investigations of madness, discipline, and sexuality not as studies of those subjects in themselves, but in order to understand reason and the modern subject (for discussions see Gutting 1989; Han 1998; Elden 2002, 2003).

The question of what made National Socialism *possible*, and therefore of why certain modes of thought were *necessary*, is the central concern of this paper. Although this paper therefore takes its point of departure from this philosophical position, it focuses on the historical period at hand in order to illuminate both a particular instance of the politics of calculation, and a calculative understanding of the political. It undertakes this through an examination of the way in which mathematics and politics intertwined in National Socialist Germany, particularly in relation to the period between 1933 and 1939.

The reason for these dates is twofold. First, that it does seem to make sense to delimit two periods with Nazism. These are the periods 1933–39 and 1939–45, in other words the period before, and the period of, the war. This is not an absolute distinction. As shall become apparent, Germany was geared up for a war economy long before 1939, albeit not for the slave economy it became, and many of the mechanisms derived in the pre-1939 period continued after it. But it seems a convenient break, and this is the second reason: this distinction enables the paper to put to one side—to bracket—*Blitzkrieg* and the Final Solution. This is obviously not to say they are unimportant, but that there is an already extensive literature. Nor is it to suggest that they do not fit the analysis made, although, as the conclusion makes clear, the Holocaust is the limit case for the study and the applicability of this perspective on the question of Nazism. In any case, the period of 1933–39 offers a wide range of important and challenging issues to investigate the politics of calculation. I will concentrate here on two key instances: the understanding of race, particularly as found in the Nuremberg Laws and the spatial planning aspects of geopolitical strategy. In order to investigate these two areas, I will first spend a while investigating the ontological presuppositions behind the notion of *Gleichschaltung*, suggesting that this illuminates the very basis of calculative politics. Essentially the claim of this paper is that through a particular sense of the politics of calculation Nazism sought control of the earth in a way that both *made possible* and therefore *exceeds* its quest for world domination. This control is to render it as something understandable through number and calculation.

Gleichschaltung and the Politics of Calculation

The notion of *Gleichschaltung* is an interesting and important term in the lexicon of National Socialism. It is usually translated as coordination or synchronization, but has a sense of unification, of bringing into line or the elimination of opposition. Literally, the word means "same wiring" or "connection," the bringing of things under a common measure, subordination (see Bracher 1972; Friedlander 1980; Klemperer 2000). The two *Gleichschaltung* laws passed in 1933 coordinated the federal *Länder*—regional governments—with the Reich as whole and served two crucial purposes: to give the *Länder* the same post-Enabling Act powers as the Reich, and to ensure their conformity with the Reich they were to politically mirror the makeup of the *Reichstag* (for the text of documents, see Hofer 1957; Noakes and Pridham 2000). While reforms in early 1934 effectively abolished the *Länder*'s powers entirely, and dissolved the *Reichrat*—the second chamber of parliament, comprising regional deputies—the reforms of subnational government were models to be followed elsewhere.

Gleichschaltung was therefore effectively a mechanism for the way in which all parts of society were homogenized and organized. Some parts were suppressed, outlawed, or violently disbanded. The *Sturmabteilung* (SA), itself to be violently repressed in June 1934, was essential to this coordination (Frei 1993, 10). The *Gleichschaltung* of the *Länder* did not only mean that local unity was achieved, but was a euphemistic expression for the obliteration of independent regional government and the centralization of power (Bracher 1972, 121; Broszat 1981, chap. 4; Frei 1993, 39). In particular, the distribution of seats in the *Länder* would mirror that of the *Reichstag*, and the Reich governors would be chosen by Hitler, except in Prussia, where he would be the governor himself (Frei 1993, 41).

Universities were part of this overall organization or synchronization, with the Reich Ministry for Science, Education and Adult Education created in 1934, rebranding university rectors as "Führers of the Universities" (Frei 1993, 97). Many other organizations were *gleichgeschaltet*: there were associations of National Socialist lawyers, doctors, and so on. Racial profiling for such organizations was crucial, forming the model for and later based explicitly on the model of the Nuremberg laws to be discussed below. Equally, we can see a concern for synchronization in the reaction against particular forms of art and literature, taking the form of boycotts, book burnings, and denunciations, and in the reorganization of areas such as broadcasting. Indeed, as Frei notes, "in no other area of culture and mass communications was the grip of the new rulers comparably efficient" (ibid., 63).

The ideas behind *Gleichschaltung*, that humans, groups, and organizations can be understood as either the same or different, and in the latter case rendered the same is, ontologically important in terms of understanding what was happening in Germany. Determining things as different and seeking to render them more equivalent, or counting them the same in the first place, requires a number of important moves: most importantly recognizing things as sufficiently similar in their essence that they can be summed or evaluated against each other. Rather than a logical structure based on identity or non-identity, *Gleichschaltung* requires a notion of difference, in that things are not so irreconcilably distinct that they cannot be compared, but that they can be rendered the same, or similar, *gleich*. *Gleichschaltung* therefore requires a disruption of the binary of identity or non-identity, where something is either "A" or "not-A," but in thinking that the difference can be changed, or reduced to "A" it is similarly distinct from the notion of "difference" advanced in recent European thought.

This is not to suggest that this thought worked on simple binaries, of Aryan or not-Aryan, for example. In practice, all sorts of gradations could be possible, codified, and represented. Although these differences could be qualitative, in that actions could be coordinated to be more similar, or a homogenization could be put through in education or the arts, these differences could also be quantitative. In fact, as the establishment of quotas in the Four Year Plan shows, coordination could be around numerical targets, but in terms of an overall determination *Gleichschaltung* depends on the understanding that underpins calculative politics.

Despite his own direct involvement in the regime, Heidegger's work can be useful here. In the mid-1930s, as his disillusion with the movement he had once enthusiastically supported waned, he developed an analysis of what he called "machination," a term that later becomes the more famous question of technology. For Heidegger, machination depends upon a particular notion of metaphysics, a particular casting of being, that is, to be is to be calculable. Heidegger traces this shift to developments in what is sometimes called the scientific revolution in the sixteenth and seventeenth centuries, focusing on the roles of Galileo, Descartes, Leibniz, and Newton, especially the second. To explore this transition in detail is beyond the scope of this paper, but a number of issues can be taken from his work to inform an analysis of National Socialism.

For Heidegger there are three ways in which the question of being—his key concern—has come to be forgotten in the modern age. These are calculation (*die Berechnung*), massiveness (*Massenhaften*), and acceleration (*die Schnelligkeit*). In all three we see interlinked themes of measure (*Maß*) and calculative thinking, grounded on a particular way of reckoning (*Rechnung*), based on number and the celebrative of quantitative enhancement (Heidegger 1989, 120–21). Throughout many works of the mid-1930s on, Heidegger is regularly concerned with the way notions of measure and calculation or reckoning are used (see Elden 2005a).

The idea of massiveness is not merely hugeness or enormity, although it may superficially take such forms. Rather it is grounded in the very notion of measure (*Maß*), and is related, Heidegger contends, to the notion of the gigantic. But this too he understands in a very particular way.

> The gigantic was determined as that through which the "quantitative" is transformed into its own "quality," a kind of magnitude or greatness [*Größe*]. The gigantic is thus not something quantitative that begins with a relatively high number (with number and measurement)—even though it can appear superficially as "quantitative." The gigantic is grounded upon the decidedness and invariability of "calculation" and is rooted in a prolongation of subjective representation unto the whole of beings. (Heidegger 1989, 441)

For Heidegger what is central to understanding the modern world, of which Nazism increasingly becomes its most acute symptom, is therefore not the mere size, scope, or ambition of its various projects, but the very issues underpinning this. The politics of calculation that emerges from this is thus an important tool for an understanding of modernity, and is a way in which Heidegger can be used politically, and indeed, for political purposes he would not himself have condoned (for discussions, see Janicaud 1990; Ott 1993; Wolin 1993; Bambach 2003; Elden 2006). While to my mind therefore Heidegger offers us the opportunity of a powerful critique of National Socialism, and—within certain tightly circumscribed limits—a useful tool for analysis more generally, it is important to note right at the outset that Heidegger avoids any kind of moral position. This is the limit case, a point to which I shall return in the final section of this paper.

The Nuremberg Laws

A particularly stark example of the problematic reduction of qualitative differences to quantitative ones can be found in the Nuremberg laws of 1935, based on a calculation of race. An earlier tradition within anthropology had given the twofold distinction of *Volkskunde* as the study of traditional German customs and *Völkerkunde* as studying dark-skinned races. The second term is the German title of Friedrich Ratzel's anthropological study translated into English as the much more neutrally sounding *The History of Mankind* (1896–98). In this work and others of its kind (Chamberlain 1911; Gobineau 1967 [1853–55]), there was an explicit attempt to link cultural practices to physiological differences, in other words to equate qualitative differences with quantitative ones that they felt could be measured in more mathematical ways. While in part this may first rely on observation of the qualitative, once models have been created, the quantitative can be used to make judgments in itself (see Pick 1989; Borneman 1997). In this there is a conflation of a number of elements. Race and the *Volk* became increasingly related, and the nation seen as the congregation of a racially pure *Volk*. Neumann helpfully explains their understanding:

> Race is an entirely biological phenomenon: the concept of "the people" contains an admixture of cultural elements. . . . The concept of a *racial* people, a term the Germans are fond of, is, however, based primarily on biological traits; the cultural elements serve only to distinguish various groups within one race. In contrast the nation is primarily a political concept. (Neumann 1966 [1944], 99; see Hitler 1939, 229)

The Reich Citizenship Law of September 15, 1935, attempted to clarify matters. In Article 2.1 it declared that "a citizen of the *Reich* is that subject only who is of German or kindred blood and who, through his conduct, shows that he is both desirous and fit to serve the German people and *Reich* faithfully." What comes through here is both a measure of blood, that is race, but also one of disposition, a more cultural element. The same day saw the "Law for the Protection of German Blood and German Honour," which prohibited marriage and sexual relations "between Jews and nationals of German or kindred blood" (sections 1 and 2). Although these build on the October 18, 1933, laws for the "protection of the hereditary health of the German people," as Agamben (1998) has shown, what gets introduced here is an important distinction between those of German blood, defined as four German grandparents, who are citizens, and those who are not German but still fall under the jurisdiction of the state. This distinction between state subjects (*Staatsangehörige*) and citizens (*Reichsbürger*) was one that Hitler had made back in *Mein Kampf*, where he had suggested a three-part division of citizens, subjects of the state, and aliens, that is, subjects of another state (1939, 247). It was something that had first been put into practice on July 14, 1933, when all people naturalized between the end of the First World War and the Nazi takeover had their citizenship removed (see Nathans 2004, 219–21; on the drafting see Kershaw 2000, 107–8).

In order to clarify the gradations of state subjects the First Supplementary Decree of November 14, 1935, created charts which demonstrated how different numbers of Jewish grandparents created both Jews (three or four grandparents) or *Mischlinge*, those of mixed race. What we find here, lest there be infinite regress in determining the racial characteristics of the grandparents that determine the race of the grandchildren, is the shift from cultural traits to race (supposed to be scientific and non-cultural). For the definition is that "full-blooded Jewish grandparents are those who belonged to the Jewish religious community" (Article 2.2). These indications were supposed to be scientifically proven (see Stolting, 1987, 192–93; Borneman 1997, 101; Aly and Roth 2004; more generally Gould 1996). The charts created included ones where they showed the various permutations of unions between "Aryans," *Mischlinge*, and Jews as combinations of circles cut into quarters colored black or white for Jewish or Aryan grandparents. (It is important to note here that Aryan was defined negatively, as not being of Jewish descent.) These potential pairings were defined as either allowed, forbidden, or allowed in various circumstances. Other charts indicated this in more graphic form with either outline or silhouette figures in family trees of three generations, showing the race of the offspring, and then diagrams showing various pairings allowed into buildings together or not: the buildings being registry offices. What we have here is the rational, calculative, and bureaucratic as a gloss on the "irrational" anti-Semitism.

Inevitably, given the construction of these tables, Jews could neither be citizens nor aliens, rather they would fall into the categories of subjects of the state: that under the jurisdiction of laws, but denied rights and a role in the legislative process. As Sax notes:

> From the Nazi point of view, the Jews were an anomaly and therefore impure. The Jews did not fit into received classifications, being neither nationals nor foreigners. . . . Other groups persecuted by the Nazis were also ambiguous.

> The handicapped, almost by definition, are anomalies; so are homosexuals. (Sax 2000, 155)

As Foucault has shown in numerous works (e.g., Foucault 1999), models of classification that are used to classify sexualities bear close comparison to those of race and other forms of supposed abnormality. It is worth noting that the word Foucault uses—*anormaux*—can equally mean "anomalies" or "irregularities" as it can "abnormals." Trading on pioneering work by Canguilhem (1978), Foucault is interested in how a statistical notion used to catalog and codify behavior or traits becomes one that categorizes individuals. As one of Foucault's lecture courses notes, the categorizations of race are behind both reactionary forms of politics and, transformed into notions of class and class enemy, more "progressive" forms (1997; see Enoch 2004). Foucault's analysis is useful in a number of ways, but perhaps most on this topic when he suggests that there is a break or a cut, a *coupure*, that is fundamental to racism, one which fractures the "biological continuum of the human species." This continuum, like the continuum of geometry divided by Descartes in the seventeenth century, is fragmented by "the apparition of races, the distinction of races, the hierarchy of races" and their qualification as good or inferior. The "first function of racism is to fragment, to make caesuras in the biological *continuum* biopower addresses" (Foucault 1997, 227). The human species is subdivided into subgroups that are thought of as races, in a manner which is akin to the modern, orderable, calculable view of space.

Indeed, when the territories of the East were seized, there was a question about how useful foreign races (i.e., Balts and Ukrainians) could be separated from the "undesirable tribes." The anthropologists of the Kaiser Wilhelm Institute were involved in such decision making (see Frei 1993, 126), and as Burleigh (2002 [1988], 8) notes, the *Ostforscher*—academic researchers on the East—"had a distinctive contribution to make to the accurate 'data base'—the statistical and cartographical location of persons—upon which *all* aspects of Nazi policy in the East, as elsewhere, ultimately rested."

Geopolitics and Spatial Planning

One of the standard complaints in National Socialist rhetoric was that the map-making of the Versailles treaty had excluded German *Volk* from the borders of the country (Articles 27 and 28). Although this was not new, in that German territory and that occupied by German-speaking peoples have rarely been the same (see Urwin 1982; Herb 1997), in the Weimar period it did raise a number of issues. An expansionist foreign policy could come in many forms, but included those which wanted a return to 1914 borders, those which sought to incorporate all German-speaking people within the confines of the Reich, and even more aggressive attempts to expand even beyond these confines (see Herb 1997). Hitler, for example, declared the idea of a return to 1914—i.e., a straightforward overturn of Versailles and a return to the borders following German unification under Bismarck and the Kaiser in 1871—as "a glaring political absurdity that is fraught with such consequences as to make the claim itself appear criminal" (1939, 357); 1871—the second Reich—could not be looked at as a lost ideal.

The notion of *Lebensraum*—living space or room for the German *Volk*—was an important issue for Hitler in *Mein Kampf*, where he talked about the restrictions on the German living space, and that the solution to this was not to be sought in merely colonial acquisitions—the "place in the sun" claims of previous generations since unification. Germany, because of its late emergence as a unified nation-state, lagged behind its European neighbors in colonization. But the aim was not simply to dispossess and catch them up. Rather the living space was to be attained through an expansion of the German territory itself, a greater magnitude. For Hitler this was essential: "the right to territory may become a duty when a great nation seems destined to go under unless its territory be extended. . . . Germany will either become a World Power or will not continue to exist at all" (1939, 360).

What is striking is that not only does Hitler clearly outline the need for expansion, as is well known, but that he continues to use measures of quantitative extent: magnitude, greatness, proportion. Russia, with her extensive landmass, and the buffer countries between Russia and Germany, are the prime targets: eastward expansion.

> Germany is not at all a World Power today. Even though our present military weakness could be overcome, we still would have no claim to be called a World Power. What importance on earth has a State in which the proportion between the size of the population and the territorial area is so miserable as in the present German *Reich*? At an epoch in which the world is being gradually portioned out among States many of whom embrace whole continents one cannot speak of a World Power in the case of a State whose political motherland is confined to a territorial area of barely five-hundred-thousand square kilometres. (1939, 354)

As is well known, ideas of *Lebensraum* have a long pedigree. The crucial initial source is Friedrich Ratzel's *Politische Geographie* (1897) and his later *Der Lebensraum* (1901), which discusses the laws of the spatial growth of states. The term itself came from a review of Darwin's *On the Origin of*

Species by Oscar Peschel (see Borneman 1997, 100). It is worth noting that Ratzel was influenced by his trips to the USA, seeing the vast landmasses available (1988 [1876]; Kamenetsky 1961, 28–29). Indeed, Hitler used comparison with America as a justification of an expansionist aim in his response to Roosevelt's peace telegram of April 28, 1939, and earlier in *Mein Kampf* (Hitler 1939, 351; Neumann 1966 [1944], 130–31).

In Ratzel's mind, space was the most vital requirement of the state; the struggle for it was dependent on a natural law, in which strength prevails, and the most likely victors were racially pure peoples, rooted in the soil (Kamenetsky 1961, 29), buying into notions of autochthony (see Bambach 2003). Ratzel felt that colonies were only a potential partial solution (Agnew 2002, 60). There were other important works in the pre-Nazi period, notably Friedrich Naumann's *Mitteleuropa* (*Central Europe*) (1915) advocating a union of countries under German dominance, and Hans Grimm's novel *Volk ohne Raum* (*People without Space*) (1926). The ideas were passed on to the National Socialist leadership particularly through Karl Haushofer, who was in contact with Rudolf Hess while he was in prison in Landsberg in the 1920s following the failed *Putsch*, and through Hess, only indirectly, to Hitler (see Heske 1987). As Fest summarizes:

> The original idea of a political geography under the catchword "geopolitics" had undergone an imperialist transformation into a "pseudo-scientific expansionist philosophy." It offered the humiliated national spirit the idea that the destiny of Germany would be decided in the East and thus added a fundamental ideological category of National Socialism, that of "space," to that of "race." These two ideas, linked by that of struggle, constituted the only more or less fixed structural elements in the intricate tactical and propagandist conglomerate of the National Socialist *Weltanschauung*. (1972, 289)

It is important, of course, to understand that while National Socialism sought to contest the particular drawing of boundaries and the amount of "space" Germany had, mainstream thought did not challenge the ways in which these were conceived. The calculations might be inaccurate or unjust, but calculation itself was not seen as the problem. Indeed much emphasis was given to ideas that Versailles and the other treaties of the Peace of Paris had been based on erroneous maps and data (see Herb 1997). While Heidegger and others sought to challenge a Cartesian notion of space (*spatium*, *Raum*) as calculative, bounded, and exclusive, and replace it with a more originary sense of place (*locus*, *Platz*, *Ort*), and certain legislation such as Himmler's decree "On the Treatment of the Land in the Eastern Territories" saw a particular ecological sense of the land at stake (see Staudenmaier 1995, 16), this was not the dominant critique. Let me take three examples of the more dominant strand in Nazi thinking about space.

The first is found in an important 1940 book by Kopp and Schulte which provided a rereading of the seventeenth-century Peace of Westphalia that was in effect a surrogate attack on the Peace of Versailles. The two treaties signed in Westphalia in 1648 had effectively ended the coherence if not the existence of the Holy Roman Empire—the "First *Reich*"—and are often, if somewhat misleadingly, seen as the birth of both the modern state and the sense of territory (see Osiander 2001; Teschke 2003; Elden 2005b). Kopp and Schulte's work, to which the party philosopher Alfred Baeumler contributed a "Foreword," challenged the substance but not the essence of these spatial politics. Similarly, as Neil Smith notes, Alexander Supan's *Leitlinien der allgemeinen politischen Geographie* (*Geometries of German Political Geography*) analyzed the "colonial quotient" as a "ratio between the area and population of a country's colonial possessions with that of its homeland." This argued that Germany deserved more territory (Smith 2003, 282). Paul Mombert also used mathematical formulations to look at the relation between population size (*Volkszahl*) and "feeding capacity" or economic resources (Aly and Heim 2003, 60). These were, effectively, challenges to the distribution and separation of power, the drawing of boundaries and the partition of territory, not to the determination of these ideas.

Second, there was the establishment of a number of research initiatives that show that far from criticizing such a modern notion of space, much National Socialist thought embraced it and sought to understand it better. There was, for example, the establishment of the Reichsstelle für Raumordnung (Office for Spatial Planning or Ordering), which was a revised version of the Reich Office for the Regulation of Land Claims, on June 28, 1935. Both of these offices planned and organized space, but also studied the concept of space in general, effectively looking at how ideas of "blood and soil" could be put into practice. This shows that the organic *völkisch* ideas could work with, rather than against, science, in spite of the rhetorical resistance to "reason" and "science." The Reichsstelle produced a review called *Raumordnung und Raumforschung* (*Spatial Planning and Spatial Research*) under the editorial hand of Hans Boehm and one called *Zeitschrift für Geopolitik* (Kamenetsky 1961, 46; Neumann 1966 [1944], 136ff.; Heske 1986). It is worth noting that the procedures used in the East were dependent on the "central place" theory of Walter Christaller, where geometrically drawn regions are orientated around a central town, a hub that provides more efficient services. As Aly and Heim note, the "degree of centrality" was "calculated originally in terms of the number of telephone connections" (2003, 97).

Third, and finally, the jurist Carl Schmitt, whose legal work on the foundations of the Nazi rule is often discussed, also analyzed ideas of space in their legal context in the service of the regime. For example, in 1939 lectures (Schmitt, 1991 [1939]) he analyzed the American Monroe Doctrine and its potential counterpart for Germany, put forward a notion of *Großraum* of an expansionist territory, a kind of German-dominated *Mitteleuropa*, particularly tied to notions of the *Volk*, and analyzed the legal basis for such claims. While this traded on both *völkisch* and calculative elements, it was not based on the idea of *Blut und Boden*, but rather on a more juristic notion of land law, deriving in part from Locke's famous argument about the cultivation of the land (see Bendersky 1983; Balakrishnan 2000). Schmitt (1997, 58) relates his work to Heidegger's view of space, but this seems to underplay the differences, particularly around the notion of calculation. Equally, Schmitt attempted to distance his ideas from notions of *Lebensraum*, *Mitteleuropa*, and a greater German Reich (1995). Rather, his idea was closer to a German Monroe Doctrine, with Germany the dominant power and other world powers not involved in the affairs of "Central Europe." Nonetheless, Schmitt's ideas here were hardly critical of Hitler's expansionist policies to the East.

As Burleigh has shown (2002 [1988]), this academic work was important in putting these ideas in practice. In particular it was realized that if land was conquered it needed to be settled, as the most effective way of ensuring its continued possession. The racial profiling discussed above became extremely important here. So, not only was the space calculated in terms of area, and to a level of some detail, but it was also racially profiled. While Bassin (1987) has argued that we should be careful in seeing too direct a relation between *Geopolitik* and Nazi policy, as there are some important differences, there is unquestionably a relation, and despite inflections of *völkisch* ideas, at root a calculative understanding of territory is at play in both (see Raffestin, Lopreno, and Pasteur 1995). In part, as Bassin recognizes (1987, 122), it was the crossing of ideas of *Volk* with scientific racism that was so important. Just as the racism is given a calculative coding, so too is it spatialized, with a stress on the distribution of races. Equally, the spaces to be conquered are extensively mapped, measured, and assessed. It is for both of these reasons that the notion of calculation is central to Nazi concepts of *Lebensraum*. And this idea was, of course, crucial to what happened after 1939.

> We must realise, that the whole sense of this war rests in a natural enlargement of *Lebensraum* for our people. (Governor Hans Frank, August 1, 1942, Nuremberg Trial Doc. 2233-PS, cited in Kamenetsky 1961, 25)

Indeed, in a reversal of Grimm's formulation, in 1937 Hitler characterized the east as *Volkloser Raum*, space without people, space empty of people. For Agamben, this is not "a desert, a geographical space empty of inhabitants," but "the driving force of the camp . . . understood as a biopolitical machine that, once established in a determinate geographical space, transforms it into an absolute biopolitical space, both *Lebensraum* and *Todesraum*, in which human life transcends every assignable biopolitical identity" (2002, 86, see 156; 1998, 169–71, 173–74). Particularly in the lands of the East the calculation of race and the calculation of space come together.

Heidegger, Ethics, and Calculation

Although analyzing National Socialism through the lens of calculation is important in a range of ways, and challenges any straightforward equation of the movement with irrationalism and vitalist modes of thought, there are, of course, limits to this analysis. In practice, the National Socialist state was often disorganized and not nearly as hyperefficient as it might sometimes have appeared. Indeed, Neumann has noted that "it is doubtful whether Germany can be called a state. It is far more a gang, where the leaders are perpetually compelled to agree after disagreements" (Neumann 1966 [1944], 522). A more nuanced argument can be found in the work of Ian Kershaw, where he analyses the different power blocs operating as competing interests (Kershaw 2000, 2001a, 2001b; see Broszat 1981). In this we might draw on Lefebvre's analysis of technocracy under the French Fifth Republic, where he notes that the technocrats, for all their plans and mechanisms, are often very poor administrators, that their efficiency is often mythical, and that they fail to use technology in the service of everyday, social life (see, in particular, 1967). Perhaps the exception, and the archetypal technocrat of the Third Reich, was Albert Speer (see Fest 1972, 300; more generally Speer 1970; Sereny 1995). Speer is an interesting case, for not only was he the most "efficient" and "calculative" in many ways, moving from being Hitler's architect to control of arms production, but is also seen as important in that he was the Nazi who admitted at least some measure of guilt at Nuremberg.

Alongside numerous other sources I have made use of the pioneering analysis of Franz Neumann, a member of the Institut für Sozialforschung (Institute for Social Research—also known as the Frankfurt School). Neumann's work *Behemoth* was written in America in exile in the last years of the war, and was a powerful critique and analysis of Nazism, which, given the time of its original publication in 1944, was of course more than a dispassionate academic work. Neumann's analysis is revealing for a number of reasons, but it is worth noting the way in which he thinks allied war policy should proceed.

> A military defeat of Germany is necessary. Whether National Socialism can be crushed without a military defeat, I do not know. But of this I am certain: a military defeat will wipe it out. The military superiority of the democracies and of Soviet Russia must be demonstrated to the German people. The philosophy of National Socialism stands and falls with its alleged "efficiency." This must be proved untrue. The stab-in-the-back legend of 1918 must not be allowed to arise again. More and better planes, tanks and guns and a complete military defeat will uproot National Socialism from the mind of the German people. (Neumann 1966 [1944], xiii)

What is perhaps revealing in this is the suggestion that rather than compete at the level of ideas, in a sense brute force and *greater* efficiency and efficacy is what is needed. Neumann calls for "more and better planes, tanks and guns," greater quantities and better quality. In 1940, following France's defeat, Heidegger makes a related point in reverse: the country of Descartes has been beaten on its own terms (Heidegger 1991, 116–17).

What is, of course, neglected in such an analysis—made here through an appropriation of various thinkers, but crucially Heidegger—is the question of ethics. It is undoubtedly the case that without some sense of ethics it is difficult to take a position on the topics analyzed here. Can the actions of the Allies— including of course the firebombing of Dresden and the atomic bombs dropped on Japan—not be seen as founded on the same techniques and rationalities? Does this render all of this equivalent? The point about these issues is, of course, not just that they were mathematical, but about how mathematics was utilized in particular ways. We find this neglect most explicit in Heidegger's two comments on the Holocaust: a dispassionate equation of this with other forms of quantitative excess of death and calculative disregard for life.

Heidegger's comments are found in two 1949 lectures. The first states that, "Agriculture is now a motorised food industry, the same thing in its essence [*im Wesen das Selbe*] as the production of corpses in the gas chambers and extermination camps, the same thing as blockades and the reduction of countries to famine, the same thing as the production of hydrogen bombs" (Heidegger 1994, 27). This has been extensively discussed in the Heidegger literature, although rarely with much thought (for two notable exceptions, see Lacoue-Labarthe 1987; and Kisiel 2002). It is essential to note that Heidegger is claiming that these are the same *in their essence*, and not in essence the *same*. This is important, not only because a straightforward equation of agriculture and the Holocaust is abhorrent, but because he hints here at the way that ontologically there may be an equivalence. They rest on the same view of the world as something orderable, measurable, controllable, and ultimately destroyable. The second asks:

> Hundreds of thousands die *en masse* [*in Massen*]. Do they die? They perish. They are cut down. Do they die? They become items in an inventory for the business of manufacturing corpses [*Sie werden Bestandstücke eines Bestandes der Fabrikation von Leichen*]. Do they die? They are liquidated without ceremony in extermination camps. And even without such a machinery, millions of poor souls are now perishing from hunger in China. (Heidegger 1994, 56)

There is clearly a brutal detachment from the events here. In fact it is the *process*, with the emphasis on the reduction of humans to *Bestand*, standing reserve or stock, that is emphasized. In Agamben's analysis, this actually brings Heidegger close to writers like Primo Levi, who declared that "one hesitates to call their death death" (Levi 1986, 82), and Hannah Arendt (1963; see Agamben 2002, 70–76).

In a sense the most generous reading of the equation of agriculture as a "motorized food industry" to "the production of corpses in the gas chambers and extermination camps . . . blockades and the reduction of countries to famine . . . the production of hydrogen bombs" is to see it as an attempt to suggest an equivalence between the crimes of Stalin, Mao, and Communism and the National Socialists, a version of the *Historikerstreit* thirty years in advance (on the *Historikerstreit* see Baldwin 1990; Kershaw 2000, chap. 10). Heidegger delivered the lecture in early December 1949, shortly after the USSR declared itself a nuclear power, and a few months after the end of the Berlin Blockade. To follow this through would require that the reference to farming is intended to suggest the collectivization and mechanization of Soviet agriculture in the 1920s and 1930s, with the attendant famines and mass slaughter and deportation of the Kulaks; although it could equally refer to the killing of landlords in the reorganization taking place in China following the revolution. Indeed, although Heidegger refers to famine in China in the second passage, he does not make either of these anything like explicit.

To view the Holocaust as an example of modern technology has occasioned critical responses, such as that of Berel Lang, who hubristically describes this as "a formulation that seems as dramatically revisionist as any other in Holocaust historiography" (Lang 1996, 98). Heidegger's analysis is not unique, and elements of it, albeit not penetrating to the ontological level, can be found elsewhere. We could point, for example, to Boria Sax's suggestion that "Heinrich Himmler, who founded the SS and oversaw the Nazi death camps, was initially a chicken farmer. Many of his ideas for both systematic breeding and slaughter of human beings were simply the extension of mechanised farming to people" (Sax 2000, 150). Alternatively, in a much more mainstream analysis, it has been suggested that "much of the correspondence of Rudolph Hoess, the commandant of Auschwitz, was concerned with quotas and rates of disposal and sounded as if he were the manager of a synthetic firm of a factory for the conversion of waste materials" (Craig 1982, 327). Finally, as some recent work on animal rights has pointed out, the parallels between the meat production industry and the organization of the extermination camps are not merely coincidental (Patterson 2002; Davis 2005). Generally though, and certainly in Heidegger's case, this is a critique of the ontological casting and not the ethical crime. For Kershaw, "dealing with the problem of explaining Nazism, historical-philosophical, political-ideological, and—above all—moral issues remain inescapable" (Kershaw 2000, 262).

Conclusion

Moral issues are of course crucial, but it is worth remembering that while it is not difficult to condemn National Socialism today, to analyze what made it possible is necessarily a more complex and challenging issue. This is not, in terms of the Holocaust, merely in terms of the "historical, material, technical, bureaucratic, and legal circumstances in which the extermination of the Jews took place," which, as Agamben has argued, has been detailed extensively (see e.g., Bauman 1989; Black 2002; Aly and Heim 2003; and particularly Sofsky 1997). Nor is it merely, as he suggests, that the "ethical and political significance" of these events is much less examined (Agamben 2002, 11). Rather, it is that ontologically this is even more the case. A particular grasping of the politics of calculation is a contribution to this ongoing analysis. As Frei notes, the regime modeled itself on particular scientific and industrial theories. It may have been research on the limits—such as eugenics—but it testified to a fascination with the rational, the calculative and the hyperefficient. Indeed Frei suggests that a younger breed of technocrats, coming in the wake of Hitler's old generation, envisaged "a post-war order, informed by the Nazi *Weltanschauung* but founded on science, which was supposed to be much more 'rational' and efficient than the Hitler State of its quasi-revolutionary beginnings" (Frei 1993, 154–55).

> From this perspective—which admittedly moves into the realm of historical speculation—the question arises, once again, of the structural capabilities of National Socialism as a system and the modernity of the "Führer State": the

> many "modern" elements of Nazi rule then appear, not simply as unintended or even dysfunctional side effects of a basically reactionary, atavistic political philosophy, but rather as the harbingers of an attempt to complete the project of the modern in the particular variant of a racial order. Technical rationality and efficiency were the absolute values in this dead, technocratic world. Barbarism wore the clothes of modernity. (Frei 1993, 155)

Yet this is not something that we can feel is safely consigned to the history books. Many of the most pressing issues of today can be seen as distinctly related: the simplistic division of "with us or against us," and the coercion of the "willing" into a coalition against the "axis of evil"; biometrics and offender profiling, and the denial of basic rights to selected individuals in extraterritorial spaces; the ever greater expansion of markets and the requirement of the state to make its companies competitive; along with the geopolitical strategies required to underpin and support such policies. Once again the moral question arises, both as a distinction and as a cause. As Aly and Roth conclude their study of statistics and the Nazis:

> Many of the bureaucratic and scientific techniques used by the Nazis have not been taken in[to] account. This is probably because they are, in many respects, considered normal techniques of the modern state—used, to be sure, in extreme cases, but by no means considered shady. (Aly and Roth 2004, 149; see Aly and Heim 2003, 295)

But it is not merely the techniques, the technologies of the state, that are parallel. It is the essence of these technologies, their conditions of possibility. This is why Heidegger is useful, because he enables us to see how the politics of calculation displayed in Nazism showcases a much broader calculative understanding of the world, a calculative understanding of the political. This ontological determination necessarily exceeds any ontic politics. This particular grasp of the political can be approached through a historical analysis, a historical ontology of the politics of calculation.

References

Adorno, Theodor W., and Max Horkheimer. 1973. *The Dialectic of Enlightenment*. London: Allen Lane.

Agamben, Giorgio. 1998. *Homo Sacer: Sovereign Power and Bare Life*. Stanford, CA: Stanford University Press.

———. 2002. *Remnants of Auschwitz: The Witness and the Archive*. New York: Zone Books.

Agnew, John. 2002. *Making Political Geography*. Oxford: Oxford University Press.

Aly, Götz, and Susanne Heim. 2003. *Architects of Annihilation: Auschwitz and the Logic of Destruction*. London: Phoenix.

Aly, Götz, and Karl H. Roth. 2004. *The Nazi Census: Identification and Control in the Third Reich*. Philadelphia: Temple University Press.

Arendt, Hannah. 1963. *Eichmann in Jerusalem: A Report on the Banality of Evil*. London: Faber and Faber.

———. 1973. *The Origins of Totalitarianism: New Editions with Added Prefaces*. San Diego, CA: Harcourt Brace Jovanovich.

Balakrishnan, Gopal. 2000. *The Enemy: An Intellectual Portrait of Carl Schmitt*. London: Verso.

Baldwin, Peter, ed. 1990. *Reworking the Past: Hitler, the Holocaust, and the Historians' Debate*. Boston: Beacon Press.

Bambach, Charles. 2003. *Heidegger's Roots: Nietzsche, National Socialism, and the Greeks*. Ithaca, NY: Cornell University Press.

Bassin, Mark. 1987. "Race contra Space: The Conflict between German Geopolitik and National Socialism." *Political Geography Quarterly* 6 (2): 115–34.

Bauman, Zygmunt. 1989. *Modernity and the Holocaust*. Ithaca, NY: Cornell University Press.

Bendersky, Joseph W. 1983. *Carl Schmitt: Theorist for the Reich*. Princeton, NJ: Princeton University Press.

Black, Edwin. 2002. *IBM and the Holocaust*. London: Time Warner.

Borneman, John. 1997. "State, Territory, and National Identity Formation in the Two Berlins, 1945–1995." In *Culture, Power, Place: Explorations in Critical Anthropology*, edited by Akhil Gupta and James Ferguson, 93–117. Durham, NC: Duke University Press.

Bracher, Karl D. 1972. "Stages of Totalitarian 'Integration' (Gleichschaltung): The Consolidation of National Socialist Rule in 1933 and 1934." In *Republic to Reich: The Making of the Nazi Revolution: Ten Essays*, edited by Hajo Holborn, 109–28. New York: Random House.

Broszat, Martin. 1981. *The Hitler State: The Foundation and Development of the Internal Structure of the Third Reich*. London: Longman.

Burleigh, Michael. 2002 [1988]. *Germany Turns Eastwards: A Study of Ostforschung in the Third Reich*. London: Pan.

Canguilhem, Georges. 1978. *On the Normal and the Pathological*. Translated by Carolyn R. Fawcett. Dordrecht: D. Reidel.

Chamberlain, Houston S. 1911. *The Foundations of the Nineteenth Century*. Translated by John Lees. London: J. Lane.

Craig, Gordon A. 1982. *The Germans*. Harmondsworth: Penguin.

Davis, Karen. 2005. *The Holocaust and the Henmaid's Tale: A Case for Comparing Atrocities*. New York: Lantern Books.

Elden, Stuart. 2002. *Mapping the Present: Heidegger, Foucault, and the Project of a Spatial History*. London: Continuum International Publishing Group.

———. 2003. "Reading Genealogy as Historical Ontology." In *Foucault and Heidegger: Critical Encounters*, edited by Alan Milchman and Alan Rosenberg, 187–205. Minneapolis: University of Minnesota Press.

———. 2005a. "Contributions to Geography? The Spaces of Heidegger's *Beiträge*." *Environment and Planning D: Society and Space* 23 (6): 811–27.

———. 2005b. "Missing the Point: Globalization, Deterritorialization, and the Space of the World." *Transactions of the Institute of British Geographers* 30 (1): 8–19.

———. 2006. *Speaking against Number: Heidegger, Language, and the Politics of Calculation*. Edinburgh: Edinburgh University Press.

Enoch, Simon. 2004. "The Contagion of Difference: Identity, Bio-Politics, and National Socialism." *Foucault Studies* 1: 53–70.

Fest, Joachim C. 1972. *The Face of the Third Reich*. Translated by Michael Bullock. Harmondsworth: Penguin.

Foucault, Michel. 1997. In *"Il faut défendre la société": Cours au Collège de France (1975–1976)*. Edited by Mauro Bertani and Alessandro Fontana. Paris: Seuil/Gallimard.

———. 1999. In *Les Anormaux: Cours au Collège de France (1974–1975)*. Edited by Valerio Marchetti and Alessandro Salomani. Paris: Seuil/Gallimard.

Frei, Norbert. 1993. *National Socialist Rule in Germany: The Führer State, 1933–1945*. Translated by Simon B. Steyne. Oxford: Blackwell.

Friedlander, Henry. 1980. "The Manipulation of Language." In *The Holocaust: Ideology Bureaucracy, and Genocide*, edited by Henry Friedlander and Sybil Milton, 103–13. Millwood: Krause International Publishers.

Gasman, Daniel. 2004. *The Scientific Origins of National Socialism*. New Brunswick, NJ: Transaction Publishers.

Gobineau, Arthur, compte de. 1967 [1853–55]. *The Inequality of Human Races*. Translated by Adrian Collins. New York: Howard Fertig.

Gould, Stephen J. 1996. *The Mismeasure of Man*, revised and expanded. Harmondsworth: Penguin.

Grimm, Hans. 1926. *Volk ohne Raum*. Munich: Albert Langen.

Gutting, Garry. 1989. *Michel Foucault's Archaeology of Scientific Reason: Science and the History of Reason*. Cambridge: Cambridge University Press.

Han, Béatrice. 1998. *L'ontologie manquée de Michel Foucault: Entre l'historique et le transcendantal*. Grenoble: Jérôme Millon.

Heidegger, Martin. 1989. *Beiträge zur Philosophie (vom Ereignis), Gesamtausgabe Band 65*. Frankfurt am Main: Vittorio Klostermann.

———. 1991. *Nietzsche*. 4 vols. Translated by David F. Krell, Frank Capuzzi, and Joan Stambaugh. San Francisco: Harper Collins.

———. 1994. *Bremer und Freiburger Vorträge, Gesamtausgabe Band 79*. Frankfurt am Main: Vittorio Klostermann.

———. 1997. *Kant and the Problem of Metaphysics*. Translated by Richard Taft. Bloomington: Indiana University Press.

Herb, Guntram H. 2002. *Under the Map of Germany: Nationalism and Propaganda, 1918–1945*. London: Taylor and Francis.

Heske, Henning. 1986. "Political Geographers of the Past III German Geographical Research in the Nazi Period: A Content Analysis of the Major Geography Journals, 1925–1945." *Political Geography Quarterly* 5 (3): 267–81.

———. 1987. "Karl Haushofer: His Role in German Geopolitics and in Nazi Politics." *Political Geography Quarterly* 6 (2): 135–44.

Hitler, Adolf. 1939. *Mein Kampf: Unexpurgated Edition (Two Volumes in One)*. London: Hurst and Blackett.

Hofer, Walther. 1957. *Der Nationalsozialismus Dokumente, 1933–1945*. Frankfurt: Fischer Taschenbuch.

Janicaud, Dominique. 1990. *L'ombre de cette pensée: Heidegger et la question politique*. Grenoble: Jérôme Millon.

Kamenetsky, Ihor. 1961. *Secret Nazi Plans for Eastern Europe: A Study of Lebensraum Policies*. New York: Bookman Associates.

Kant, Immanuel. 1988. *Critique of Pure Reason*. Translated by Paul Guyer and Allen W. Wood. Cambridge: Cambridge University Press.

Kershaw, Ian. 2000. *The Nazi Dictatorship: Problems and Perspectives of Interpretation*. London: Arnold.

———. 2001a. *Hitler, 1889–1936: Hubris*. Harmondsworth: Penguin.

———. 2001b. *Hitler, 1936–1945: Nemesis*. Harmondsworth: Penguin.

Kisiel, Theodore. 2002. "Heidegger's Apology: Biography as Philosophy and Ideology." In *Heidegger's Way of Thought: Critical and Interpretive Signposts*. Edited by Theodore Kisiel, Alfred Denker, and Marion Heinz, 1–35. London: Continuum.

Klemperer, Victor. 2006. *Language of the Third Reich: LTI, Lingua Tertii Imperii: A Philologist's Notebook*. Translated by Martin Brady. London: Continuum.

Kopp, Friedrich, and Eduard Schulte. 1940. *Der Westfälische Frieden*. Munich: Hoheneichen.

Lacoue-Labarthe, Philippe. 1987. *La fiction du politique: Heidegger, l'art et la politique*. Paris: Christian Bourgois.

Lang, Berel. 1996. *Heidegger's Silence*. Ithaca, NY: Cornell University Press.

Lefebvre, Henri. 1967. *Position: Contre les technocrates en finir avec l'humanité-fiction*. Paris: Gonthier.

Levi, Primo. 1986. *Survival in Auschwitz and the Reawakening: Two Memoirs*. New York: Summit Books.

Nathans, Eli. 2004. *The Politics of Citizenship in Germany: Ethnicity, Utility, and Nationalism*. Oxford: Berg.

Naumann, Friedrich. 1915. *Mitteleuropa*. Berlin: Georg Reimer. (Published in English as: Naumann, Friedrich. 1916. *Central Europe*. London: P. S. King.)

Neumann, Franz L. 1966 [1944]. *Behemoth: The Structure and Practice of National Socialism, 1933–1944*. New York: Harper and Row.

Nietzsche, Friedrich. 1966. *Beyond Good and Evil*. Translated by Walter Kaufmann. New York: Vintage.

Noakes, Jeremy, and Geoffrey Pridham, eds. 2000. *Documents on Nazism, 1919–1945*. 4 vols. Exeter: Exeter University Press.

Osiander, Andreas. 2001. "Sovereignty, International Relations, and the Westphalian Myth." *International Organization* 55 (2): 251–87.

Ott, Hugo. 1993. *Martin Heidegger: A Political Life*. London: Harper Collins.

Patterson, Charles. 2002. *Eternal Treblinka: Our Treatment of Animals and the Holocaust*. New York: Lantern Books.

Pick, Daniel. 1989. *Faces of Degeneration: A European Disorder, c. 1848–1918*. Cambridge: Cambridge University Press.

Raffestin, Claude, Dario Lopreno, and Yvan Pasteur. 1995. *Géopolitique et histoire*. Lausanne: Éditions Payot.

Ratzel, Friedrich. 1896–98. *The History of Mankind*. 6 vols. New York: Macmillan.

———. 1897. *Politischen Geographie*. Munich: Oldenbourg.

———. 1901. "Der Lebensraum: Eine Biographische Studie." In *Festgaben für Albert Schäffle zur Siebenzigsten Wiederkehr Seines Geburtstages am 24. Februar 1901 Dargebracht von K. Bücher [et al.]*, edited by Karl Bücher, 103–89. Tübingen: Laupp.

———. 1988 [1876]. *Sketches of Urban and Cultural Life in North America*. New Brunswick, NJ: Rutgers University Press.

Sax, Boria. 2000. *Animals in the Third Reich: Pets, Scapegoats, and the Holocaust*. London: Continuum.

Schmitt, Carl. 1991 [1939]. *Völkerrechtliche Großraumordung mit Interventionsverbot für Raumfremde Mächte: Ein Beitrag zum Reichsbegriff im Völkerrecht*. Berlin: Duncker and Humblot.

———. 1995. In *Staat, Großraum, Nomos: Arbeiten aus den Jahren, 1916–1969*. Edited by Günter Maschke. Berlin: Duncker and Humblot.

———. 1997. *Land and Sea*. Washington, DC: Plutarch Press.

Sereny, Gitta. 1995. *Albert Speer: His Battle with Truth*. London: Macmillan.

Smith, Neil. 2003. *American Empire: Roosevelt's Geographer and the Prelude to Globalization*. Berkeley: University of California Press.

Sofsky, Wolfgang. 1997. *The Order of Terror: The Concentration Camp*. Princeton, NJ: Princeton University Press.

Speer, Albert. 1970. *Inside the Third Reich*. London: Sphere.

Staudenmaier, Peter. 1995. "Fascist Ideology: The 'Green Wing' of the Nazi Party and Its Historical Antecedents." In *Ecofascism: Lessons from the German Experience*, edited by Janet Biehl and Peter Staudenmaier, 4–30. London: AK Press.

Stolting, Edward. 1987. "Anthropologie." In *Disziplinen: Wissenschaft in Berlin*, edited by Tilmann Buddensieg, Kurt Düwell, and Klaus-Jürgen Fembach, 128–33. Berlin: Gebr. Mann.

Teschke, Benno. 2003. *The Myth of 1648: Class, Geopolitics, and the Making of Modern International Relations*. London: Verso.

Urwin, Derek W. 1982. "Germany: From Geographical Expression to Regional Accommodation." In *The Politics of Territorial Identity: Studies in European Regionalism*, edited by Stein Rokkan and Derek W. Urwin, 165–249. London: Sage.

Wolin, Richard, ed. 1993. *The Heidegger Controversy: A Critical Reader*. Cambridge, MA: MIT Press.

8

Applied Geography and Area Research in Nazi Society: Central Place Theory and Planning, 1933–1945

MECHTILD RÖSSLER

Introduction

In recent years the volume of literature on the history of science and, particularly, social sciences in Nazi Germany has been growing rapidly (Tröger 1984; Lundgreen 1985; Klingemann 1987). Additionally, historical studies on geography outside the realm of geopolitics in the 1930s have appeared; Sandner (1988), for example, provides an account of the gradual development of the history of geography. In the context of recent research, "Central Place Theory" (Christaller 1933c) and the implications of applied geography, as well as area research and planning in Nazi society, have emerged as important themes.

Because of the complex, highly industrialized state of the National Socialist dictatorship that existed in Germany in the 1930s, social research and planning were regarded by the Nazis as essential to society's growth.[1] Unfortunately, however, few modern studies on National Socialism emphasize the highly relevant interaction between social science and politics, thereby too often neglecting the importance of empirical research and planning in the 1930s. In contrast to geographers, general historians of National Socialism are beginning to recognize this significant interrelationship. A current investigation suggests that the Nazi state was based on "reactionary modernism" (Herf 1984), a blending of traditional elements and modern technical development:

> Reactionary modernism was a specifically German response to a universal dilemma of societies facing the consequences of the industrial and French revolutions: How can national traditions be reconciled with modern culture, modern technology, and modern political and economic institutions? (Herf 1984, 217)

Thus, in trying to understand the social and political history of National Socialism we have to address the role of scientists and scientific disciplines as well as the integration of tradition and modernization. The present study, therefore, is directed toward placing the production of theory and its application in the context of social and political processes.

Area Research, Social Theories, and Practice in Nazi Institutions

The "German spatial reorganization" (*die deutsche Raumordnung*) began in 1935 with the first Reich's Office for Spatial Planning (Reichsstelle für Raumordnung) headed figuratively by Hitler himself (that is, it was subordinated directly under Hitler). Its tasks encompassed the centralization of all planning and research activities. At that time, planning was legally regulated by the domicile settlement law (*Wohnsiedlungsgesetz* 1933), the settlement organization law (*Siedlungsordnungsgesetz* 1934), and the Reich's environment protection law (*Reichsnaturschutzgesetz* 1935).

As the economist Gerhard Hesse wrote in 1944, "area research and spatial planning" were confronted with "mighty tasks" at this time:

> Great expectations have been placed before the German area planning program. If German *Raum* formation is to measure up to the new economic, logistical, social and demographic-political objectives, then changes in the composition of towns on a huge scale, the dispersal of major cities and industrial centres, the relocation of factories and the resettlement of people and other transformations will have to be tackled. Further examples of similar changes are the new city planning of Berlin, Munich and Nuremberg, the building of the *Reichsautobahn* [highway], the emergence of the central German industrial region in the North of the Harz mountains . . . the town border and industry workers settlements, and the building up of rearmament production companies. (Hesse 1944, 118)

In view of the far-reaching scope of these plans, it is hardly surprising that the "anarchy of competences" (conflicts over competence and power), or bureaucratic infighting, which was so characteristic of the Nazi regime, was soon in evidence in this area too. Although officially the Reichsstelle für Raumordnung was responsible for this field of research, its significance was soon attenuated both in regard to its own subordinate research organizations as well as in regard to the new research and planning agencies created by Himmler, Rosenberg, Göring, and the German Labor Front.

A great number of geographers worked in the Reich Association for Area Research (Reichsarbeitsgemeinschaft für Raumforschung). It was founded in 1936 and had branches at all universities. The first leader of this organization was the agronomist Konrad Meyer,[2] whose ideas determined the nature of area research in its initial stages from 1936 to 1939. He combined two elements of National Socialism into one concept: the traditional "folk" or "ethnic" sciences and modern social research.

Meyer institutionalized his new concept during the so-called revolutionary *Umbruch* (radical change) at the universities.[3] Meyer directed scientists from different faculties to work in a community on one subject and one region. Topics of their research included, for example, *Volkstums* (ethnic) studies on the border areas of Germany and Alsace (at Freiburg University) and economic evaluations of industrial growth in the Hamburg harbor region (at Hamburg University). The leader of the Hamburg group, Professor Andreas Walther, developed a "social cartography" showing class structure, political orientation, and quality of health in districts and town quarters (Pahl-Weber and Schubert 1987; Rössler 1988b). Walther's approach, significantly, paved the way for the Reich's urban population policy.

In general, scientists had to work for the extension of German "living space" (*Lebensraum*) and in the "interest of the Aryan race." This concept was influenced by Ernst Krieck's ideas about the National Socialist takeover of the universities. At the core of this ideological stance were the notions of the German Nation as one "*Volk*" (*Volksgemeinschaft*), and anti-Semitic and racist visions.

The German *Altreich* (that is, Germany in 1937) was divided into twenty-two different planning areas, and geographers made studies of various subjects, including area inventories (maps), the development of transportation and communication, comparisons of population densities, and the intensity of agricultural and industrial production. These research and planning efforts were geared toward the formulation of population policies and rearmament production for the Four Year Plan. Insofar as the study of space (*Raum*) was interdisciplinary and concerned with concrete ethnic-political goals, it broke new ground and represented a departure from the traditional organization of a discipline like geography at the universities.

Central Place Theory in Nazi Germany

Apart from research associations at the universities in 1937, special study groups were also founded, one of which was the Arbeitskreis Zentrale Orte (Study Group on Central Places). Other groups studied subjects such as mineral resources, labor service and mobilization, and distressed areas.[4] The foundation for the Study Group on Central Places was laid by Walter Christaller in his famous PhD dissertation "Central Places in Southern Germany: An Economic-Geographical Investigation into the Regularities of the Distribution and Development of Settlements with Urban Functions"

(Christaller 1933c; see appendix). Essentially he argued that there is a hierarchically structured network of settlements, and a hierarchically structured order to regional economies. His model was dependent upon economic, administrative, and political circumstances.

After finishing his thesis in 1932, Christaller received research and travel grants from the Penck Foundation, the German Research Foundation (Deutsche Forschungs-gemeinschaft), and the Reich's Association for Area Research for various projects, including the *Atlas for German Living Space* (*Atlas des deutschen Lebensraumes*).[5] During that time he initiated the study group on "Central Places" (see Meyer 1970).

Christaller himself, Hans Bernhard von Grünberg, Walter Geisler (the former vice chancellor of the Reich's University Posen), Gottfried Feder (the founding member of the National Socialist Party and Reichssiedlungskommissar [Reich's Settlement Commissioner] up to 1935), and other scientists converged to discuss the implications of central place theory for planning in Nazi Germany.

My research efforts have revealed two phases in the development, institution, and application of central place theory:[6] first, the initial reception of central place theory, 1933–39; and, second, development of the theory in planning the occupied eastern areas (1939–45). Significantly, Christaller's theory was used for the "Secret Nazi Plans for Eastern Europe." For example,

> Plans were made to build completely new German cities in the East, with approximately 15,000–20,000 population, at important railway and highway junctions. Such cities were to be surrounded by a 5–10 mile broad belt of German villages. (Kamenetsky 1961, 67)

Christaller's thesis on central places in southern Germany (completed in 1932) was published in Jena in 1933. There is a persistent myth and misconception that the importance of his famous work was only recognized in the 1950s and 1960s (see, for example, Klöpper 1970). Lichtenberger (1978, 9) pointed out the enormous time lag between the publication of Christaller's thesis and the earliest attempts by empirical social scientists to make appropriate quantitative studies (Burton 1963). Von Böventer (1968, 111) stated that the "application of Christaller's model is limited to cases, like southern Germany around 1930. . . . Thus the model remains valid as one element of a more comprehensive theory of spatial structures." Wirth (1982, 293) mentioned the critical reception of Christaller's concept by geographers, and Meyer (1970, 405) wrote that Christaller's model was ignored for a long time. James pointed out: "German geographers in the 1930's paid little attention to Christaller's work. It was tested by Otto Schlier in 1937 on the basis of census data . . . and it was applied to the pattern of settlement in Estonia by Edgar Kant" (1972, 519).[7] It is true that only a few geographers noticed Christaller's thesis in the 1930s (see Wunderlich 1933; Dörries 1934; Bobek 1935), but in planning and research organizations it was soon discussed as a new planning theory. The proceedings of the Study Group on Central Places revealed that the theory was recognized as a model for the spatial reorganization of Germany.[8] With this model, it seemed possible to plan the distribution of goods and services in an economic manner. Efficiency and economic viability of the spatial order were essential for building a rational society, especially in the prewar period when industry was being brought into line for rearmament.

Christaller's theory was not positively received by all scientists. Some scholars rejected his theory on the grounds that it was too "theoretical," and they called for a more "organic" model which would correspond with the "organically" conceived National Socialist "ethnic community." A case in point is the debate between Christaller and the economist Friedrich Bülow about the problems National Socialists had with the concept of central place theory. The planning of the eastern occupied areas gave rise to the debate, which began in the mid-1930s and appeared in print a few years later in a new journal called *Archiv für Wirtschaftsplanung* (*Archive of Economic Planning*). Its editor articulated Christaller's theoretical orientation as opposed to the practical realization of Bülow's ideas. Bülow, for example, favored a more applied approach and intended planning for the creation of a real Nazi society and community. Christaller conceptualized a *hierarchy of settlements* on a theoretical level. His model assumed a uniform populated plain, the proportional distribution of services, and a hexagonal market area. Bülow, on the other hand, presumed a *hierarchy of settlements* as a projection of the political structure and reorganization of the German nation and *Volk*. Bülow criticized Christaller's deductive methods and theoretical focus. Yet by that time Christaller had already modified his theory and had included the creation of the Nazi *Volk*-community (*Volksgemeinschaft*) as part of his concept (Christaller 1941b). The distinction between his theory and Bülow's model was the difference between a deductive and an inductive method. Practically, the debate was of little consequence because both models remained in the planning stage and their application was prevented by circumstances of war.

Applied Geography—Christaller's Approach to Local Government Science

In 1937 Walter Christaller's book *Rural Settlements in Germany in Their Relation to Community Administration* was published by the Kommunalwissenschaftliches Institut (Institute for the Study of Local Government) in Berlin (Christaller 1937b). Local government science was a new research field at that time, developed for the instruction of municipal officers and public servants and for local planning studies.

In the same year, Christaller went to Freiburg University and began a short career as a lecturer. He founded and expanded the Kommunalwissenschaftliches Institut under Professor Maunz, the author of the "concentration camp legal code" (*KZ-Recht*). Jointly financed by the university and the town of Freiburg-Breisgau, the institute was closely connected to the University Law School. As a renowned public law specialist (*Lehrstuhl für Staatsrecht, Verwaltungsrecht und Finanzrecht* [Chair of Constitutional, Administrative and Financial Law]), and a member of the law faculty, Maunz was a logical choice for the directorship of the institute.[9] It was there that Christaller developed a "new form of applied geography" called "local government geography" (*Kommunalgeographie*), for which he did not win any recognition and which was largely ignored by other geographers (1937a, 1937b).

His approach brought central place theory together with theories of local government and the different political organization of towns, villages, and small places. He gave lectures on industrial planning and small settlements. In Freiburg he was supported by the geographer Friedrich Metz,[10] who was the leader (*Rektor*) of Freiburg University from 1936 to 1938.[11] Metz promoted the acceptance of Christaller's habilitation dissertation (*Rural Settlements in Germany in Their Relation to Community Administration*) in 1938. Metz was also the head of the Freiburg University branch of the Reich Association for Area Research, through which he and Christaller collaborated. Nevertheless, Metz's approach to human geography was quite different from Christaller's systematic one. Christaller's time in Freiburg was short: 1937 to the early 1940s (see Rössler 1983). However, "local government geography" was only one of his first attempts at applied geography. At the same time he organized the aforementioned study group on "Central Places" and was involved in debates on planning theory for Nazi Germany.

"Secret Nazi Plans for Eastern Europe": The General Plan for the East

With the beginning of the Second World War and the war program entitled "The German East," the studies of the Reich's Association for Area Research and the *Reichsstelle* were increasingly directed towards fundamental plans for the development of the conquered East.[12] During this period most geographers were active in empirical research on the East within different scientific organizations (see Rössler 1988a).

A noteworthy fact is that the Polish government exiled in London knew about the connections between area planning and German settlement policy in the annexed countries of the East. In 1942 the Polish Ministry of Information published a text entitled "The German New Order in Poland," which referred to the Reich's Office for Spatial Planning and its research.[13]

In 1940 Konrad Meyer brought Christaller to Berlin. Thirty years later, Meyer wrote an overview of Christaller's life that included a glimpse into that phase:

> When in the beginning of the 1940s the writer of this short biography was authorized with the planning of the eastern areas, he invited Christaller to his planning office. There Christaller worked (until the dissolution of this authority) as a scientific colleague on formulating the conception of the projected new order after the war of the German Reich. He was able then to verify (economically supported and protected from several political problems and disagreements) the theoretical understanding of his dissertation through the reality of culturally and geographically different settlements. (Meyer 1970, 405)

Meyer was in charge of the head office of Planning and Soil in Himmler's Reich's Commissariat for the Strengthening of Ethnic Germandom (*Reichskommissariat für die Festigung deutschen Volkstums*). This new organization, which was created by Himmler at the beginning of the war, was given the task of area planning in the conquered East:

> What is significant and worthwhile to note is the fact that the Nazi colonization activities were not entrusted to the regular German administration and occupation authorities in the East but to special agencies under the authority of SS leader Heinrich Himmler. (Kamenetsky 1961, 70)

Under this organization, Konrad Meyer, the former leader of the area research groups, had selected a few social scientists to work using innovative methods and theories. The primary social scientists he employed were the sociologist Herbert Morgen, and the geographers Angelika Sievers and Walter Christaller in the Planning and Soil Office and in Meyer's university institute. At that time, Christaller was financed by several research foundations and institutes—the Forschungsdienst (Research Service), the German Research Foundation (DFG), and the Institute of Agriculture and Agricultural Politics (Institut für Agrarwesen und Agrarpolitik)— and worked as a freelance scientist in the main office of Planning and Soil.

During his time in Berlin, Christaller produced many studies, which to date are unknown. For example, "*Die zentralen Orte in den Ostgebieten und ihre Kultur- und Marktbereiche*" ("The central towns of the eastern countries and their cultural and market areas," 1941c) and others were published for official use only. Hottes, Hottes, and Schöller (1980, 12) wrote:

> Several of Christaller's works from the years 1940–1944 reveal his co-operation with National Socialist regional planning, including the redevelopment of the settlement pattern in the areas then under German military occupation in Eastern Europe. These plans, under the conditions of totalitarian power though to their author scientifically sincere, had political consequences which made the acceptance of his views by the academic community cautious, and by some hostile, not only in Eastern Europe.

This last statement is very problematic, because we have to differentiate between the geographic academic community and the scientific community of regional planning and space organization which was developed in Nazi society. The geographic academic community was, on the one hand, partly involved in research on the eastern countries during this period (for example, universities such as Posen and the Institute for German Work in the East in Kraków had geographical branches). On the other hand, they did not employ new theories such as the central place model because they used a traditional regional geographical approach based on inductive methods.

The acceptance by the wider academic community was positive and multifaceted. This community, to which Christaller belonged in the 1930s, used his theory in their planning studies. He initiated study groups and, after 1939 when new conflicts emerged over competence and power between the different planning organizations of Himmler, Alfred Rosenberg (Hauptämter für Raumordnung—Main Offices for Space Planning in the occupied areas) and the *Reichsstelle*, he undertook studies for Himmler's office. This planning institution, the Planning and Soil Office (Stabshauptamt für Planung und Boden) was comparatively speaking the most powerful. The head office and the Institute for Agriculture and Politics under Konrad Meyer invented the well-known *Generalplan Ost* (General Plan for the East) (see Heiber 1958; Eichholtz 1982; Benz 1985; Roth 1985). In broad terms this plan was made for Himmler, who needed a detailed concept for the settlement policy and administration in the new areas in the East for the next twenty-five years. The plan included a proposed settlement pattern, the construction of new central places, and a time scale. It presumed such population policies as the Germanization of native Poles and Russians, the resettlement of German peasants to the East, and the liquidation of Jews and most of the native population. The aim of the General Plan for the East was to build up a "truly German and Aryan community" through settlement construction. This plan

> mentions the future creation of special villages and village complexes. Such an ideal future village was supposed to cover from 2,500–3,500 acres and to secure a livelihood for 300–400 people. . . . The groups of villages were not to be centered around a town, as might naturally be expected, but around one large village called *Hauptdorf* (main village). Such a *Hauptdorf* was to be the unifying core of the surrounding villages. (Kamenetsky 1961, 77)[14]

This final stage of planning in Hitler's state was conducted by the staff of Meyer's institute at Berlin University. Several scientists were involved in the General Plan for the East. Some of them were members of the SS (SS guard detachments), others (including Walter Christaller) were members of the Nazi Party.[15] Whether or not the scientists were fascists themselves is, however, a moot point. Bunge failed to recognize the contradictions inherent in his statement "Walter Christaller was not a Fascist" when he wrote:

> The internal content of Christaller's work was obviously that of a man of science. Science is something that Fascists are short of. Engineers yes, but scientists no. Every major theoretician, no matter from what country or what religion opposed Hitler. (Bunge 1977, 84)

In fact, the Nazi State clearly relied on the input of pure and social scientists, including planners and area organizers. Thus social scientists and geographers found themselves in "favorable" circumstances for research and implementation of their ideas, which were involved in the political situation during the Nazi dictatorship. For many of them, the political context of their work hardly came into question. Their research, however, became inseparable from the Nazi's general attempts toward modernization of the society.

The Discussion of Planning in the Nuremberg Trials

The chief of the head office of Planning and Soil, Konrad Meyer, was brought to the Nuremberg trials in 1946 because of the General Plan for the East and his SS membership. In his memoirs[16] he gave detailed information on his relationship to Himmler, his ideology, and a commentary on the postwar period. Although an analysis of the manuscript lies outside the scope of this paper, I want to draw attention to some interesting facts on the planning debate in the archival material of the Nuremberg Trials (BAK: AIL Proz. 3, Behling).

Meyer was accused in "case 8," which was called *Volkstumsprocess*. His defender at the trial was Dr. Behling, who collected material from the Nazi era (papers from Konrad Meyer, planning laws, etc.) together with "whitewash papers" (*Persilscheine*) from many well-known people such as the physicist Werner Heisenberg, Walter Christaller, and other members of his former staff. Walter Christaller pointed out in his "whitewash paper" that Meyer had employed several scientists of antifascist political inclination. The line of defense was to show that the work of Himmler's planning office was only to produce scientific planning studies, which never were realized in any form. Konrad Meyer emphasized that the *Generalplan Ost* was a totally independent scientific study, even though it prescribed deportation and destruction of the non-German people and their land. Further, he argued that he had supported critical scientists like the former member of the Social Democratic Party, Walter Christaller, who was a member of the Communist Party after the war.[17]

As the Allies could not demonstrate that the General Plan had been realized in any way, Konrad Meyer was freed in 1948 and continued his scientific career as a professor for Land Planning in Hannover. The problem of the trial was twofold: the judges at the Nuremberg Trial failed to recognize the flexible manner in which scientific research was conducted within the Nazi State as well as the relationship between social science, and organization and planning, especially in the Eastern occupied areas.

Conclusion

Historically, social scientists have been influenced in a specific way by contemporaneous social and political conditions. Germany in the 1930s was a highly industrialized and modernized society, which needed planning authorities and organizations. With the Reichsstelle für Raumordnung in 1935, a centralized planning organization was created which had research branches at every university.

At that time a new scientific community grew up—academics working in planning organizations. They adopted Walter Christaller's central place theory as a new planning model in two stages: first, the initial reception of the theory, 1933–39; and second, its application in planning institutions.

Christaller's theoretical model had already assumed considerable significance in the context of Nazi planning activities. This was soon in evidence in the endeavors of the various planning agencies responsible for the occupied East.

In 1940 Konrad Meyer summoned Christaller to his Institute for Agricultural and Agricultural Political Studies in Berlin where he was to work within a research team including geographers, rural sociologists, and landscape planners. In this context, Christaller researched the "cultural and market centers" of the Warthegau, and hence endeavored to apply his theories to occupied Poland. Concretely, this means that he established the basic plans for German settlement in occupied Poland.

Significantly, Christaller was simultaneously employed on the staff of the Main Office for Planning and the Soil, an agency within Himmler's Reich's Commissariat for the Strengthening of Ethnic Germandom. Himmler had such a planning office because Hitler had entrusted him with "the reorganization of ethnographic relations and the creation of settlement areas in the East." Himmler's planning agency was also responsible for the overall plan for *Lebensraum* in the occupied East, which became notorious later as the General Plan for the East. Here one also encounters Christaller's theory of a hierarchy of places, referred to in this context as "settlement pearls," which appear in graduated zones of settlement. Academics on the staff of Meyer's planning office worked on the General Plan, making calculations about cost, available settlers, and the spatial transformation of the area concerned. Therefore, Walter Christaller worked in high-profile Nazi organizations from 1940 to 1944. There he modified his theory together with other geographers, town planners, and economists.

In my analysis, I have tried to give an overview of the different fields in which area research and planning was practiced during the Nazi period. The use of the concept of "space" under this regime shows the social significance of and the demand for geographical and area research during the Nazi era. This was confirmed by the financial support this field of research received from Nazi agencies. Area research and space planning, as pursued in the Third Reich, continued after 1945. Their institutional and organizational conceptions then became paradigmatic for institutions in the Federal Republic of Germany. I have tried to point out the complex relationships between biographical, internal academic, and external factors, which in reality are inseparable. Political systems, social developments, and academic considerations are bound together in a specific manner which illustrates the clear interrelationship between the production of social science and society.

Acknowledgments

This chapter is based on a presentation given at the 1988 AAG Conference in Phoenix, Arizona, with the title "Area Research and Planning in Nazi Germany: Walter Christaller and the Central Place Theory." I wish to thank

Guntram Herb (University of Wisconsin, Madison), Michael Burleigh (London School of Economics), and Joanna Carter (New York and Hamburg) for helpful comments.

Notes

1. For the situation in the United States during this period see (for example) the institutionalization and the development of the Chicago School of Sociology (Bulmer 1984).

2. Konrad Meyer was succeeded by Professor Paul Ritterbusch, who headed the *Kriegseinsatz der Geisteswissenschaften* (application of humanities for the war effort) in 1939–44, and by the geographer Kurt Brüning (1944–45), the first president of the Academy of Area Research and Regional Planning after the war.

3. For example, see the well-known speech of Martin Heidegger as the leader of Freiburg University in 1933, "Die Selbstbehauptung der deutschen Universität." Heidegger's role in National Socialism has induced a lot of discussion at Freiburg University. An interesting example of the political involvement of geographers is seen in geomorphologist Hans Mortensen, Heidegger's "Senator zur besonderen Verwendung" (Senator for Special Employment) in 1933 and a lecturer in Heidegger's series of speeches on "German Socialism" in the summer term 1933 (Rössler 1983).

4. See the archival material in the Bundesarchiv Koblenz (BAK) (Rl13/93).

5. Research in the Hoover Archive at Stanford revealed political judgments and efficiency reports (by the *NS-Dozentenbund* [National Socialist university teachers organization], München) in regard to Christaller's former activities. This archival material was found in the German Research Foundation Collection, which was brought to the United States after the war (during the denazification period).

6. I used material and primary sources from the archives in Freiburg, the Bundesarchiv in Koblenz, the Zentrales Staatsarchiv of the German Democratic Republic, the Public Record Office in Kew, London, and the Hoover Archive in Stanford, California.

7. Kant represents a typical example of the complex interrelationship between biography and social science in a political context. Born in the Russian city of Dorpat, he studied at the Estonian University of Tartu (later renamed Dorpat by the Nazis). After lecturing there during the war, he went to Sweden in 1944, where he founded the Lund school of urban geography. For further information on Kant, see Buttimer (1988).

8. Source of the proceedings: BAK R113 and R164.

9. Maunz was the official in charge of "Jews in the law" (Referent für Judentum in der Rechtwissenschaft) in the National Socialist Rechtswahrerbundes (NSRB—lawyers organization). For more information on Maunz's role in the Third Reich and postwar era see Seeliger (1964, 43–45).

10. Metz supported Christaller's trip to the International Congress of Geographers in 1938 in Amsterdam. He got travel expenses (240 Reichsmark) from the Ministry of Science and Education (see *Zentrales Staatsarchiv Potsdam*, 49-01/2818).

11. Metz was, after Heidegger, the third *Rektor* of Freiburg University during National Socialism. Nothing is known about any connection between Heidegger and Christaller.

12. There were many organizations and institutions of eastern research, such as the Reich's University Posen, opened on Hitler's birthday in 1941, and the Institut für Deutsche Ostarbeit (Institute for German Work in the East) in Kraków. Both institutions had geographical branches.

At the Reich's University Posen, Walter Geisler was the head of the geographical institute, and Hans Graul headed the department of *Landeskunde* (regional geography) at Kraków.

13. "Even while consolidating his conquests . . . in accordance with the high-sounding slogan of National Socialism: *Lebensraum für das deutsche Volk* (Living space for the German people) Hitler has created a new state organization called Reichsstelle für Raumordnung (Reich's Office for Space Planning). The Kölnische Zeitung of November 21, 1940, writes that the task of the new office, which is to cooperate with the Reich's Commissariat for the Strengthening of Ethnic Germandom, is: 'to fill the unpopulated (*menschenleere*) areas in the East by settling German peasants, German business men and workers, so that as a result a country truly German shall arise'" (Polish Ministry 1942, 199).

14. On the creation and development of the *Hauptdorf* concept, see Christaller (1942b).

15. Christaller's membership number was 8.375.670 from July 1, 1940 (see Berlin Document Center, personal files Walter Christaller).

16. Private property of Mrs. Dina Meyer, Salzderhelden.

17. Christaller rejoined the Social Democratic Party in 1959.

References

Primary Literature

*These references have not been cited in the text, but are of particular interest to this paper.

Bobek, H. 1935. "Eine neue Arbeit zur Stadtgeographie: Rezension von Walter Christaller, Die zentralen Orte in Süddeutschland." *Zeitschrift der Gesellschaft für Erdkunde zu Berlin*: 125–30.

*Bülow, F. 1937a. "Gedanken zu einer volksorganischen Standortslehre." *Raumforschung und Raumordnung* 1: 385–90.

*———. 1937b. "Zur Standortstheorie des Wirtschaftsliberalismus." *Raumforschung und Raumordnung* 1: 353–58.

*———. 1941. *Großraumwirtschaft, Weltwirtschaft und Raumordnung: Raumforschung und Raumordnung: Volks- und Raumpolitische Reihe.* Heft 1. Edited by P. Ritterbusch. Leipzig: K. F. Koehler.

*Christaller, Walter. 1924. "Beispiele, wie der abgebaute Beamte sich durch Garten, Heimstätte oder Siedlung wirtschaftlich besser stellen wird." *Die Heimstätte* supplement of "Die Bodenreform," August 22.

*———. 1933a. "Grundsätzliches zu einer Neugliederung des deutschen Reiches und seiner Verwaltungsbezirke." *Geographische Wochenschrift* 1: 913–19.

*———. 1933b. "Kommunale Elektrizitätswirtschaft in Württemberg." *Zeitschrift für Kommunalwirtschaft*: 640.

———. 1933c. "Die zentralen Orte in Süddeutschland: Eine ökonomisch-geographische Untersuchung über die Gesetzmaßigkeit der Verbreitung der Siedlungen mit städtischen Funktionen." PhD diss., Erlangen. Jena: Gustav Fischer.

*———. 1934a. "Allgemeine geographische Voraussetzungen der deutschen Verwaltungsgliederung." *Jahrbuch für Kommunalwissenschaft* 1: 48.

*———. 1934b. "Baut eine Stadt!" *Kreuz-Zeitung* 86 (195): 1–2.

*———. 1934c. "Zur französischen geographischen Wissenschaft und Kartographie." *Geographische Wochenschrift* 2: 1000–1001.

———. 1937a. "Kommunalwissenschaft und Siedlungsgeographie." *Jahrbuch für Kommunalwissenschaft* 4: 492–500.

———. 1937b. *Die ländliche Siedlungsweise im Deutschen Reich und ihre Beziehungen zur Gemeindeorganisation Einzelschriften des Kommunalwissenschaftlichen Institutes der Institutes der Universität Berlin 7*. Stuttgart: Kohlhammer.

*———. 1938a. "Rapports fonctionnels entre les agglomérations urbaines et les campagnes." In *Comptes Rendus du Congrès Internationale de Géographie Amsterdam 1938*, 123–27. Leiden: F. J. Brill.

*———. 1938b. "Siedlungsgeographie und Kommunalwissenschaft." *Petermanns Geographische Mitteilungen* 84: 49–52.

*———. 1938c. "Zweite Reichstagung der Reichsarbeitsgemeinschaft für Raumforschung in Graz vom 17.–20. Oktober 1938." *Petermanns Geographische Mitteilungen* 84: 372–73.

*———. 1940a. "Grundgedanken zum Siedlungs- und Verwaltungsaufbau im Osten." *Neues Bauerntum* 32: 305–12.

*———. 1940b. "Die Kultur- und Marktbereiche der zentralen Orte im deutschen Osten und die Gliederung der Verwaltung." *Raumforschung und Raumordnung* 4: 498–503.

*———. 1941a. "Das Land in der Verwaltungsgliederung des Altreichs." In *Landvolk im Werden: Material zum ländlichen Aufbau in den neuen Ostgebieten und zur Gestaltung des dörflichen Lebens*, edited by Konrad Meyer, 2nd ed., 1942, 283–88. Berlin: Deutsche Landbuchhandlung.

———. 1941b. "Raumtheorie und Raumordnung." *Archiv für Wirtschaftsplanung* I: 116–35.

———. 1941c. *Die zentralen Orte in den Ostgebieten und ihre Kultur- und Marktbereiche. Bd. I des Gemeinschaftwerkes der Reichsarbeitsgemeinschaft für Raumforschung: Struktur und Gestalt der zentralen Orte des deutschen Ostens*. Leipzig: K. F. Koehler.

*———. 1942a. "Land und Stadt in der deutschen Volksordnung." *Deutsche Agrarpolitik* 1: 53–56.

———. 1942b. "Die Verteilung der nichtlandwirtschaftlichen Bevölkerung im Hauptdorfbereich." *Neues Bauerntum* 34: 139–45.

*———. 1942c. "Die Verteilung der nichtlandwirtschaftlichen Bevölkerung im Hauptdorfbereich." *Neues Bauerntum* 34: 169–76.

*———. 1944. "Die Landstadt als Stufe des Siedlungsaufbaus." *Neues Bauerntum* 36: 300–306.

*———. 1948. "Bildung einer Arbeitsgemeinschaft für funktionale Landes- und Verwaltungsgliederung." *Berichte zur deutschen Landeskunde* 5: 35–37.

*———. 1968. "Wie ich zu der Theorie der zentralen Orte gekommen bin." *Geographische Zeitschrift* 56: 88–101.

*Christaller, Walter, and Johannes Lubahn, eds. 1922. *Praktische Heimstättenarbeit: Heimstättenamt der deutschen Beamtenschaft*. Berlin: Eichkamp.

Dörries, Hans. 1934. "Rezension von Walter Christaller, Die zentralen Orte in Süddeutschland." *Geographische Zeitschrift* 40: 233–34.

Heidegger, Martin. 1933. "Die Selbstbehauptung der deutschen Universität." Reprinted in 1983 in *Die Selbstbehauptung der deutschen Universität: Das Rektorat 1933/34*. Frankfurt: Tatsachen und Gedanken.

Hesse, Georg. 1944. *Die Deutsche Wirtschaft: Volkswirtschaftslehre*. Hannover: Meyer.

*Meyer, Konrad. ca 1970. "Über Höhen und Tiefen: Ein Lebensbericht." Mimeo. Salzderhelden; now in the Bundesarchiv, Koblenz.

Polish Ministry, ed. 1942. *The German New Order in Poland*. London: Hutchinson.

Wunderlich, H. 1933. "Rezension von Walter Christaller, Die zentralen Orte in Süddeutschland." *Geographische Wochenschrift* 1: 957–58.

Secondary Literature

Benz, Wolfgang. 1985. "Der Generalplan Ost: Zur Germanisierungspolitik des NS-Regimes in den besetzten Ostgebieten, 1939–1945." In *Die Vertreibung der Deutschen aus dem Osten: Ursachen, Ergebnisse, Folgen*, edited by Wolfgang Benz, 39–48. Frankfurt: Fischer.

Bulmer, Martin. 1984. *The Chicago School of Sociology*. Chicago: University of Chicago Press.

Bunge, William. 1977. "Walter Christaller Was Not a Fascist." *Ontario Geography* 37: 84–86.

Burton, Ian. 1963. "Quantitative Revolution and Theoretical Geography." *Canadian Geographer* 1: 151–62.

Buttimer, Ann. 1988. "Edgar Kant: Baltic Pioneer." Paper presented at Session D3 of the XXVI International Geographical Congress held at Sydney, August 1988. Published in *Geografiska Annaler: Series B, Human Geography* 87 (3): 175–92, September 2005.

*Carol, Hans. 1970. "Walter Christaller—A Personal Memoir." *Canadian Geographer* 14: 67–69.

Eichholtz, Dietrich. 1982. "Der 'Generalplan Ost': Uber eine Ausgeburt imperialistischer Denkart und Politik." *Jahrbuch für Geschichte* 26: 217–53.

Freeman, Michael J. 1987 *Atlas of Nazi Germany*. Beckenham, Kent: Croom Helm.

Gröning, Gert, and Joachim Wolschke-Bulmahn. 1987. *Die Liebe zur Landschaft Teil III: Der Drang nach Osten: Zur Entwicklung von Landespflege im Nationalsozialismus wdhrend des Zweiten Weltkrieges in den "eingegliederten" Ostgebieten*. Munich: Minerva.

*Hantschel, R. 1978, "Geographie Heute—Zwischen Wissenschaftstheorie und Technokratie." In *Quantitative Modelle in der Geographie und Raumplanung*, edited by Gerhard Bahrenberg and Wolfgang Taubmann, 129–34. Bremer Beiträge zur Geographie und Raumplanung 1. Bremen: Bremen University Press.

Heiber, Helmut. 1958. "Der Generalplan Ost." *Vierteljahreshefte für Zeitgeschichte* 6: 281–325.

Herf, Jeffrey. 1984. *Reactionary Modernism: Technology, Culture, and Politics in Weimar and the Third Reich*. Cambridge: Cambridge University Press.

Hottes, Karlheinz, Ruth Hottes, and Peter Schöller. 1980. "Walter Christaller, 1893–1969." In *Geographers: Biobibliographical Studies*, edited by Thomas W. Freeman, 11–16. London: Mansell.

Hottes, Karlheinz, and Peter Schöller. 1968. "Werk und Wirkung Walter Christallers." *Geographische Zeitschrift* 56: 81–87.

Hottes, Ruth. 1981. "Walter Christaller—ein Überblick über Leben und Werk." In *Geographisches Taschenbuch, 1981/82*, edited by Eckhart Ehlers and Emile Meynen, 59–69. Wiesbaden: Steiner.

———. 1983. "Walter Christaller." *Annals of the Association of American Geographers* 73: 51–54.

James, Preston E. 1972. *All Possible Worlds: A History of Geographical Ideas*. Indianapolis: Odyssey Press.

Kamenetsky, Ihor. 1961. *Secret Nazi Plans for Eastern Europe: A Study of Lebensraum Policies*. New York: Bookman Associates.

Klingemann, Carsten, ed. 1987. *Rassenmythos und Sozialwissenschaften in Deutschland*. Opladen: Westdeutscher Verlag.

Klöpper, Rudolf. 1970. "Zentrale Orte und ihre Bereiche." In *Handwörterbuch der Raumforschung und Raumordnung*, 3851–62. Hannover: Akademie für Raumforschung und Landesplanung.

Lichtenberger, Elisabeth. 1978. "Klassische und theoretisch-quantitative Geographie im deutschen Sprachraum." *Berichte zur Raumforschung und Raumplanung* 22: 9–20.

Lundgreen, Peter, ed. 1985. *Wissenschaft im Dritten Reich*. Frankfurt: Suhrkamp.

Meyer, Konrad. 1970. "Walter Christaller." In *Handwörterbuch der Raumforschung und Raumordnung*, 403–9. Hannover: Akademie für Raumforschung und Landesplanung.

Pahl-Weber, Elke, and Dirk Schubert. 1987. "Großstadtsanierung im Nationalsozialismus: Andreas Walthers Sozialkartographie von Hamburg." *Sozialwissenschaftliche Informationen* 2: 108–17.

Rössler, Mechtild. 1983. "Die Geographie an der Universität Freiburg, 1933–1945: Ein Beitrag zur Wissenschaftsgeschichte des Faches im Dritten Reich." Reprinted in 1989 in *Urbset Regio 51. Geographie und Nationalsozialismus*, edited by P. Jüngst, K. P. Fromm, and H.-J. Schulze-Göbel. Kassel: Urbs et Regio.

*———. 1987a. "'Blut und Boden—Volk und Raum': Thesen zur Geographie im Nationalsozialismus." In *23. Deutscher Soziologentag 1986*, edited by J. Friedrichs, 741–44. Opladen: Westdeutscher Verlag.

*———. 1987b. "Die Institutionalisierung einer neuen Wissenschaft im Nationalsozialismus: Raumforschung und Raumordnung, 1935–1945." *Geographische Zeitschrift* 75: 177–94.

———. 1988a. "Géographie et National Socialisme: Remarques à une relation problématique." *L'Éspace Géographique* 17: 5–14.

———. 1988b. "Wissenschaft und Lebensraum: Geographische Ostforschung im Nationalsozialismus." PhD diss., Department of Economic Geography, University of Hamburg; published in 1990 as *Hamburger Beiträge zur Wissenschafts-geschichte* 8. Berlin: Dietrich Reimer.

Roth, Karl H. 1985. "Erster Generalplan Ost." In *Mitteilungen der Dokumentationsstelle zur NS-Sozialpolitik* Heft 4: 45–52. Hamburg: Dokumentationsstelle.

*Sandner, Gerhard. 1983. "Die 'Geographische Zeitschrift,' 1933–1944: Eine Dokumentation über Zensur, Selbstzensur und Anpassungsdruck bei wissenschaftlichen Zeitschriften im Dritten Reich." *Geographische Zeitschrift* 71: 65–87; 127–49.

———. 1988. "Recent Advances in the History of German Geography, 1918–1945: A Progress Report for the Federal Republic of Germany." *Geographische Zeitschrift* 76: 120–33.

Seeliger, Rolf. 1964. *Braune Universität: Deutsche Hochschullehrer gestern und heute.* Munich: Hine Dokumentation.

*Smit, Jan G. 1983. *Neubildung Deutschen Bauerntums: Innere Kolonisation im Dritten Reich.* Kassel: Urbs et Regio.

Tröger, Jörg, ed. 1984. *Hochschule und Wissenschaft im Dritten Reich.* Frankfurt: Campus.

von Böventer, Edwin. 1968. "Walter Christallers zentrale Orte und periphere Gebiete: Ruckblick nach 35 Jahren—anlässlich des 75 Geburtstages." *Geographische Zeitschrift* 56: 102–11.

Wirth, Eugen. 1982. "Fünfzig Jahre Theorie der zentralen Orte." *Geographische Zeitschrift* 70: 293–97.

Archival Material

Berlin Document Center

Bundesarchiv, Koblenz

Hoover Archives, Stanford, California

Private correspondence with Professor Theodor Maunz (Munich), Mrs. Dinn Meyer (Salzderhelden), Professor Herbert Morgen (Wiesbaden), Professor Josef Umlauf (Munich)

Public Record Office, Kew, London

Stadtarchiv, Freiburg

University Archive, Freiburg

Zentrales Staatsarchiv, Potsdam

9

A Morality Tale of Two Location Theorists in Hitler's Germany: Walter Christaller and August Lösch

TREVOR J. BARNES

Introduction

Hitler's Nazi project was fundamentally geographical, with space and place pressed into horrific service. That geographical agenda was articulated in such terms as: *Lebensraum*,[1] *Großraum*,[2] *Entfernung*,[3] *Blut und Boden*,[4] *Volk ohne Raum*,[5] and *Drang nach Osten*.[6] Each of these expressions, and others like them, implied heinous acts as regions were annexed, or countries were invaded, or people were dispossessed, or rounded up, or forced marched, or ghettoized, or imprisoned, or murdered in death camps (Rössler 1993, 204). Space and place were never neutral for the Nazis. They were an affirmation and means of enforcement of the regime's deadly authority.

Consequently, the Nazis were concerned with managing, planning, organizing, and contorting space and place, making geography conform to and realize their larger ideological ends. Specific bureaucratic units were established to undertake those operations. Some were based in Berlin, such as Konrad Meyer's Department of Planning and Soil in Dahlem, others in the spaces and places most affected: at the Institut für deutsche Ostarbeit (Institute for Research on the German East) in Breslau or Kraków (Burleigh 1988); or at the research program on "Ethnic Studies, Frontier and Ethnic Germandom" at the Reich University, Posen; or, yet again, at the Institute of Geographical Studies in Kiev (Rössler 2001, 69).

This goes to a second feature of Nazi operations, the use of academic labor, in this case the labor of geographers, urban and rural planners, landscape architects, and agronomists. Each of these academic specialties possessed expert knowledge about geography, as well as theories, concepts, and practical methods potentially applicable to transforming space and place for the purposes of National Socialism. Michael Burleigh (1988, 9) contends that such "scholars . . . put their knowledge at the service of the government . . .

willingly and enthusiastically. There was virtually no resistance." That is not quite true, though, and this is why I believe it is important to engage with a larger theme in the recent literature and explored in this chapter: the extent to which non-Jewish German academics collaborated with the regime, as in Burleigh's suggestion, and the extent to which they resisted.[7] This chapter contends there were a range of responses by German academics to their situation, and a range of motivations. Easiest was to do nothing, but if explicitly asked to participate, some felt they could not refuse—any alternative would be worse, from serving on the "Eastern Front" to imprisonment in Dachau or Buchenwald. Some believed that collaborating would enhance their career, giving them power, position, resources, and an audience. Others thought it was their national duty to participate, especially in time of war. Yet others believed in the Nazi project, thinking it could best be realized through contributing their academic expertise.[8] But there were those who balked, who defied Nazism. Some relinquished their academic positions as a form of protest.[9] Others joined opposition groups bent on removing Hitler from power.[10]

There is also another related issue here that again runs throughout the chapter. It stems from Hannah Arendt's book *Eichmann in Jerusalem* (1977), her famous account of the trial and conviction of the German Nazi bureaucrat of the Holocaust, Adolf Eichmann. Equally famous was her book's subtitle: *A Report on the Banality of Evil.* Arendt's thesis was that Eichmann was not fundamentally evil, although he committed unbelievably evil acts. He joined the Nazis less out of conviction than because he was a joiner. As a bureaucrat he was then unthinkingly swept along. While "terrifyingly normal" he committed terrifying acts (Arendt 1977, 273). Arendt's thesis provoked immediate controversy, with Lionel Abel (1963, 211) famously charging that Eichmann "comes off so much better in her book than do his victims." Subsequent historical research, for example, by Lozowick (2002), contended that Eichmann, and other so-called bureaucrats of the Holocaust, were not so unthinking, that their evil was not so banal. The larger question here, and the subject of this chapter, is: were German academic bureaucrats who signed up to do the Nazis' bidding only banally evil or really truly evil?

All these issues are discussed in relation to the lives and works of two contemporaneous German academics with expert geographical knowledge, Walter Christaller (1893–1969) and August Lösch (1906–1945), but who each had a very different relationship with the Nazis. First, the two are such a useful comparison because they developed a similar spatial conceptual framework, *central place theory*, to explain the geographical pattern of different sized settlements across a theoretical landscape. Both deployed the same hexagonal geometry as their basic spatial grid, both undertook supporting empirical work, and both believed that state planning could improve reality, bringing it closer to their theories. Christaller was the first to complete his version of the theory in 1932. It was his doctoral thesis written in only nine months at the University of Erlangen's Department of Geography (and subsequently published in 1933 in Jena as *Die zentralen Orte in Süddeutschland*; Christaller 1966).[11] Lösch's version of the theory was published seven years later in 1940 as *Die räumliche Ordnung der Wirtschaft*,[12] and made explicit reference to Christaller's prior work (Lösch 1954, xiv).

Second, Christaller and Lösch each had a radically different relationship with the Nazis, and that is the basis for the morality tale at the center of this chapter. Initially taking fright because of his membership in the German Social Democratic Party (SPD), banned once Hitler took power in 1933, Christaller from the mid-1930s increasingly worked for the Nazis. He formally joined the party in 1940, and was employed by Konrad Meyer (see note 8). Christaller used central place theory to redesign the landscape of Germany's newly acquired eastern territories. Lösch, in contrast, never worked for the Nazis. He was strongly critical of Nazism, and Hitler. He worked as a researcher at Kiel University's Institut für Weltwirtschaft (Institute of World Economy), under its director, Andreas Predöhl (1893–1974). He never swore allegiance to Hitler, and despite the strong normative character of his central place theory, he never implemented it under Nazism.

The chapter is divided into four sections. The first section reviews the Nazi commitment to planning, especially spatial planning.[13] It was affected by enlisting German academics into the bureaucratic machinery of the National Socialist government. Treated as holders of expert knowledge, academics were essential to the larger Nazi regime, enabling its obscene ends to be realized ever more efficiently and rationally. The second section examines one of those Nazi academic experts, the geographer Walter Christaller. Effectively given the annexed province of Warthegau, western Poland, as his drawing board, Christaller was charged with "re-planning" its landscape, to make it look like Germany's. Not only would it look like Germany, but also those who lived there would be exclusively ethnic Germans. Existing residents of Warthegau—Slavs, Romani, Jews—would be removed, replaced by *Volksdeutsche* from elsewhere in Europe, and possibly not having spoken German for decades if not centuries (Berger 1994).[14] The third section is on August Lösch. Also an academic specialist of spatial planning, he actively resisted interpellation within the Nazi regime while continuing to undertake his own geographical work. The culmination of that preoccupation, *Die räumliche Ordnung der Wirtschaft* (Lösch 1954), was published in two separate editions during the war. Seemingly having a theory ripe for spatial practice, as was

Christaller's, Lösch nonetheless escaped notice, holding an unassuming research position at the University of Kiel, and never being recruited by the Nazis.[15] The conclusion uses the lives and works of Christaller and Lösch, and their respective connection with the Nazi regime, to reflect on the complicated relation among knowledge, power, collaboration, and morality.

The Rule of Experts: Nazi Spatial Planning, and German Academics

SPATIAL PLANNING

Spatial issues were integrated into the very Nazi project, inseparable from its fulfillment. To be realized they required enormous energy and resources, the coordinated efforts of myriad people and institutions, and a directed instrumental rationality. Once the ends of Nazism were defined, however ghastly or perverted, a technocratic army of experts and bureaucrats would get to work to plan their most efficient realization. Zygmunt Bauman (1989) argues that the Holocaust was the perfect example.[16] Given the Nazi characterization of a perfect society, defined by members who were able-bodied and -minded, heterosexual, and Aryan, techniques and technologies forged by the regime's planning technocracy would then be set in motion to realize it. In this case, it was to be achieved by exterminating massive numbers of non-Aryans, Jews, but other groups and peoples too, including Aryans who did not come up to scratch on other criteria (Gregory 2009). Furthermore, it required the Nazis to establish an infrastructure and logistics for mass, Fordist-scale, assembly-line factories of murder, based on a meticulous, functional division of labor, scientific management, exact timing, and bureaucratic efficiency.

The first German spatial planning (*Raumordnung* or *Raumplanung*) had begun sporadically during the first decade of the twentieth century, associated with implementing Ebenezer Howard's garden city concept within large cities (Fehl 1992; Rössler 1994, 2001). With the rise of Nazism and centralized state control, spatial planning became comprehensive and systematic (Rössler 1994, 130). Robert E. Dickinson (1938, 1943, postscript), a British geographer, spent six months in Germany in 1936–37 and was concerned precisely with the emergence of the Nazi spatial planning apparatus.[17] He quoted a January 1936 speech made by Reichminister Kerri, head of the National Board of Spatial Planning (Reichstelle für Raumordnung), made nine months after the Nazis established that board:

> Fundamentally it is our endeavour to direct all changes in the German State . . . on a basis of planning and foresight, in the common interests of the

people and the State; and also to place all such developments under the guidance and security of the State. (Quoted in Dickinson 1938, 623–24)

The Nazi State and its ideological agenda were thus everything. Even local "community planning," as Konrad Meyer, the arch-Nazi academic spatial planner wrote in 1938, "can only take place in the closest connection with the state and Reich planning, which formulates more general ideas and plans" (quoted in Rössler 1994, 130).

The National Board of Spatial Planning had its own separate research office, the Reichsarbeitgemeinschaft für Raumforschung (the Reich study group for area research, and initially headed by Konrad Meyer; Rössler 2001, 67–69). There was a branch at every major university, as well as a house journal, *Raumforschung und Raumordung* (Spatial Research and Spatial Planning; Dickinson 1938, 625n34; Rössler 1994, 130–32). It was in this journal that academics such as Walter Christaller published. That work, and germane to Christaller, was partly concerned with devising new techniques to provide greater spatial efficiency in the distribution and use of economic resources, but also to allow "the rich regional variations of the nation's cultural heritage . . . [to] reach their fullest expression" (Dickinson 1938, 626).

The dual expectation within Nazi spatial planning to realize both economic and cultural objectives reflected a deeper binary shaping the regime's beliefs and practices, what Jeffrey Herf (1984) labeled "reactionary modernism." While the larger ends of Nazism were reactionary, based on a mythic history, racial essentialism, dogma, bigotry, and much worse, those ends were to be achieved through modernist means: science, rationality, new technology, and cutting-edge management, logistical, and organizational techniques. Spatial planning cut across both worlds. It involved a set of shiny new theories, concepts, methods, and algorithms to obtain "geographical efficiency," to squeeze out the most from the least. But the cultural ends that planning set itself could be horrendous, involving spilling blood on the soil to achieve blood and soil.

The culmination of this double trajectory was a geographical master plan, *Generalplan Ost* (General Plan for the East). The desire for Germany to expand its borders to the East was deeply entrenched, going back centuries (Burleigh 1988, 6n6). Under the Nazi version of that desire, fulfillment would be through a comprehensive spatial state plan that came with specific goals, strategies, cost estimates, maps, figures, and organizational charts, along with dedicated offices, bureaucrats, and experts. It culminated over a period of more than three years from 1939. Associated with Himmler, it began as a new research program within the Reichsarbeitgemeinschaft für Raumforschung, deutsche Osten (the German East). There, scholars were charged with calculating "how many German settlers should live [in the German East] as well as . . . construct[ing] new industrial sites and agrarian regions" (Rössler 1994, 132). Christaller later did some of that work, along with planning the conversion of "the East" into the look and feel of settled Germanic space (Rössler 1994, 134; Preston 2009, 11). By 1942, these efforts folded into the last version of *Generalplan Ost*, and became linked to Hitler's "Final Solution."

ACADEMICS, GEOGRAPHY, AND PLANNING

The National Socialist project relied crucially on academic labor. While Hitler's ends were malevolently irrational, they were implemented by a scrupulous logic and knowledge provided by experts and advisors from the German academy (Renneberg and Walker 1994; Szöllösi-Janze 2001). Admittedly, some Nazi projects that involved the professoriat, such as at Himmler's Das Ahnenerbe ("ancestral heritage") institute, could be barking mad (Szöllösi Janze 2001, 1–3). But generally, the work of ordinary, everyday academics—scientists, social scientists, and assorted technocrats—was "largely rational, and result oriented . . . [and] not ideologically dogmatic" (ibid., 12).

The Nazi reliance on academics stemmed from the modernist half of Herf's reactionary modernism label, bound up with rationality, instrumentalism, and technocracy. Modernism was made possible only by drawing on a pool of experts and specialists. For the Nazis there was no better pool than that found in German universities. German scholars led the world. If anyone could achieve the regime's instrumental goals it would be German academics. Aly and Heim (2002, 3) write, "the National Socialist leadership sought to maximize the inputs for scientific policy advisors and used their research findings as an important basis for their decisions." But, and this is where the reactionary part of reactionary modernism becomes relevant, that "includ[ed] the decision to murder millions of human beings" (ibid., 3). Going to my earlier comments about the motivations of academics to collaborate with the Nazis, and the character of the evil committed, Zygmunt Bauman (1989, 109) writes:

> With relish, German scientists boarded the train drawn by the Nazi locomotive towards the brave, new, racially purified and German-dominated world. Recent projects grew more contributions by the day, and research institutes grew more populous and resourceful by the hour. Little else mattered.

Not all academic knowledge was the same, though. For the Nazis it was ranked according to its practical usefulness in achieving the regime's ends.

According to Renneberg and Walker (1994, 9), at the top were the disciplines of "biology, chemistry, geography, and engineering." The prominent position of geography is particularly germane. Henning Heske (1986, 279) claims that "geography more than any other established science in Germany, consciously and unconsciously, supported National Socialism."

Geography certainly supported National Socialism's plans for its newly conquered territories in the German East (along with other kindred spatial sciences). Konrad Meyer, professor of agronomy at Berlin University, one of the Nazi key academic movers and shakers, Walter Christaller's boss, and "the Father of Regional Planning" (Christaller quoted in Preston 2009, 11), was in no doubt about the crucial role of academic knowledge for Germany's new territorial acquisitions. On October 23, 1941, in Posen, Poland, Meyer said in a speech inaugurating an exhibition, *Planning and Reconstruction in the East*:

> Our planning enterprise can only prosper with the sure collaboration of, and in close enduring association with, the university establishment. I am not talking a value-free science, with no preconceptions . . . but a science that sees its *raison d'être* in serving the people and embracing the forces of blood and soil. For the work of planning we need a scientific approach that seeks not so much to abstract and generalize its knowledge as to relate its findings to concrete situations . . . a forward looking science that plays a constructive part in shaping events as they unfold. (Meyer quoted in Aly and Heim 2002, 94)

Michael Burleigh (1988) provided a brilliant study of German academics who supplied the right kind of "concrete" scientific knowledge for planning the East, for "shaping events" as much as describing them. Known as *Ostforschung* (Eastern research), it involved scholars working in one of many institutes spread across the newly occupied German East. Burleigh (ibid., 10) writes:

> As scholarly experts in the East, the *Ostforscher* had a distinctive contribution to make to the accurate "data base"—the statistical and cartographic location of persons—upon which all aspects of Nazi policy in the East, as elsewhere, ultimately rested. Deportations, resettlements, repatriations and mass murder were not sudden visitations from on high, requiring the adoption of some commensurate inscrutable, quasi-religious, meta-language, but the result of the exact, modern, "scientific" encompassing of practices with card indexes, card sorting machines, charts, graphs, maps and diagrams. . . . This was why [*Ostforschung*] received generous funding.

This work was made for geographers. That is why they ranked so highly on the Nazi academic hierarchy. As Rössler (2001, 67–69) documents, area-based and regional geographical institutes followed the frontier of the easterly expanding German Reich. Kiev became the easternmost outpost, with the Institute of Geographical Studies established in 1941. After that it was retreated, although the geographical sections of such institutes were often the last to leave (Burleigh 1988, 284–85).

Walter Christaller, Central Place Theory, and Collaboration

Walter Christaller was another *Ostforscher* who "voluntarily and enthusiastically put [his] knowledge at the disposal of the Nazi regime" (Burleigh 1988, 8). He certainly fixated on "charts, graphs, maps and diagrams." In October 1936, Robert E. Dickinson went on a five-day car trip with him through the Erzgebirge to Dresden, and noted Christaller's "passion for figures, and his constant use of pencil and paper and reams of statistics."[18] He was more than just a numbers man, though. He had a larger scientific theory, central place theory, and this theory was perfect for the Nazi purpose of re-planning the East, satisfying both economic and cultural ends.

Christaller's central place theory had a long gestation period. In an autobiographical essay he said it began when he was only eight (Christaller 1972, 601). But interrupted by active service in World War I, it took him seventeen years before, in 1930, he finally received his diploma in economics at the University of Erlangen (Hottes, Hottes, and Schöller 1980, 11). Christaller's intention was to carry on with a PhD in economics, but because he "found no response from the economists," he went back to his childhood interests and asked the biogeographer Robert Gradmann in the Geography Department to supervise his dissertation (ibid.). He did. The dissertation, "Die zentralen Orte in Süddeutschland," was written in less than a year, and was published the same year Hitler became Germany's chancellor.

There are three points to make about Christaller's central place theory. First, it was a theory about the organization of space. In this case, how different sized cities (central places) that ranged from traditional rural hamlets to modern metropolitan regions were locationally related to one another. The underlying spatial pattern, Christaller argued, was hexagonal, and was discerned from the central place landscape of southern Germany he studied (Christaller 1972, 610). Second, Christaller believed he propounded a *modern* scientific theory predicated on underlying spatial laws. "My goal was staked out for me: to find laws according to which number, size and distribution of cities are determined," he wrote (ibid., 607). Consequently, this was not traditional regional geography, but the expression of modern science. Finally, central place theory was a planning tool, a technology for practicing instrumental rationality. That was there in his doctoral thesis, laid out as three planning principles (K = 3 [marketing], K = 4 [transportation], and K = 7 [administrative]). Later they were further refined in his 1938 *Habilitation* at the University of Freiburg, and then from 1940 put into practice when he worked for Konrad Meyer.

How he ended up in Meyer's office is instructive. After initially fleeing Germany in 1934, he was lured back by work, including from the Reichsarbeitsgemeinschaft für Raumforschung (Rössler 1989, 422). Between 1937 and the early 1940s he was a lecturer at the University of Freiburg. There he established the Kommunalwissenschaftliches Institut (Institute for the Study of Local Government) under Professor Maunz of the law faculty, and author of the "concentration camp legal code" (ibid., 423). Christaller's Institut following his central place theory was about devising and practicing a new kind of applied geography.

Traditional regional German geographers were not interested, but Konrad Meyer was. He had been the original director of Reichsarbeitsgemeinschaft für Raumforschung. In October 1939, he was appointed chief of the Planning and Soil Department (Hauptabteilung Planung und Boden) within Himmler's Reichskommissariat für die Festigung deutschen Volkstums (RKFDV) (Reich's Commission for the Strengthening of Germandom). One of RKFDV's mandates was "the whole planning of settlement and development of Germany and in the territories under German supremacy as well as the realization of that planning" (quoted in Koehl 1957, 253). Meyer's Planning and Soil Department was crucial for planning the Germanization of conquered land. It employed an interdisciplinary group of academics, including the planner Joseph Umlauf, the rural sociologist Herbert Morgen, and from 1940 onward the geographer Walter Christaller (Rössler 1989, 425).

Christaller's central place theory was ideal for the Nazis. It was fundamentally about spatial relations, speaking to key aspects of the Nazi project. It embodied the contradictory reactionary modernism characterizing the larger regime. On the one hand, it was seemingly modernist (rational, law-seeking, scientific), but, on the other, it acknowledged cultural tradition and the past by making its beginning point the smallest urban unit, the village (*Dorf*), emphasizing rural community (*Volksgemeinschaft*) and land. And it came as a ready-made planning tool. Christaller's detailed maps, figures, and plans needed only to be unfurled, the bulldozers brought in, and the East became "central places in Southern Germany." As Rössler (1994, 134) writes, the "aim was the transformation of the East into German land and as German landscape." That is what Christaller's model did.

According to Richard Preston (2009, 6), by working at Meyer's office, Christaller "contributed directly to plans facilitating German *Lebensraum* (search for living space) policy, on the one hand, and Himmler's RKFDV [Germanization], on the other."

The first of these roles took the form of re-planning the now expanded German East, the annexed province of Warthegau, Poland. Warthegau was the "workshop" for the Reich as Joseph Umlauf, Christaller's colleague, put it (quoted in Fehl 1992, 96).

This was Christaller's view too. Writing in 1940, he said:

> Because of the destruction of the Polish state and the integration of its western parts into the German Empire, everything is again fluid. . . . Our task will be to create in a short time all the spatial units, large and small, that normally develop slowly by themselves . . . so that they will be functioning as vital parts of the German Empire as soon as possible. (Translated and quoted in Preston 2009, 23)[19]

A year later Christaller was more strident and specific.

> The aim of regional planning . . . is to introduce order into impractical, outdated and arbitrary urban forms or transport networks, and this order can only be achieved on the basis of an ideal plan—which means in spatial terms a geometrical schema. . . . [C]entral places will be spaced an equal distance apart, so that they form equilateral triangles. These triangles will in turn form regular hexagons, with the central place in the middle of these hexagons assuming a greater importance. (Quoted in Aly and Heim 2002, 97)[20]

Parts of Warthegau were thus redesigned, "completely changing the face of the countryside," as Himmler demanded in 1940 (quoted in Aly and Heim 2002, 74). But to do so, a lot of work was required. Christaller wrote in the same 1941 planning document quoted above: where "it seemed absolutely essential . . . that a new town of at least 25,000 inhabitants" be built then a new town would be "created from scratch" (ibid., 97). If Upper Silesia needed "a Dusseldorf or Cologne" of 450,000 people "to provide a cultural centre" then so be it (ibid.). If "Posen . . . has the power and potential to develop into a town of 450,000 [from 350,000]," it should (ibid.). Christaller planned for Warthegau thirty-six new *Hauptdorfs*.

Before this could happen, though, and this is where Christaller is joined to *Generalplan Ost*, many of the non-Aryan existing residents had to go—560,000 Jews and 3.4 million Slavs. Only 1.1 million of the existing population were thought to be Germanized enough to stay. Given the large expulsion, 3.4 million *Volksdeutsche* settlers needed to be brought in. This was Christaller's second role, to assist in the migration of ethnic Germans from various places in Europe so as to strengthen Germandom. As Christaller put it, this was another reason to construct a new central place system: "to give settlers roots so they can really feel at home" (quoted and translated by Preston 2009, 23).[21]

August Lösch, Central Place Theory, and Resistance

Christaller left no account of why he collaborated with the Nazis. His earlier SDP affiliation, and that he joined the German Communist Party after the war, made it unlikely he was politically on board. He was scared by the regime, which accounts for his getaway to France in 1934. But he needed a job (and given the unorthodox kind of geography he practiced that was unlikely at a German university). He also wanted recognition, to know that someone found his work interesting, that it was relevant, and as a practitioner for his ideas to be realized on the ground. Maybe he also wanted to be close to power, which working under Konrad Meyer gave him. In contrast, August Lösch, the other inventor of central place theory, resisted all such blandishments, opposing Hitler and his geographies.

Also unlike Christaller, Lösch was a star student, winning prizes and scholarships, rattling through his higher education. From 1927 he studied at the Universities of Freiburg, Kiel, and finally Bonn, receiving at the latter his diploma in economics in 1931, his PhD in 1932, and his *Habilitation* in 1936. Bonn University's Economics Department was one of the best in the country. It included among its faculty Walter Eucken, Albert Spiethoff, and Joseph Schumpeter (the *Meister*). Schumpeter left for Harvard the year Lösch completed his doctoral dissertation on the economics of declining population, albeit not before he took on the role of "Doctor Father" (Stolper 2007, 381).[22]

Hitler becoming chancellor on January 30, 1933, and the Enabling Act in March of the same year that made the Nazi government in effect a legal dictatorship, deeply disturbed Lösch. Unflinchingly principled, and an archrationalist, he believed in absolute truth and freedom of expression (Riegger 2007; Stolper 2007). "Dogmatism was death and ultimately untruth" (Stolper 2007, 380). Hitler and Nazism repulsed him. He wrote in his diary in April 1933:

> A black Friday. *Imperium fuit* [his reign was]
>
> I am still raging in pain and fury: the empire has gone, and everybody dances now to the whistle of that one fool. In this Germany, only creatures are tolerated. But I am hoping and working for *that* Germany which will come thereafter, so God willing, it comes.[23]

His first act, though, was to leave Germany, facilitated by a Rockefeller Fellowship for twelve months for 1934–35, and a second award for 1936–37.

As he wrote in his diary in September 1934: "I go to America. Not because I am not up for a fight or that I want to leave home, but I'm suffocating in this atmosphere of violence and I want to breathe. Freedom!" (Riegger 1971, 86). He was based at Harvard on both occasions, reunited with *Der Meister*, and also with Wolfgang Stolper (2007, 384–85), a student with him at Bonn, but Jewish, and for whom an arrest warrant had been issued.

It was at a class seminar at Harvard run by Edward Chamberlain, the economic theorist of monopolistic competition, that Lösch gave the first public presentation of his spatial theory. There had been a long-standing German interest in location theory from the early nineteenth century with Johann von Thünen (1783–1850) (Blaug 1979). That spatial sensibility might always have been in Lösch's intellectual marrow. But it was never explicit until Chamberlain's class, when, according to Stolper, Lösch "went to the blackboard and filled its whole expanse" (Stolper 2007, 385) with equations and diagrams to illustrate "the structure of an economic region" (published as Lösch 1938). Lösch (1954, xiii) later wrote in the preface to the 1940 edition of *The Economics of Location* that he had "conceived the plan of the present work ten years ago as a young student." That seminar presentation in Chamberlain's class was thus still very much a work in progress.

On the two trips to America, Lösch collected empirical data for what was to become *The Economics of Location*. He went to central Canada, as well as to the US South, and to what later became central place theory's principal postwar testing ground, the American Midwest. In Iowa he met Harold McCarty, later the first chair of Geography at the University of Iowa, who modeled his department explicitly on Lösch's theories and methods. In 1937 at Harvard on the eve of his departure from America he also met Edward Ullman. A graduate student in geography at the University of Chicago, Ullman was told to see Lösch because of Ullman's interest in the formal, spatial distribution of different sized cities. In an interview, Ullman said, "I described my idea to [Lösch] and a strange light came in his eyes—that's the only way I can describe it. And he wrote down on a piece of paper, 'You should see Walter Christaller in Deutschland'" (Dow 1972, 3). Ullman (1941) never saw Christaller in Deutschland, but he did read his book, which led him in 1941 to write a well-known review article introducing North American scholars to central place theory. Ullman's story also speaks to Lösch's knowledge of Christaller's work. Lösch in fact explicitly acknowledges him in his 1940 preface: to "Dr. Christaller in Freiburg," Lösch (1954, xiv) writes.

But by the time that preface appeared in print in 1940 (written in "Heidenheim [Württemburg] Autumn 1939," xvi), Christaller was no longer in Freiburg, but in Konrad Meyer's office in suburban Berlin. And Lösch was not in Heidenheim, but at the Institut für Weltwirtschaft at the University of Kiel.

Kiel's institute was founded at the beginning of World War I, its researchers providing military and political assistance to the German Reich. When the Nazis took it over, its director, Barnhard Harms, was initially compliant, but when Jewish academics were forced to leave, he resisted, and stepped down. In July 1934 Harms's former student, Andreas Predöhl, was made director of the institute, a position he kept until November 1945. Predöhl, a decorated World War I pilot, had studied theoretical and substantive issues in economic geography at Kiel, receiving both his doctorate (1921) and *Habilitation* (1924) from there. While still director of the institute, Predöhl, was made pro-rector of the University in 1935. Joining the Nazi Party in 1937, he became university rector in 1942 (Klee 2001, 471).

Clearly, Predöhl had at least one foot in the Nazi world.[24] Under his directorship, the institute conducted international economic research important for Germany's war planning and its economy. For example, it investigated the Reich's access to natural resources and the geopolitical significance of areas that Germany had annexed, invaded, or resettled. But Predöhl was not a patsy. He prevented the institute's library from being cleansed of books written by Jews. He also ensured that the library could buy foreign literature, and that the academics who worked in the institute could disseminate their research to the English-speaking world. That included Lösch's book, which before the German invasion of Russia was sent via the Soviet Union to North America (Stolper 2007, 386). Microfilm versions were also made of it, and it was translated into English (Lösch 1954, xvi). The founder of American regional science, Walter Isard (2000), remembers translating Lösch during the war "in the wee hours of the night, when it was quiet," when he was a conscientious objector assigned to the graveyard shift at a Philadelphia mental asylum. And in 1943 Stolper wrote a review of the book for the *American Economic Review*. His copy was sent to him by Lösch himself, and included on the flyleaf as part of the dedication a hand-drawn "picture of a sinking steamer" (Stolper 2007, 386). Clearly, Lösch believed he and Germany were going under.

But not immediately. Perhaps the most important act Predöhl performed was keeping Lösch at Kiel. Moreover, his duties were such that he had time to revise *The Economics of Location*, with the second edition published in 1944. Not that Lösch was grateful. In Lösch's (Riegger 1971, 102–22) diary entries, Predöhl is presented as the bad boss, referred to as *sie*, "they," who prevented him from fulfilling his potential. In a January 1945, diary entry Lösch complains that the "Director" made him undertake a "completely useless report," which "at most would have academic interest," and which then robbed him of the time to do what was really important, his own work (Riegger 1971, 116). But Predöhl was his protector, keeping him safe. A colleague at the institute, Erich Schneider, said to Lösch, the same month he wrote his diary entry: "you are the last 'liberal' living at the brink of the concentration camp" (quoted in Funck 2007, 408n1).[25]

What Predöhl also managed was insuring Lösch did not turn his theory into a spatial planning model for the Nazis. Here it was better that Lösch authored a "completely useless report" than to contribute as Christaller did to *Generalplan Ost*. Lösch's version of central place theory bore striking resemblances to Christaller's. Both provided a theoretical account of the spatial distribution of different sized central places based on a hexagonal geometrical grid. Both were empirically tested against actual city distributions. And both came with a planning mandate: a belief that implementation of the theory on the ground would improve the world, not just explain it. Given his spatial sensibility, Lösch's version of central place theory could have been deployed by the Nazis. He could have been working in the *Hauptabteilung Planung und Boden* with Christaller. He could have been drawing up plans for Warthegau or any of the newly conquered territories of the German East. But he didn't. He may have been frustrated, bored, felt he was wasting his talent, under the beck and call of a bad boss, but he held fast, he did the right thing. Unlike many German academics, rather than collaborating, enjoying Nazi largesse, furthering their career, becoming interpellated within the regime, he resisted, living within the beast but not being beastly.

Conclusion

Because it is easy to think of Nazism as irrational, even lunatic, one can forget the enormous effectiveness and efficiency of its bureaucratic and logistical apparatus that facilitated both its military successes and its management and control of large occupied expanses of territory. What the Nazis did in those occupied territories, and especially to the people who lived there, was repulsive and abhorrent. But once the ends were determined, an army of academic experts and civil servants descended, fulfilling Nazi objectives in the most rational and expeditious manner, using leading-edge knowledge, ideas, technology, and devices.

My particular concern was with those academic experts employed to reconfigure space and place. For the Nazis, geography was not merely a stage on which to mount their operations, but was an integral part of their ideological mandate, inextricably joined to their dogma. To practice Nazism *was to do geography*. The use of academic experts by the Nazis—geographers, planners, architects, agronomists—was to ensure geography met in the best possible way ideological ends of the worst possible kind.

As discussed in the introduction, following Arendt's book *Eichmann in Jerusalem* there has been debate around the extent to which German academic experts cum bureaucrats were morally culpable in the acts they committed on behalf of the Nazis. Were they, as Arendt contends, merely assiduous civil servants who could not but join, following orders? Or, as Lozowick (2002, 279) suggests, did they "work hard, took the lead, over many years . . . [becoming] Alpinists of Evil?"

My chapter's account of two German academics, Walter Christaller and August Lösch, both with expertise in planning space, shows that neither Arendt's nor Lozowick's position quite fits. Lösch was able to assert moral agency, never joining the Nazis, never swearing allegiance to Hitler, and being subversive when possible.[26] He didn't simply fall in line, as Arendt suggests, practicing the banality of evil. He made a deliberate moral choice. But he was able to do that in part because others, in this case his "Director," Andreas Predöhl, chose to work with the Nazis, to become one, to join the party, but which then allowed him to protect some of his dissenting staff, including Lösch; that is, to allow Lösch to make his moral choice. Was Predöhl "an Alpinist of Evil?" It seems more complicated than that simple designation. And the same is true for Christaller. One can see how he was drawn into Nazism, which as a party and a set of beliefs was at first anathema to him. But Christaller seemingly couldn't help himself, ineluctably he is interpellated, ending up saying and doing evil things. But calling him an "Alpinist of Evil" also is too extreme. But saying that he had no choice, that his evil was banal, is not right either given that August Lösch, who developed a similar theory, exercised moral choice. This morality tale of two location theorists in Hitler's Germany reveals shades of gray, not stark black and white.

Acknowledgments

I am grateful to Paolo Giaccaria, Ron Johnston, and Claudio Minca for excellent comments and suggestions that greatly improved the chapter.

Notes

1. *Lebensraum* meant living space, or the space necessary for a people to conduct their way of life. It was first set out by the nineteenth-century Swedish political scientist Rudolf Kjellén and the German geographer Friedrich Ratzel (Abrahamsson 2013).

2. *Großraum* was conceived by the Nazi legal theorist Carl Schmitt, and meant "greater

space." It was used to justify the division of the world into a few large regional blocks, each with a dominating nation (see Barnes and Minca 2013).

3. *Entfernung* meant expulsion, or removal, or just distance. It was an essential part of Nazi racial politics, implying spatial separation of Aryans from non-Aryans (Clarke, Doel, and McDonough 1996).

4. *Blut und Boden*, blood and soil, was a phrase invented in the nineteenth century but popularized by the Nazi propagandist Richard Walther Darré to link racial politics (blood) and place (the soil).

5. *Volk ohne Raum*, literally people without space, was the title of Hans Grimm's 1926 novel that suggested because of boundary redrawing enacted by the 1919 Treaty of Versailles German people had lost their space. Recouping that space became one of Hitler's geopolitical ends.

6. *Drang nach Osten* meant yearning for the East, and a touchstone of German nationalist politics. Even as late as 1944, as German military forces were pushed back ever westward by Soviet troops, Hitler exhorted: "It is eastwards, only and always eastwards, that the veins of our race must expand" (in Hauner 1983, 197).

7. Noakes's (1993) paper is a useful overview, and for the details about what academics did for the Nazis, and to a lesser extent why, see: Renneberg and Walker (1994); Szöllösi-Janze (2001); Cornwell (2003); Sherratt (2013). I say non-Jewish German professors because Jews and those deemed politically unreliable were expelled from the German civil service, which included university employees, in April 1933 following a new Nazi law, *Gesetz zur Wiederherstellung des Berufsbeamtentums* (The Law for the Restoration of the Professional Civil Service). Noakes (1993, 378) estimated that 3,120 university academics lost their positions.

8. Konrad Meyer (or sometimes Meyer-Hetling; 1901–1973), professor of agronomy at the University of Berlin, was a key Nazi academic cum planner, a believer in the National Socialist agenda. He joined the Nazi Party in 1932, and the SS a year later working under Himmler. From 1942 he was one of the architects of *Generalplan Ost* that aimed to resettle "the East" with Aryans, and in the process to murder over sixty million Slavs, Romani, and Jews. He was brought into *Generalplan Ost* precisely because he could be trusted with academic details, particularly numerical estimates, unlike some others in the Nazi upper echelon of command (Burleigh 2000, 546–48).

9. Alfred Weber (1868–1958), the location theorist at the University of Heidelberg, and brother of sociologist Max Weber (1864–1920), was critical of Hitler. Following a tussle in April 1933 between Weber and Nazi administrators at Heidelberg about the flying of the swastika over Weber's Institute for Social and State Sciences, Weber resigned his position, asking to be transferred to emeritus status (Loader 2012, 113). He remained part of a group of German intellectuals who kept alive a critique of Hitler and the Nazis throughout the 1930s and into the war (Loader 2012, epilogue). Weber was reappointed to his chair in 1945, assisting in the subsequent denazification of Heidelberg University.

10. For example, the political geographer Albrecht Haushofer, after initially working for von Ribbentrop at the German Foreign Office, turned against the Nazis. He participated in the failed von Stauffenberg attempt to assassinate Hitler on July 20, 1944. Fleeing Berlin, Haushofer was later captured by the Nazis in December 1944, and executed by the SS a few days before the war's end (Herwig 1999, 2010).

11. An American student, Carlisle W. Baskin, translated Christaller's book into English as his 1957 PhD thesis at the Department of Economics, University of Virginia, and it was later published by Prentice-Hall as *Central Places in Southern Germany* (Christaller 1966).

12. In 1954 Lösch's book was translated into English by a retired American cancer researcher formerly at the Rockefeller Institute, William H. Wolgom (Stolper 2007, 386). According to Wolfgang Stolper (ibid.), an economist and friend of Lösch's at Bonn, Wolgom's initial translation was "unintelligible to economists" so it was retranslated by Stolper. He made the English title of Lösch's book *The Economics of Location* (Lösch 1954). Some suggested that the literal translation of *Die räumliche Ordnung der Wirtschaft*, "The spatial order of the economy," was more appropriate (Funck and Kowalski 2007, 15).

13. It is useful to distinguish spatial planning as a field different from geography. German geography had been typically practiced as a regionally descriptive, ideographic science, with few practical, instrumental ends or techniques other than cartography and map interpretation. Spatial planning, in contrast, was intensely practical and instrumental, geared to realizing specific geographical objectives using formal methods. Christaller's central place theory blurred the distinction between the two fields. Although he completed his PhD thesis in a Department of Geography, the dissertation deployed formal methods, showing how those methods could be used for spatial planning. It was that blurring of geography and spatial planning that led to his difficulties in getting a job in German academic geography before the war, but it also explains why he got a job with the Nazis during the war, and why from the mid-1950s he was hailed by proponents of the "quantitative revolution" in North America (Barnes 2012). The regional descriptive tradition of German geography was useful to the Nazis, but its value increased when Christallerian formalism and spatial planning were added.

14. The *Volksdeutsche* were defined as people whose racial and ethnic heritage was German even though they did not live in Germany.

15. That Lösch escaped the attention of the Nazis was perhaps even more remarkable if as McCraw (2007, 586) asserts he was Jewish.

16. Bauman's position has been labeled functionalism, a position also attributed to another Holocaust theorist who I discuss below, Götz Aly (Aly and Heim 2002). A functional explanation of the Holocaust means that it was caused by deep historical internal contradictions within the social structure of German society, making individuals, even Adolf Hitler, only bearers of their functional role. The opposite position, and represented also by someone I discuss below, Jeffrey Herf (1984), is intentionalism. Here the Holocaust is explained by the anti-Semitic intentions of a single person who comes to exercise total power, Hitler. Both positions are criticized by Yehuda Bauer (2001, chaps. 2 and 4), who offers yet another explanation, a consensus-based one.

17. Dickinson wrote about the six months he spent in Nazi Germany, 1936–37, in his "Field Diaries II: Europe 31–66," a copy of which is held by Ron Johnston at the School of Geographical Science at Bristol University. See Johnston's (2001) paper about Dickinson's life.

18. From "Field Diaries II: Europe 31–66," 4–5.

19. The quote is from an article Christaller published in *Raumforschung und Raumordnung*, 4, 498–503 (1940), "*Die Kultur- und Marktbereiche der zentralen Orte im Deutschen Osteraum und die Gliederung der Verwaltung*" (Cultural and market segments in central places in the German East and the breakdown of administration).

20. Taken from Walter Christaller, "*Die Zentralen Orte in den Ostgebieten und ihre Kultur- and Marktbereiche*" (Central places in the Eastern territories and their cultural and market segments), Leipzig (1941).

21. Taken from Walter Christaller, *Land und Stadt in der Deutsche Volksordung* (Country and city in the German national order), *Deutsche Agrarpolitik* 1, 53–56 (1942).

22. The rise of Nazism during the late 1920s and early 1930s did not bear on Schumpeter's decision to move to America. Schumpeter was Catholic not Jewish. In fact, Schumpeter was investigated in 1941 by the FBI as a potential Nazi sympathizer, although he was cleared of the suspicion (McCraw 2007, 337–43).

23. Selected diary entries are found in German in Riegger (1971, 67–120). They begin in March 1925, and finish in May 1945. The quote is from page 78 (my translation).

24. Predöhl was dismissed by the Allies from his position as director of the institute in November 1945 but was allowed to retain his position as professor of economics.

25. Rolf Funck (2007, 407–8), who became a colleague of Predöhl's in the 1950s, says that Predöhl's "main objective" was "to save, [and] to perpetuate the existence and as much as possible the international reputation of the Institute of World Economics; and for this purpose thorough reports had to be presented, harsh confrontations with the Nazi regime had to be avoided, political compromises accepted, the 'slave language,' as he called it, had to be spoken."

26. In 1944 Lösch wrote in English a critical article about Hitler, "Hitler and Heidenheim," which he tried to publish in an American newspaper (reprinted in German in Riegger 1971).

References

Abel, Lionel. 1963. "The Aesthetics of Evil: Hannah Arendt on Eichmann and the Jews." *Partisan Review* 30 (2): 211–30.

Abrahamsson, Christian. 2013. "On the Genealogy of *Lebensraum*." *Geographica Helvetica* 68: 37–44.

Aly, Götz, and Susanne Heim. 2002. *Architects of Annihilation: Auschwitz and the Logic of Destruction*. Translated by Allan G. Blunden. First published in German in 1991. London: Weidenfeld and Nicolson.

Arendt, Hannah. 1977. *Eichmann in Jerusalem: A Report on the Banality of Evil*. Harmondsworth: Penguin.

Barnes, Trevor J. 2012. "Notes from the Underground: Why the History of Economic Geography Matters: The Case of Central Place Theory." *Economic Geography* 88: 1–26.

Barnes, Trevor, and Claudio Minca. 2013. "Nazi Spatial Theory: The Dark Geographies of Carl Schmitt and Walter Christaller." *Annals of the Association of American Geographers* 103 (3): 669–87.

Bauer, Yehuda. 2001. *Rethinking the Holocaust*. New Haven, CT: Yale University Press.

Bauman, Zygmunt. 1989. *Modernity and the Holocaust*. Ithaca, NY: Cornell University Press.

Berger, Doris L. 1994. "The Nazi Concept of '*Volksdeutsche*' and the Exacerbation of Anti-Semitism in Eastern Europe." *Journal of Contemporary History* 29: 569–82.

Blaug, Mark. 1979. "The German Hegemony of Location Theory: A Puzzle in the History of Economic Thought." *History of Political Economy* 11: 21–29.

Burleigh, Michael. 1988. *Germany Turns Eastwards: A Study of Ostforschung in the Third Reich*. Cambridge: Cambridge University Press.

———. 2000. *The Third Reich: A New History*. New York: Hill and Wang.

Christaller, Walter. 1966. *Central Places in Southern Germany*. Englewood Cliffs, NJ: Prentice-Hall. Originally published in German as *Die zentralen Orte in Süddeutschland* in 1933.

———. 1972. "How I Discovered the Theory of Central Places: A Report about the Origin of Central Places." In *Man, Space, and Environment: Concepts in Contemporary Human Geography*, edited by Paul Ward English and Robert C. Mayfield, 601–10. New York: Oxford University Press.

Clarke, David B., Marcus A. Doel, and Francis X. McDonough. 1996. "Holocaust Topologies: Singularity, Politics, and Space." *Political Geography* 15 (6–7): 457–89.

Cornwell, John. 2003. *Hitler's Scientists: Science, War, and the Devil's Pact.* Harmondsworth: Penguin.

Dickinson, Robert E. 1938. "The Economic Regions of Germany." *Geographical Review* 28: 609–26.

———. 1943. *The German Lebensraum.* Harmondsworth: Penguin.

Dow, Maynard W. 1972. "Geographers on Film" interview of Edward L. Ullman by Preston E. James, April 23, 1972, Kansas City, Missouri. Available also as an "Online Geographers on Film" transcription, 1–7. Accessed August 15, 2013. http://oz.plymouth.edu/~gof/home.html.

Fehl, Gerhard. 1992. "The Nazi Garden City." In *The Garden City: Past, Present, and Future,* edited by Stephen Ward, 88–105. London: E. and F. N. Spon.

Funck, Rolf H. 2007. "Concluding Remarks." In *Space-Structure-Economy: A Tribute to August Lösch,* edited by Ulrich Blum, Rolf H. Funck, and Jan S. Kowalski, 407–12. Baden Baden: Nomos.

Funck, Rolf H., and Jan S. Kowalski. 2007. Foreword to *Space-Structure-Economy: A Tribute to August Lösch,* edited by Ulrich Blum, Rolf H. Funck, and Jan S. Kowalski, 13–19. Baden Baden: Nomos.

Gregory, Derek. 2009. "Holocaust." In *The Dictionary of Human Geography,* edited by Derek Gregory, Ron Johnston, Geraldine Pratt, Michael Watts, and Sarah Whatmore, 337–40. Chichester: Wiley-Blackwell.

Hauner, Milan. 1983. *Hitler: A Chronology of His Life and Time.* London: Palgrave Macmillan.

Herf, Jeffrey. 1984. *Reactionary Modernism: Technology, Culture, and Politics in Weimar and the Third Reich.* Cambridge: Cambridge University Press.

Herwig, Holger H. 1999. "Geopolitik, Haushofer, Hitler, and Lebensraum." *Journal of Strategic Studies* 22: 218–41.

———. 2010. "The Daemon of Geopolitics: Karl Haushofer, Rudolf Hess, and Adolf Hitler." Accessed June 2013. http://www.usafa.edu/df/dfh/docs/Harmon53.pdf.

Heske, Henning. 1986. "Political Geographers of the Past III. German Geographical Research in the Nazi Period: A Content Analysis of the Major Geography Journals, 1925–1945." *Political Geography Quarterly* 5: 267–81.

Hottes, Karlheinz, Ruth Hottes, and Peter Schöller. 1980. "Walter Christaller, 1893–1969." In *Geographers: Biobibliographical Studies,* edited by Thomas W. Freeman, 11–16. London: Mansell.

Isard, Walter. 2000. "Interview with the Author." Ithaca, NY: Cornell University.

Johnston, Ronald J. 2002. "Robert E. Dickinson and the Growth of Urban Geography: An Evaluation." *Urban Geography* 22: 702–36.

Klee, Ernst. 2001. *Deutsche Medizin im Dritten Reich: Karrieren vor und nach 1945.* Frankfurt: Fischer.

Koehl, Robert L. 1957. *RKFDV: German Resettlement and Population Policy, 1939–1945.* Cambridge, MA: Harvard University Press.

Loader, Colin. 2012. *Alfred Weber and the Crisis of Culture, 1890–1933.* New York: Palgrave Macmillan.

Lösch, August. 1938. "The Nature of Economic Regions." *Southern Economic Journal* 5: 71–78.

———. 1954. *The Economics of Location.* New Haven, CT: Yale University Press. Originally published in German as *Die räumliche Ordnung der Wirtschaft* first in 1940, and in a revised second edition in 1944.

Lozowick, Yaacov. 2002. *Hitler's Bureaucrats: The Nazi Security Police and the Banality of Evil.* London: Continuum.

McCraw, Thomas K. 2007. *Prophet of Innovation: Joseph Schumpeter and Creative Destruction.* Cambridge, MA: Belknap Press of Harvard University Press.

Noakes, Jeremy. 1993. "The Ivory Tower under Siege: German Universities in the Third Reich." *Journal of European Studies* 23 (4): 371–407.

Preston, Richard E. 2009. "Walter Christaller's Research on Regional and Rural Development Planning during World War II." *METAR– Papers in Metropolitan Studies*, 52, Freie Universität Berlin, Institut für Geographische Wissenschaften: Gerhard O. Braun.

Renneberg, Monika, and Mark Walker. 1994. "Scientists, Engineers, and National Socialism." In *Science, Technology, and National Socialism*, edited by Monika Renneberg and Mark Walker, 1–35. Cambridge: Cambridge University Press.

Riegger, Roland. 2007. "Three Decades of August Lösch Days in Heidenheim." In *Space-Structure-Economy: A Tribute to August Lösch*, edited by Ulrich Blum, Rolf H. Funck, and Jan S. Kowalski, 401–6. Baden Baden: Nomos.

———, ed. 1971. *August Lösch in Memoriam*. Heidenheim: Meuer.

Rössler, Mechtild. 1989. "Applied Geography and Area Research in Nazi Society: Central Place Theory and Planning, 1933–1945." *Environment and Planning D: Society and Space* 7: 419–31.

———. 1993. "Secret Nazi Plans for Eastern Europe: Geography and Spatial Planning in the Third Reich." *Treballs de la Societat Catalana de Geografia* 35: 202–10.

———. 1994. "'Area Research' and 'Spatial Planning' from the Weimar Republic to the German Federal Republic: Creating a Society with a Spatial Order under National Socialism." In *Science, Technology, and National Socialism*, edited by Monika Renneberg and Mark Walker, 126–38. Cambridge: Cambridge University Press.

———. 2001. "Geography and Area Planning under National Socialism." In *Science in the Third Reich*, edited by Margit Szöllösi-Janze, 59–78. New York: Berg.

Sherratt, Yvonne. 2013. *Hitler's Philosophers*. New Haven, CT: Yale University Press.

Stolper, Wolfgang F. 2007. "August Lösch: A Record of a Friendship." In *Space-Structure-Economy: A Tribute to August Lösch*, edited by Ulrich Blum, Rolf H. Funck, and Jan S. Kowalski, 379–90. Baden Baden: Nomos.

Szöllösi-Janze, Margit. 2001. "National Socialism and the Sciences: Reflections, Conclusions, and Historical Perspectives." In *Science in the Third Reich*, edited by Margit Szöllösi-Janze, 1–35. New York: Berg.

Ullman, Edward L. 1941. "A Theory of Location for Cities." *American Journal of Sociology* 46: 853–64.

10

Social Engineering, National Demography, and Political Economy in Nazi Germany: Gottfried Feder and His New Town Concept

JOSHUA HAGEN

Introduction

An engineer by training, Gottfried Feder was one of the earliest and more enigmatic members of the Nazi Party. Feder's exact role remains debated, but he was clearly a significant figure within the nascent movement (Ludwig 1974; Tyrell 1978; Schubert 1986; Schenk and Bromley 2003; Meyer 2004; Guse 2010). Feder would serve as one of the first Nazis elected to the Reichstag, was editor of the National Socialist Library book series, and was the semiofficial guardian of the party's 25 Point Program. These positions helped earn Feder the moniker of the *Programmatiker*, or theoretician of the movement, yet Feder was not in Hitler's inner circle. This ambiguity continued after the seizure of power as Feder was appointed to the relatively inconsequential post of state undersecretary in the Reich Economics Ministry in 1933. It appeared Feder would serve as little more than a mid-level functionary in the new regime.

The situation changed dramatically in March 1934 when Hitler appointed Feder to the new post of Reichskommissar for Settlement Affairs. Feder soon launched a series of "programmatic speeches" signaling a new approach to housing policy and settlement planning (Feder 1934a). Feder recounted how continued industrial and urban agglomeration fostered various economic and social ills, such as the erosion of traditional culture, declining fertility rates, and the concentration of wealth. As Feder summed in his stock phrase, "the metropolis is the death of the nation" (Feder 1934b, 434). Instead, Feder called for the "dispersal of large cities, the re-rooting and the sedentarization of the population in the native soil." By dispersing people from urban centers, Feder promised to advance several ideological and practical goals, including protecting traditional values, countering communism, increasing fertility rates, and ultimately producing a healthier, stronger, and racially pure nation.

Raising many of the standard tropes of agrarian romanticism and cultural conservatism, Feder explained:

> Beautiful German cities, as are available to us in a few rare specimens from the organic and harmonious German Middle Ages, shall rise again in a new form. These new settlements, these new rural towns amid a rich peasantry, shall be created anew according to the best aspects of urban planning and architecture. They shall not exceed a certain manageable size and are to be a reflection and expression of social spirit and the best public mindedness. (Feder 1934a, 186)

Feder estimated that this goal might entail the establishment of 1,000 new cities of 10,000 to 15,000 people each over a fifty-year period. Feder claimed the effort would "create the foundation for a new state structure, for a new economic structure . . . and give back to the people the most priceless thing that they have lost in the gears of the metropolis: homeland and the sense of homeland." Feder's speeches were replete with dramatic rhetoric but few details on how this would be accomplished. In at least one speech though, Feder did hint at what would become a central organizational principle in his later book: "There may not be a settlement without a settlement core!" ("Kundgebung" 1934, 550).

Feder's ambitious program entailed wide-ranging consequences. It is unclear if Hitler intended his recent appointee to proffer such a radical program, but it galvanized opposition among leading industrialists whose fortunes were endangered, among military leaders who feared any disruption to rearmament, and among other senior officials pursuing their own agendas. Feder was overmatched, and his sudden ascent was matched by his rapid downfall. By the end of 1934, Feder was dismissed from his ministerial office, and Hitler dissolved the post of Reichskommissar (Münk 1993, 181–86; Harlander 1995, 53–66). The Nazi press characterized Feder's departure as voluntary and temporary with every indication that Feder remained in good standing. Feder was eventually appointed as a professor at the Berlin Technical University in 1936. Hitler was personally involved in arranging the appointment, and the Labor Ministry provided financial support for his research. Further, Feder was allowed to continue using the title of state undersecretary and retained his Reichstag post until 1936 (Schubert 1986, 201; Schenk and Bromley 2003, 116; Meyer 2004, 94–95). Undoubtedly a demotion, the title of professor was still highly respected and offered financial security. It also afforded Feder the opportunity to pursue his interests in settlement planning from an academic perspective. The remainder of this chapter will examine Feder's efforts to promote his vision of a new settlement paradigm for Nazi Germany by focusing on his book *The New Town.*

Feder and *The New Town*

Beginning in 1936, Feder resurfaced in professional trade journals with a brief article asking "what is the ideal form for the city of the future?" Feder answered, "undoubtedly, it is the small and medium town, systematically planned, with the opportunity to grow organically." Feder outlined a general approach for planning new towns that would be "racial-hygienic," meaning that single family homes with ample gardens would raise living standards and fertility rates; "state political" meaning that every town would be a "political unit"; economical meaning that they would be relatively self-sufficient; and finally, "military-strategic" meaning that towns would be spread out to avoid an "all too compact target for attacks by air forces" (Feder 1936, 193–94).

Feder expanded his ideas the following year. The task was to approach "spatial planning and spatial research [as] a secular task, which must be handled clearly and systematically without haste and nervousness. It is an area of responsibility that in this originality and totality appears for the first time in history" (Feder 1937, 71). To achieve this, the state would serve as the central conduit ensuring that German territory was used for the maximum benefit of the nation. From Feder's perspective, the "great state policy goal of the reorganization of German living space is a *well-planned and preferably complete utilization of German living space for the preservation and requirements of the German people*" (ibid., original emphasis). Feder announced that he was leading a team to develop guiding principles for a comprehensive spatial reordering of society encompassing geography, sociology, demography, and economics. Feder largely eschewed strident party rhetoric, instead presenting the effort as rational, technocratic, and scientific. In calling for a comprehensive, state-centric approach to spatial planning, Feder's thinking was widely shared among many architects, engineers, and planners across Germany. In short, the time was ripe for Feder to publish his ideas.

In March 1939, Feder published the results of his efforts in a book titled *The New Town: An Attempt to Establish a New Art of Town Planning Out of the Social Structure of the Population* [orig.: *Die neue Stadt: Versuch der Begründung einer neuen Stadtplanungskunst aus der sozialen Struktur der Bevölkerung*]. *The New Town* is hefty at nearly five hundred pages total, including about three hundred different illustrations, tables, and building plans. It begins with a short introduction outlining the project's objectives and ideological foundation intertwined with a cursory history of urban development. This introduction most resembles Feder's other writings. The book's main body consists of four parts, each including five to eight chapters or subsections. The first part covers general considerations concerning the overall size, composition, and layout of the ideal new town. This section seems to build on Rechenberg's dissertation while also drawing from ideas in broader circulation at the time among planners and policy makers. The second and third parts detail the public services and businesses that should be included in these new towns. In the foreword, Feder credited his research assistant W. Knoblich for researching the businesses covered in the third section but stopped short of crediting him with any level of authorship. The fourth part provides the final analysis, conclusions, and practical considerations, generally paralleling the work acknowledged to Max Mueller in the foreword. The book ends with a brief discussion of the estimated costs for building one of these new towns and a one-page conclusion, which shares some similarities with the introduction, suggesting Feder as the main author. Since it appears impossible to resolve authorship definitively, this chapter will simply refer to Feder and Feder's team interchangeably. Despite these questions, there is little doubt that Feder was the supervisor of the work, even if substantial portions were drafted or even written by others.

Feder claimed to offer a scientific foundation for reconciling and realizing many disparate strands of Nazi ideology through a spatial reordering of the German nation. The main organizational principle of this new Germany would be the "new town." As Feder explained in the introduction:

> These new towns of a new worldview will be the most visible and lasting expression of a new community will. They will and must organically grow out of the social structure of the population. . . . The towns of the future must be in planning and construction, in their harmonious integration in the landscape and surroundings, in their relationship to district, region, and Reich, a living expression of the new spirit of the age and the will to live and work of the new Greater Germany created by Adolf Hitler. (Feder 1939b, 1–2)

Despite beginning with such strong ideological tropes, Feder emphasized that this project derived from a "strictly scientific foundation" aimed at discerning the norms, patterns, and practices of healthy settlements (Feder 1939b, 2). To bolster this claim, Feder offered a cursory history of urban development from antiquity to the present with Germany's medieval towns garnering special praise as organic, cohesive, and authentic expressions of social order.

This glorification of premodern cities contrasted with the vilification of contemporary developments. Feder repeatedly identified chaos, decline, and "the lack of authoritarian supervision of the building industry" as the defining features of modern cities (Feder 1939b, 9). This was, according to Feder, the logical extension of political and economic liberalism to urban policy and residential planning. As a result, "absolute chaos resulted. No guiding hand provided for order. . . . Speculative building and real estate reveled in orgies. There was no regulated urban planning" (ibid., 12). This had dire consequences for the cultural and physical health of the nation, or as Feder bluntly concluded: "The degenerate metropolis is the death of the nation" (ibid., 14). Feder's rhetoric was extreme but also reflective of anxieties permeating German society concerning the cultural, demographic, and economic impacts of unbridled urbanization.

Feder conceived of the team's efforts as central to the broader program of building a new National Socialist state. As the subtitle suggests, *The New Town* claimed to provide a basic blueprint for a new settlement ideal where "the social structure of the population must emerge as the new structuring principle in the field of vision of future town builders" (Feder 1939b, 479). The book would be a "complete tool for the artistic design of the new towns of the Third Reich," capturing "totality in settlement thought" (ibid., 14, 30). This undertaking would far surpass what Feder proposed in 1934. "Our research projects were equally the 'work on the old' as the 'shaping of the new,'" Feder explained, with the "reorganization of German living space" as the ultimate objective (ibid., 14). Feder argued that these new towns should have a population of approximately 20,000, or about 4,400 households. This size was "large enough to lead an independent social, cultural, and economic life," yet small enough that everything could be reached in less than fifteen minutes without mass transit (ibid., 14, Tafel IV). These towns would avoid the disadvantages of both the city and the village while retaining the advantages of both.

> Disadvantages of City: low fertility rates; lack of settledness; traffic fatalities.
>
> Disadvantages of Village: inadequate utilities; no middle point of cultural life; no administrative organization of life; incomplete business and commercial life.
>
> Advantages of City: strong centralization; involvement in public life; larger labor pool and market.
>
> Advantages of Village: direct connection with nature; economic self-sufficiency for basic needs; healthier surroundings. (Feder 1939b, 24–26)

Feder's team aimed to determine the optimal socioeconomic and spatial structure needed to achieve relative independence and self-sufficiency by studying existing towns with comparable populations. They eventually identified about 150 such towns (Feder 1939b, 28, 75). Given their modest size, Feder assumed that the recent waves of industrialization and urbanization had largely bypassed these towns, and therefore, their basic socioeconomic and spatial characteristics would reflect the organic, natural, and healthy communities believed to exist in preindustrial times. "We are confronted with the best and most organic expression of communal life in the medieval towns, some of which have retained their character almost pure," Feder explained. "Here the town lives from itself, one works for others, and so naturally forms a true living and working community" (ibid., 23). Feder hoped his new settlement model would revive this idealized past.

Feder's team compiled information on these towns from government ministries, business directories, trade associations, and other sources. Feder's team also conducted a series of surveys and inquiries focusing on a sample of seventy-two towns (1939b, 55, 75). One survey examined government institutions and public facilities. The results were difficult to interpret, because some municipal authorities were apparently reluctant to respond or simply did not bother. Nonetheless, Feder's team dutifully recorded, for example, how many towns reported having a courthouse, the number of courthouse

FIGURE 10.1. Map showing the seventy-two towns that provided the bulk of Feder's empirical data. Feder (1939b, 55).

employees, and the size of the court building. This was repeated for over sixty different types of public offices, utilities, and amenities, including police and fire departments, water and electrical works, and libraries and sport facilities.

The inventory revealed that most towns had standard services, like post offices and banks, but were often lacking in cultural and leisure facilities, like swimming pools or theaters. The results also noted that party offices were lacking. Every town should have had a house of the NSDAP (Nationalsozialistische Deutsche Arbeiterpartei, or Nazi Party) for local party administration, but most chapters were operating out of converted rental apartments or commercial properties. Ideally, the new party office buildings would be centrally located but not directly in the main commercial center. As one of "the most important institutions of the future town, its importance should also be highlighted externally," perhaps by locating the building adjacent to a small square or park (Feder 1939b, 110). Other missing party institutions included a civic center for public events, parade grounds for rallies, and Hitler Youth facilities (ibid., 113, 179, 194–200). The final results were tabulated in an oversized chart at the end of the book.

Feder's team also compiled a rather detailed survey of commercial data from a sample of forty-one towns that recorded, for example, the number of bakeries, the number of bakery employees, the physical size of bakeries, and current trends in the local bakery business for each survey site (Feder 1939b, 249). This process was repeated for nearly every conceivable commercial activity to determine an average socioeconomic structure. The goal was to approximate the regular needs of the community, especially in foodstuffs, because self-sufficiency would reduce costs for transportation, fuel, and warehousing. Any settlement, Feder explained, that is unable "to produce most of the necessities of life and carry on its own administration is social-politically undesirable and economic-politically endangered. It is not crisis-proof. It remains instead an arbitrary, artificial creation compared to the healthy organism of a community whose vitality is rooted in self-sufficiency" (ibid.). The final inventory spanned sixty different occupations in food, clothing, housing, retail, and other services. Each category had a table reporting the survey results and a second table titled "Proposal for the Planning" of a new town with 20,000 people. Based on the survey results for bakeries, for example, Feder recommended that each new town have twenty-five bakeries with one hundred total employees, equivalent to four bakery workers per shop and one for every two hundred residents (ibid., 261). This was repeated for each occupation from grocers and tailors to basket makers and chimney sweeps. These results were also compiled into an oversized chart at the end of the book.

These surveys and other sources provided a great deal of data outlining

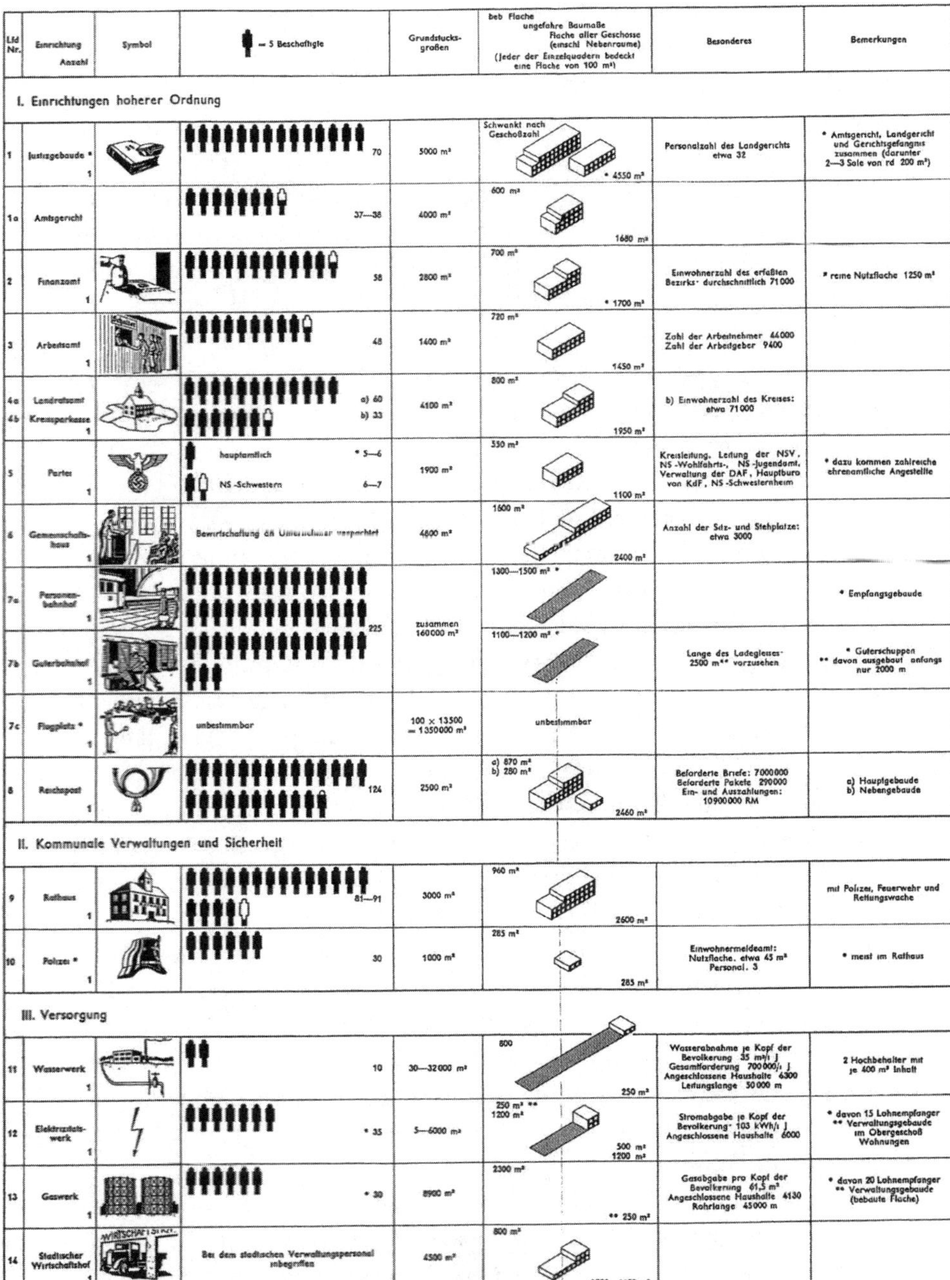

Lfd Nr.	Einrichtung Anzahl	Symbol	= 5 Beschaftigte	Grundstucksgroßen	beb Flache ungefahre Baumaße Flache aller Geschosse (einschl Nebenraume) (Jeder der Einzelquadern bedeckt eine Flache von 100 m²)	Besonderes	Bemerkungen
I. Einrichtungen hoherer Ordnung							
1	Justizgebaude * 1		70	5000 m²	Schwankt nach Geschoßzahl; * 4550 m²	Personalzahl des Landgerichts etwa 32	* Amtsgericht, Landgericht und Gerichtsgefangnis zusammen (darunter 2—3 Sale von rd 200 m²)
1a	Amtsgericht		37—38	4000 m²	600 m²; 1680 m²		
2	Finanzamt 1		58	2800 m²	700 m²; * 1700 m²	Einwohnerzahl des erfaßten Bezirks· durchschnittlich 71000	* reine Nutzflache 1250 m²
3	Arbeitsamt 1		48	1400 m²	720 m²; 1450 m²	Zahl der Arbeitnehmer 44000 Zahl der Arbeitgeber 9400	
4a 4b	Landratsamt Kreissparkasse 1		a) 60 b) 33	4100 m²	800 m²; 1950 m²	b) Einwohnerzahl des Kreises: etwa 71000	
5	Partei 1		hauptamtlich * 5—6; NS-Schwestern 6—7	1900 m²	330 m²; 1100 m²	Kreisleitung, Leitung der NSV, NS-Wohlfahrts-, NS-Jugendamt, Verwaltung der DAF, Hauptburo von KdF, NS-Schwesternheim	* dazu kommen zahlreiche ehrenamtliche Angestellte
6	Gemeinschaftshaus 1		Bewirtschaftung an Unternehmer verpachtet	4800 m²	1600 m²; 2400 m²	Anzahl der Sitz- und Stehplatze: etwa 3000	
7a	Personenbahnhof 1		225	zusammen 160000 m²	1300—1500 m² *		* Empfangsgebaude
7b	Guterbahnhof 1				1100—1200 m² *	Lange des Ladegleises· 2500 m** vorzusehen	* Guterschuppen ** davon ausgebaut anfangs nur 2000 m
7c	Flugplatz * 1		unbestimmbar	100 × 13500 = 1350000 m²	unbestimmbar		
8	Reichspost 1		124	2500 m²	a) 870 m² b) 280 m²; 2460 m²	Beforderte Briefe: 7000000 Beforderte Pakete 290000 Ein- und Auszahlungen: 10900000 RM	a) Hauptgebaude b) Nebengebaude
II. Kommunale Verwaltungen und Sicherheit							
9	Rathaus 1		81—91	3000 m²	960 m²; 2600 m²		mit Polizei, Feuerwehr und Rettungswache
10	Polizei * 1		30	1000 m²	285 m²; 285 m²	Einwohnermeldeamt: Nutzflache. etwa 45 m² Personal. 3	* meist im Rathaus
III. Versorgung							
11	Wasserwerk 1		10	30—32000 m²	800; 250 m²	Wasserabnahme je Kopf der Bevolkerung 35 m³/i J Gesamtforderung 700000/i J Angeschlossene Haushalte 6300 Leitungslange 50000 m	2 Hochbehalter mit je 400 m³ Inhalt
12	Elektrizitatswerk 1		* 35	5—6000 m²	250 m² ** 1200 m²; 500 m² 1200 m²	Stromabgabe je Kopf der Bevolkerung· 103 kWh/i J Angeschlossene Haushalte 6000	* davon 15 Lohnempfanger ** Verwaltungsgebaude im Obergeschoß Wohnungen
13	Gaswerk 1		* 30	8900 m²	2300 m²; ** 250 m²	Gasabgabe pro Kopf der Bevolkerung 61,5 m³ Angeschlossene Haushalte 4130 Rohrlange 45000 m	* davon 20 Lohnempfanger ** Verwaltungsgebaude (bebaute Flache)
14	Stadtischer Wirtschaftshof 1		Bei dem stadtischen Verwaltungspersonal inbegriffen	4500 m²	800 m²; 1000—1100 m²		

FIGURE 10.2. This is one-third of the table reporting survey data for public institutions. From left to right, each row provides the name of the institutions, number of employees, area of the facility, and additional remarks. Feder (1939b, Tafel I).

a basic socioeconomic structure but provided little guidance concerning the town layout. Reflecting his emphasis on "chaos" in critiquing contemporary cities, Feder advocated a somewhat rigid framework for these new towns while still insisting that they embody "cohesion and organic structure" (Feder 1939b, 461). Specifically, Feder conceived of a hierarchical "town organism" composed of "cells" orientated around a focal point. These basic cellular units would then be organized into a nested hierarchy extending upward, eventually reaching the Reich capital. "The integration of the individual cells must be designed so that the life of each local place orientates itself clearly toward its center and from there can flow to the next higher level up to the town center," Feder explained. "From here, the town must first be connected to the sequentially higher organism of the province and the Reich" (ibid., 19).

In practice, each cell corresponded to a primary school district of roughly five hundred to six hundred students, or about 3,500 people given contemporary fertility rates (Feder 1939b, 19, 30–431). The cells were primarily residential in nature but still contained enough businesses to meet the daily needs of residents. These cells would be arranged in a circle around a town center since, as Feder explained, "basically, the round town form is also the most organic" (ibid., 464). The town center would have the same basic services as the lower level cells but also host the government and party offices, the post office, the bank, and other services people used less frequently. The train station, utilities, and other industrial facilities would be located on the eastern edge of town, while the western edge of town would have the sports field, parade grounds, and other leisure facilities (ibid., 47). These basic cells would be linked through two boulevards, west–east and north–south, mostly lined with businesses. Despite this general schematic, Feder repeatedly assured readers that the new town concept was not rigid and "that there is not an ideal town but rather that every town must be designed anew on each location" to adapt to the local landscape (ibid., 460). To minimize excavation work, roads would follow the contours of the land, resulting in curving streets, which would also mitigate any monotony that might result from uniformity in the design of the houses and other buildings (ibid., 433–36).

Relatively little attention was paid to the architecture of individual houses or other buildings since Feder's team was concerned with broader spatial and socioeconomic structures. Given that, Feder's team merely compiled statistics on the area of the various facilities and provided a few sample blueprints. It was assumed that the bulk of housing would be some combination of detached and attached single-family homes on plots between 500 and 800 square meters. Perhaps about one-quarter of residences would be

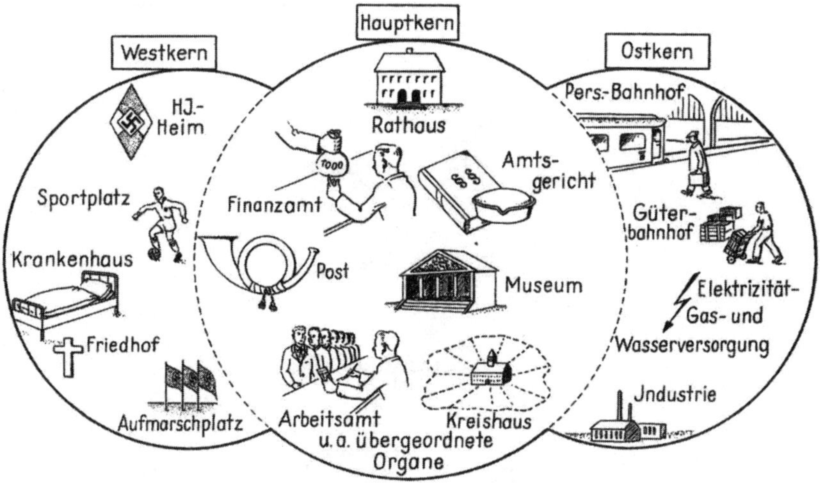

FIGURE 10.3. This simple drawing illustrates the main axis of Feder's organizational structure with government offices in the center, industry to the east, and recreation and health to the west. Residential areas would be added to the north and south. Feder (1939b, 47).

rental apartments, mostly in three-story buildings in the town center with businesses occupying the ground floor. To maintain their connection with the soil and to produce some of their own food, apartment residents would be afforded about 200 square meters of garden space along the edge of town. The relatively small plots and the concentration of apartments in the center were intended to limit sprawl so that "the settlement will appear somewhat enclosed" (Feder 1939b, 50, 461). Here, Feder was trying to balance between competing imperatives within Nazi Germany—namely, the desire to build a sense of community, enclosure, and even monumentality into the regime's new settlements as opposed to the belief that dispersed populations were necessary for meaningful connections to nature, higher fertility rates, greater economic security, and reduced danger from aerial bombardment.

The team's efforts culminated in a "crystallogramm" meant as a generic conceptual schematic instead of a literal town plan. This large foldout diagram collated the results of the public and commercial surveys into a basic spatial structure with eight subcores or subsections surrounding a main core. Each section had its own basic services, while the main core hosted the majority of government buildings. Larger and specialized retailers were spread across the main core and the east–west axis. The eastern appendage of town hosted the train station and public utilities, while the main leisure facilities and parade grounds were on the western edge. The three sections to the top and three to the bottom were framed around elementary schools and

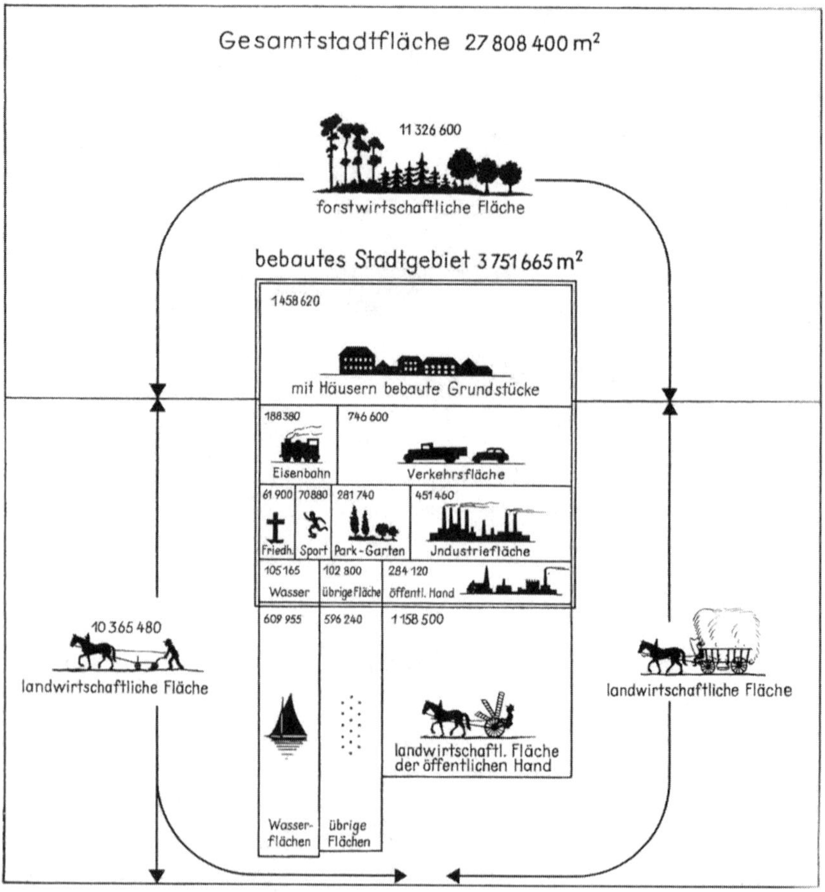

FIGURE 10.4. Map illustrating the general land use patterns with the town in the center box surrounded by fields and forests. Feder (1939b, 448).

were primarily residential in function. The final appendage to the northwest was the hospital, retirement home, cemetery, and crematorium (Feder 1939b, Tafel IV). This new town would be embedded in a landscape of farms, forests, and lakes that rooted the community in nature and provided self-sufficiency in basic necessities. Besides a total area of nearly 2,800 hectares, Feder provided no additional details concerning the location or distribution of these new towns, perhaps because Walter Christaller's central place theory—on which I will return later in the chapter—seemed to address that general question (Rössler 1989, 2001).

These new towns would not be completely autarchic but instead would adopt different specializations, such as mining or manufacturing, as determined by Reich central planners. Feder emphasized the primacy of state interests since "the ideal always remains the construction of new settlements entirely according to the new political considerations in terms of defense policy, transportation, economics, and demography as they result from the reorganization of the German living space for the Reich" (Feder 1939b, 429). Feder's team estimated that about 55 percent of each town's population, or about 11,000 people, would work outside the home. Of these, a majority of just over 6,700 would be employed in some type of "foreign industry," meaning production for nonlocal markets (ibid., 470).

Feder sponsored a small competition among his students to demonstrate the practical application of these principles. Given the rather detailed specifications, it is not surprising that the winning entry followed Feder's basic schema (1939b, 462–63). Although judged "a bit rigid," the design by Heinz Killus, one of the team members acknowledged in the foreword, was also noteworthy (ibid., 462). The Killus plan made direct linkages between daily, weekly, and monthly needs of the residents and the Nazi Party's hierarchical organization of cell, local chapter, and district. Despite their differences, both plans were based on the "foundational principle . . . that a series of smallest communities were combined into a group and this group was again united with groups of a higher order and so forth" (ibid., 467).

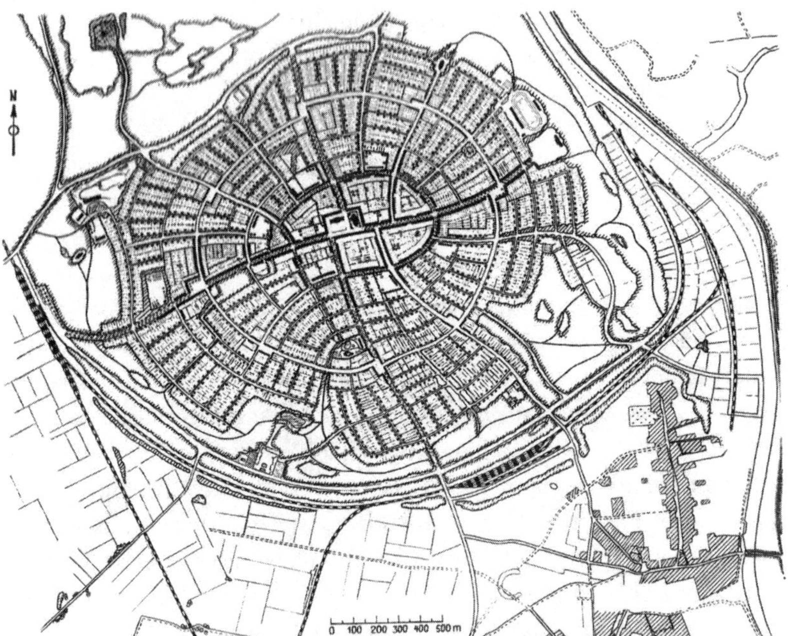

FIGURE 10.5. Feder judged this town plan by Kuhn and Fröhlich to be a reasonable approximation of his concept. Feder (1939b, 463).

From Feder's perspective, this spatial hierarchy was imperative for the reorganization of German living space. Instead of building endless, monotonous tracts of houses, a hierarchical town model based on small cells meant "that the people living there would inevitably be joined to community." This structure was "entirely organic," Feder explained, and exhibited a "striking resemblance to the structure of the party." The new town's basic spatial structure of neighborhood, sub-core, and core would correspond to the party's basic administrative structure of block, cell, and chapter. The end result would be that each new town and its adjacent hinterland would constitute a party *Kreis*, or district. "So arises here a gratifying correspondence between the structure of basic provisions, administration, organization, and transportation," Feder concluded. "One can say that our town is complete in every sense" (Feder 1939b, 468).

Taken together, Feder went beyond the idea of cells as the basic organizing principles for new towns to envision a large number of dispersed settlements nestled harmoniously into the landscape. But instead of being entirely self-sufficient homesteaders, these new communities would be interlinked and integrated into highly structured and expansive networks and hierarchies. In that sense, Feder's concept clearly drew from the earlier garden city ideas, but instead of being isolated and geometric, Feder's new towns would be interconnected and organic. Feder helped popularize the term *Stadtlandschaft*, which roughly translates as town landscape, to signify this general concept (Mantziaras 2003).

Despite this claim to totality and comprehensiveness, Feder acknowledged that some issues remained, such as the growing impact of automobiles or that air attacks may require greater provision of bunkers, shelters, or even moving some infrastructure underground (Feder 1939b, 479; 1939c). Feder also acknowledged that some larger cities would be needed, although these too would eventually be reorganized according to the same basic spatial principles (Feder 1939b, 267). Feder was also aware of the practical challenges, especially financial concerns, that often steered residential and industrial investment into ever-larger cities. Yet Feder's team estimated that each new city would cost only about fifty million Reichsmark, amounting to around 2,500 per resident or a little under 11,400 per household (ibid., 477). While certainly a considerable sum, Feder argued this was reasonable since the new towns established using his principles would be "infinitely easier and simpler, more hygienic, in every respect demographically, socially, and politically better and ultimately even cheaper than the previously used method" (ibid., 478).

Feder ignored the risks inherent in committing investment to building a new town, but the implicit assumption was that the state and its central planners would basically force people, businesses, and investment into the appropriate structures since "only superior leadership and the concentration of all power in one hand makes such building projects possible at all" (1939b, 18). Indeed, the Nazi dictatorship seemed a prerequisite for a comprehensive spatial reorganization of German society, economics, and demography. As Feder asked rhetorically at the end of the book: "But what time could be better than the turning point that National Socialism has ushered in, and what fields could possibly be more fertile and promising than the sacred ground of the Third Reich of all Germans—Greater Germany—that our *Führer* Adolf Hitler has awakened from millennium-old dreams to a power conscious, proud reality?" (ibid., 479).

Feder recognized that Germany would retain some larger population centers, so he published *Place of Work—Place of Residence* concurrently as a companion book offering a foundation for reforming existing cities. *Place of Work—Place of Residence* was much shorter than its counterpart but still aimed to provide a "principled, empirical, and methodical" study of commuting between home and work (Feder 1939a, 2). Overall, the book appears much more consistent in terms of style and format. Additionally, Feder acknowledged no other specific contributors as he did for *The New Town*. On the other hand, Feder noted that his team compiled much of the data and often used "we" when describing the work. Some of the illustrations were published previously in an article by Rechenberg (1938). On balance, it appears that Feder had a stronger claim to authorship of *Place of Work—Place of Residence*, although it is probably best to again think of him as the supervisor of the book.

Feder's basic thesis was that the mismatch between residential and work locations played a key role in the dysfunction of major cities, specifically in terms of lost time and traffic fatalities. Unsurprisingly, Feder identified liberalism and capitalism as the underlying impetus leading to deteriorating living conditions, overcrowding, class conflict, declining fertility, and ultimately support for Communism. The result was "inorganic forms of life and community" eventually leading to "the dissolution of each national community" (Feder 1939a, 11). To clarify these issues, Feder's team launched a massive survey that distributed approximately 200,000 questionnaires to private- and public-sector workers across Berlin (ibid., 21–22). Beyond some basic demographic data, the survey recorded the place of residence, place of employment, and the time and cost of commuting between those two locations. Feder's team also collected information on traffic accidents and lost free time. Based on this, Feder estimated that Berlin workers lost about 1.5 million Reichsmark daily due to transportation costs and lost free time (ibid., 50). In addition to these extra costs, cramped living conditions depressed fertility rates. Feder's team also estimated the relocation costs associated with worker migration at around thirty million Reichsmark annually (ibid., 72).

Feder's solution returned to his new town concept and reprinted two example plans included in *The New Town*. Yet Feder recognized that Berlin and other large cities would not be completely dissolved. Instead, Feder argued that these larger metropolises should be reorganized as "satellite towns" orbiting close to the main city center. Additionally, the city center should be restructured to provide greater congruence between home and work. This approach would maintain the basic concept of a cell structure, although these satellite towns would lack the agrarian character of their new town counterparts. For Feder, this was a necessary concession, but "the ideal naturally remains *the new town* that would be established, completely detached from the parasitic effects of the big city and its dangers, in open terrain from 'fresh roots' as an enclosed economic, commercial, social, and cultural unit" (Feder 1939a, 92, emphasis in original).

In his probable final publication in July 1940, Feder reiterated his vision

FIGURE 10.6. These maps of Berlin show an "incorrect" spatial structure on the top and a "correct" alternative based on a more cellular structure on the bottom. Feder (1939a, 50).

for a comprehensive reorganization of German living space but framed it within the context of war. While military and strategic considerations were obviously paramount now, Feder argued that securing Germany's future would ultimately depend on the "reorganization and expansion of this living space" in a systematic and ordered fashion: "The clarity and exactitude of technical thinking combined with the conscientiousness and thoroughness of scientific research will provide us the guarantee for the success of the reorganization of German living space." Only such a program could save Germany from the specter of "biological suicide" (Feder 1940, 14). Feder constructed a "schematic of the reorganization of German living space" to illustrate a planning process applicable to each locality and task. The process began with activities divided into two main groups. One group focused on the basic physical geography, natural resources, and social aspects (boxes numbered 1, 2, 3, in figure 10.7), while the other focused on economics, property issues, and transportation (boxes 4, 5, 6). This information provided the basis for the planning process (box 7), which would be implemented harnessing all available technologies and expertise aided by private capital and Reich laws (boxes 8, 9). This process would integrate Germany's professional builders, planners, engineers, and other scholars into an interdisciplinary cadre dedicated to the comprehensive spatial reordering of Germany, the German people, and beyond. Indeed, Feder concluded: "He [Hitler] has carved the borders of the new German living space with a brazen pen on the map of Europe; to fill this in with the flourishing of German life in his spirit is the task of German engineering" (1940, 16).

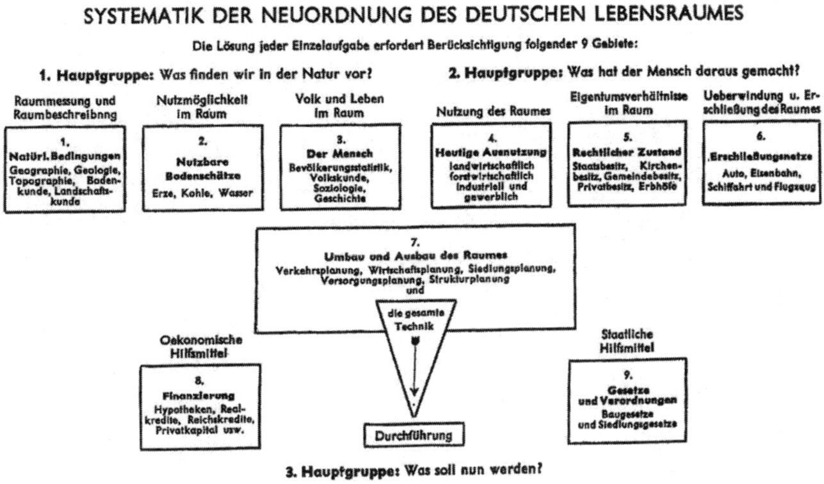

FIGURE 10.7. Feder's schematic for the reorganization of German living space. Feder (1940, 15).

Reception and Impact

Feder seemed eager to promote his work, but the Propaganda Ministry did not share that enthusiasm. Perhaps reflecting some personal acrimony from Goebbels, the ministry restricted reporting to trade publications, and there was nothing like an official party endorsement (Schubert 1986, 200–202). Yet the fact that professional-level discussion was allowed, instead of complete censorship, suggests some recognition that Feder's work contributed to the regime's objectives. Reaction to Feder's work was largely positive, and the book went into a second printing before the end of 1939. One writer declared that the book would have "lasting influence on the future of urban development and the art of urban planning." The book contained a wealth of practical information but "far beyond that will initiate a new era of urban planning" ("Unterredung" 1939, 31). Another reviewer described Feder's work as "logical," "compelling," and of "invaluable, lasting value," although acknowledging the idea would have to be tested in the real world (Wendt 1939, 126). A third writer touted the book as the "new foundation for urban planning" that was "the result of years of profound research on a strictly scientific basis" (Rühle 1939, 462). In a 1941 speech, Heinz Wetzel, professor of urban planning at the Stuttgart Technical University and an extremely influential figure in the regime's residential construction program, praised Feder's work as the "final formulation of the issues of a purposeful spatial order that in its exactitude is absolutely not to be surpassed. . . . The utilitarian organization is firmly established since Feder, now the path is free toward picturesque organization" (quoted in Voigt 1985, 240).

Feder also influenced the regime's building program through his students and assistants who contributed to various government projects at Wuppertal, Wilhelmshaven, Linz, and Stuttgart, among other places (Schmidt 1942, 14; Schubert 1986, 209). Killus also advanced Feder's basic concept in an article on the "idea of totality in new town building." Killus emphasized the importance of having the layout of neighborhoods, streets, and towns parallel the Nazi Party's organizational hierarchy to represent a "totality of thought" in the planning for a "total city" (1940, 85). Other planners also picked up these ideas as the whole topic assumed greater urgency with the prospect of vast new territories being incorporated in the Reich, especially in the East. Prominent urban planner Hans Bernhard Reichow, for example, envisioned a "new German East" characterized by the "uniform alignment of the settlement cells in the sense of the new ideological and political structure of our Reich" (Reichow 1941, 226). Building on Feder, these settlement cells would be grouped into towns that basically followed the "new town" model in terms of population size and the provision of commercial and public services, and then further organized into a hierarchical landscape of networked towns (ibid., 228–29). "As a part of this total organization, the settlement form develops from a cellular-based structure, which corresponds to the new political structure and simultaneously takes into account the perspectives of defense and air raid protection," Reichow explained. "The relationships among the cells, as well as to the higher-ranking communal facilities, can become, through optics, traffic, green areas, or otherwise, like a natural disposition" (ibid., 230).

Carl Culemann, a government planner in West Prussia, made the linkage between settlement cells and party structure most explicit: "The formation of the settlement mass through urban planning and the formation of the mass of the people through the party are concurrent and related tasks" (Culemann 1941, 123). Culemann argued that forty to sixty households constituted a block, four to eight blocks constituted a cell, and three to ten cells constituted a local party chapter. Regardless of the specific figures, the key factor for Culemann was the clear parallel between residential and party organization. Based on this, Culemann modified Feder's schema to achieve congruence between the official prescribed sizes for local chapters and the provision of basic services and amenities. Although deviating on some details, Reichow and Culemann both specifically referenced Feder's work as contributing to their overall concepts.

Feder's ideas likely influenced many planners working on Himmler's *Generalplan Ost* (General Plan for the East) and later Speer's reconstruction staff. Indeed, enthusiasm quickly spread to restructure settlement patterns around cellular models across Germany and its new territories. For example, Wilhelm Wortmann, a planner who worked for both Himmler and Speer, explained that "the city-landscape will be a new cellular-based structure of the city deliberately modeled on the political organization of our people" (Wortmann 1941, 16). Josef Umlauf, director of the urban planning section in Himmler's overall planning office, seemed to integrate much of Feder's concept into his program for restructuring Germany's newly annexed territories. Although retaining an agrarian character, cities would play a central function with the typical Kreis City having a population of 15,000 to 20,000 residents and its interior organization based on residential neighborhoods centered on schools (Umlauf 1941).

Walter Christaller, a geographer known for developing central place theory (see Barnes, this volume, chapter 9; also Barnes and Minca 2013), also contributed to Himmler's planning staff and undoubtedly saw the creation of a new East as an opportunity to turn this theory into practice. Christaller envisioned a nested hierarchy with the farmstead as the basic unit forming "the cell-like structure of the state" (Christaller 1940b, 498). These would be clustered into successively larger neighborhoods, hamlets, village groups, and finally the district centered on the Kreis City of 20,000 to 30,000 people (Christaller 1940a). Like Feder, Christaller produced general estimates of the number of various occupations needed for basic services although focusing on the village group level instead of the district city (Christaller 1941). The conceptual efforts of Feder, Christaller, and others contributed to what Umlauf declared would be a "summarized plan for the complete spatial order" of the new East (Umlauf 1942, 281–93).

The dreams of reordering the East soon faded, but allied bombardment revived the prospect of breaking up Germany's large cities. Albert Speer formed a reconstruction staff and soon began developing plans for rebuilding and reorganizing Germany's cities. Feder's concept of the cell as the basic organizational unit had already been accepted as a guiding principle by this point (Schubert 1986, 210). City planner Konstanty Gutschow, for example, advocated that Hamburg be rebuilt as settlement cells equivalent to local party chapters. This wholesale reorganization would result in a more dispersed city landscape of settlement cells interspersed amid abundant green space. Under normal conditions, there was practically no chance of realizing this vision, but Hamburg's destruction provided an opportunity to realize this new approach to city planning (Pahl-Weber 1986, 46–55). The tide of war soon overtook these plans as well, but Feder's idea of cellular-based settlements and the town landscape concept had become accepted principles among most professional planners.

Indeed, these notions easily translated into postwar planning as Feder's *The New Town* was generally interpreted as scientific, technical, and "non-ideological." There are undoubtedly a variety of reasons for this, including the heavily empirical character of the book's main sections, the lack of an official party endorsement, and the decision to link the basic cell to an elementary school instead of more directly to the party's structure. Feder also lacked an official government or party office since the early years of the regime. The continued acceptance of Feder's concept also reflected the great continuity in personnel among professional builders, planners, and engineers before and after 1945. These professionals tended to regard themselves as technocrats and aside from a handful of high-profile cases, most notably Speer, continued practicing in their fields. Rechenberg, Killus, Culemann, Wortmann, and Gutschow were all active participants in postwar reconstruction and planning projects, while books like Reichow's *Organic Urban Design: From the Metropolis to City Landscape* (1948) and Umlauf's *On the Essence of the*

City and Urban Planning (1951) helped reestablish many of these concepts outside of the Nazi period, effectively denazifying them to a large extent. These basic ideas carried beyond Germany to influence approaches to urban planning across Europe and even as far away as Japan (Hein 2008).

Conclusion

Feder died at his home in Bavaria after an extended illness in September 1941. His funeral was a relatively low-key affair. The regional party chief was the highest-ranking official in attendance and laid a wreath on Hitler's behalf. Bormann, Goebbels, Frick, and Darré also reportedly sent wreaths to the funeral. Given that the invasion of the Soviet Union was reaching what many top Nazis hoped would be its successful climax, the absence of high-ranking party leaders is not terribly surprising. Indeed, it is noteworthy that Feder's passing garnered any attention. The regime could have easily let Feder's death pass without comment. Given the limited publicity for his book, the relatively minor offices he held through 1934, and the dramatic events since 1938, few ordinary Germans were likely to remember Feder or his writings. Yet the *Völkischer Beobachter* ran his obituary, declaring that the movement had lost "a very valued old fighter" who would "forever remain unforgettable in the history of the National Socialist movement" (Meyer 2004, 79–80). Fritz Todt, Feder's one-time deputy and eventual successor who rose to become minister of armaments, offered lavish praise for Feder's contributions to the "development of a new art of town building and science of planning." Todt agreed that "for German technology, Gottfried Feder remains unforgotten" ("Gottfried" 1941, 475–76).

A more detailed obituary later appeared in an official party journal for housing. Feder was described as a "pioneer" who had "systematically researched one of the most important foundations upon which the city planning of the future must be built." Feder was portrayed as seeking a technocratic and practical approach to contemporary planning problems. The results of his efforts were "indispensable for all parties involved in city planning." Indeed, "as a loyal follower of the *Führer*, he saw his task primarily to express the foundation and objectives of National Socialism in city planning. The concept of the community settlement, in which all constituents of the population were bound together in large newly constructed communal structures, found in him an enthusiastic pioneer and promoter." Feder recognized soon after the seizure of power that "the task of National Socialism could not exhaust itself on the expansion of existing cities and the construction of new residential areas, but rather must also extend far beyond to the creation of new communities and entire cities that would become for all the times testaments to the transformative power and the transformative will of the National Socialist movement" (Schmidt 1942, 14–15).

Feder is commonly treated as a footnote in histories of the Nazi movement. Even in studies focusing on Nazi architecture and urban planning, Feder is afforded cursory attention. This is certainly justified based on his lack of official positions within the movement. Yet an assessment of Feder's work contributes to a more complete understanding of the imagined and planned spatiality of a future National Socialist state. While many of his colleagues were increasingly consumed with the immediate contingencies of war preparation and war making, Feder directed his team to develop a more comprehensive program harnessing the technocratic impulses within the movement toward the ultimate ideological objective of building a new Nazi state through a spatial reordering of Germany's society, demography, and economy.

References

Barnes, Trevor, and Claudio Minca. 2013. "Nazi Spatial Theory: The Dark Geographies of Carl Schmitt and Walter Christaller." *Annals of the Association of American Geographers* 103 (3): 669–87.

Christaller, Walter. 1940a. "Grundgedanken zum Siedlungs- und Verwaltungsaufbau im Osten." *Neues Bauerntum* 32: 305–12.

———. 1940b. "Die Kultur- und Marktbereiche der zentralen Orte im deutschen Ostraum und die Gliederung der Verwaltung." *Raumforschung und Raumordnung* 4: 498–503.

———. 1941. "Die Verteilung der nicht landwirtschaftlichen Bevölkerung im Hauptdorfbereich." *Neues Bauerntum* 33: 139–45.

Culemann, Carl. 1941. "Die Gestaltung der städtischen Siedlungsmasse." *Raumforschung und Raumordnung* 5: 122–34.

Feder, Gottfried. 1934a. "Das deutsche Siedlungswerk: Zwei programmatische Reden." *Siedlung und Wirtschaft* 16: 183–86.

———. 1934b. "Das deutsche Sieldungswerk." *Deutsche Technik* 2: 433–34.

———. 1936. "Die Zukunft der Kleinstadt und die zukünftige Kleinstadt." *Deutsche Bauzeitung* 70: 193–94.

———. 1937. "Tote und lebendige Wissenschaft." *Bauen Seideln Wohnen* 17: 71–72.

———. 1939a. *Arbeitstätte—Wohnstätte*. Berlin: Julius Springer.

———. [in collaboration with Fritz Rechenberg]. 1939b. *Die neue Stadt: Versuch der Begründung einer neuen Stadtplanungskunst aus der sozialen Struktur der Bevölkerung*. Berlin: Julius Springer.

———. 1939c. "Städtebau und Luftschutz." *Baulicher Luftschutz* supplement to *Gasschutz und Luftschutz* 9: 1–4.

———. 1940. "Reichsplanung—Stadtplanung." *Der deutsche Baumeister* 2: 13–16.

"Gottfried." 1941. "Gottfried Feder." *Deutsche Technik* 9: 475–76.

Guse, John C. 2010. "Nazi Technical Thought Revisited." *History and Technology* 26: 3–38.

Harlander, Tilman. 1995. *Zwischen Heimstätte und Wohnmaschine: Wohnungsbau und Wohnungspolitik in der Zeit der Nationalsozialismus.* Basel: Birkhäuser.

Hein, Carola. 2008. "Machi Neighborhood and Small Town: The Foundation for Urban Transformation in Japan." *Journal of Urban History* 35: 75–107.

Killus, Heinz. 1940. "Der Totalitätsgedanke im neuen Städtebau." *Monatshefte für Baukunst und Städtebau* 24: 85–88.

"Kundgebung." 1934. "Kundgebung des Reichssiedlungskommissars Staatssekretär Gottfried Feder über das deutsche Siedlungswerk." *Deutsche Technik* 2: 548–50.

Ludwig, Karl-Heinz. 1974. *Technik und Ingenieure im Dritten Reich.* Düsseldorf: Droste.

Mantziaras, Panos. 2003. "Rudolf Schwarz and the Concept of *Stadtlandschaft*." *Planning Perspectives* 18: 147–76.

Meyer, Torsten. 2004. "Gottfried Feder und der nationalsozialistische Diskurs über Technik." In *Technik und Verantwortung im Nationalsozialismus*, edited by Werner Lorenz and Torsten Meyer, 79–107. Münster: Waxmann.

Münk, Dieter. 1993. *Die Organisation des Raumes im Nationalsozialismus: Eine soziologische Untersuchung ideologisch fundierter Leitbilder in Architektur, Städtebau und Raumplanung des Dritten Reiches.* Bonn: Pahl-Rugenstein.

Pahl-Weber, Elke. 1986. "Die Ortsgruppe als Siedlungszelle." In *"Eine neues Hamburg entsteht": Planen und Bauen von 1933–1945*, edited by Michael Bose, Michael Holtmann, Dittmar Machule, Elke Pahl-Weber, and Dirk Schubert, 46–55. Hamburg: VSA-Verlag.

Rechenberg, Fritz. 1936. *Die günstigste Stadtgröße.* Berlin: Deyhle.

———. 1938. "Die Siedlung als Ausdruck der Gemeinschaft." *Bauen Siedeln Wohnen* 18: 383–90.

Reichow, Hans. 1941. "Grundsätzliches zum Städtebau im Altreich und im neuen deutschen Osten." *Raumforschung und Raumordnung* 5: 225–30.

———. 1948. *Organische Stadtbaukunst: Von der Großstadt zur Stadtlandschaft.* Braunschweig: Westermann.

Rössler, Mechtild. 1989. "Applied Geography and Area Research in Nazi Society: Central Place Theory and Planning, 1933 to 1945." *Environment and Planning D: Society and Space* 7: 419–31.

———. 2001. "Geography and Area Planning under National Socialism." In *Science in the Third Reich*, edited by Margit Szöllösi-Janze, 59–78. Oxford: Berg.

Rühle. 1939. "Neue Grundlagen für den Städtebau." *Baugilde* 21: 462–65.

Schenk, Tilman A., and Ray Bromley. 2003. "Mass-Producing Traditional Small Cities: Gottfried Feder's Vision for a Greater Germany." *Journal of Planning History* 2: 107–39.

Schmidt. 1942. "Gottfried Feder und sein Werk." *Der soziale Wohnungsbau in Deutschland* 2: 14–15.

Schubert, Dirk. 1986. "Gottfried Feder und sein Beitrag zur Stadtplanungstheorie." *Die alte Stadt* 13: 192–211.

Tyrell, Albrecht. 1978. "Gottfried Feder and the NSDAP." In *The Shaping of the Nazi State*, edited by Peter D. Stachura, 48–87. London: Croom Helm.

Umlauf, Josef. 1941. "Zur Stadtplanung in den neuen deutschen Ostgebieten." *Raumforschung und Raumordnung* 5: 100–122.

———. 1942. "Der Stand der Raumordnungsplanung für die eingegliederten Ostgebiete." *Neues Bauerntum* 34: 281–93.

———. 1951. *Vom Wesen der Stadt und der Stadtplanung.* Düsseldorf: Werner.

"Unterredung." 1939. "Unterredung mit Staatssekr. Prof. Gottfried Feder über: Die neue Stadt." *Baugilde* 21: 31–33.

Voigt, Wolfgang. 1985. "Die Stuttgarter Schule und die Alltagsarchitektur des Dritten Reiches." In *Faschistische Architekturen: Planen und Bauen in Europa, 1930–1945*, edited by Hartmut Frank, 234–50. Hamburg: Hans Christians.

Wendt. 1939. "Die neue Stadt." *Bauamt und Gemeindebau* 21: 125–26.

Wortmann, Wilhelm. 1941. "Der Gedanke der Stadtlandschaft." *Raumforschung und Raumordnung* 5: 15–17.

PART II

Geographies of the Third Reich

SECTION 3

Spatialities of the Holocaust

11

Nazi Biopolitics and the Dark Geographies of the *Selva*

PAOLO GIACCARIA AND CLAUDIO MINCA

Selva and *Città*

This chapter intends to contribute to the study of Nazi genocidal practices from a geographical perspective. In particular, it engages with the geographical metaphors of *selva* and *città*, as presented by Italian philosopher Giorgio Agamben, in order to further investigate the spatialities of the extermination practices perpetuated by the Nazis. Giorgio Agamben is today one of the most influential living critical theorists among scholars in the humanities and the social sciences. However, his contribution to the discussion of the Holocaust, of the concentration camps, of witnessing, to name a few, has been highly controversial. Dominick LaCapra, for one, while recognizing the wide impact of Agamben's philosophical speculations, argues that his take on history is "voided of specificity and counts at best as an instantiation of transhistorical theoretical concerns and post-apocalyptic apprehensions" (LaCapra 2004, 11). For LaCapra, who dedicates a key part of his influential *History in Transit* to the work of the Italian philosopher, the main problem is that Agamben's approch to Auschwitz fails to explore "the problematic, mutually questioning relation between history and theory," and ends up seeing "the historically specific, such as Auschwitz, simply as an instantiation, illustration, sign" (ibid., 111). Agamben's reference to Nazi biopolitics has been also criticized for the way in which he analyses the logic of the camp. Samuel Moyn (2010), in a recent intervention, has expressed concern for what he describes as a lack of (theoretical and historical) distinction between extermination camps and concentration camps in Agamben's description of the trasformation of the inmates into "homines sacri." At the same time, as rightly argued by Mazower, Agamben's project is "not interested in historical change but in what he sees as the deeper meaning, the potentiality, that interpretation may glean from certain historical occurences. His main concern is to find clues that will allow us to move toward redemption, to chart that course to a new politics, a way out of a fallen world" (Mazower 2008, 27).

It is with this last interpretation in mind that we suggest to move beyond Agamben's particular engagement with the camp and questions of history and to investigate instead how other elements of his philosophical endeavor can be of use for a reconsideration of genocide from an eminently spatial and geographical perspective. It is by now widely acknowledged that what Mazower (2008) describes as "Agamben's complex and sometimes obscure chain of thought" is largely based on his distinction between what he famously defines as "bare life" (*nuda vita*) and political life in the realm of biopolitics, and to the ways in which this very biopolitics is linked to a reinterpretation of Carl Schmitt's definition of the state of exception (Agamben 1998). What we will try to do here, instead, is to engage with two other, equally important, elements of Agamben's conceptual apparatus (often directly or indirectly linked to questions of space and spatial theory [Minca 2007]), drawing upon a well-consolidated methodological approach in human and political geography (see, among others: Clarke, Doel, and McDonough 1996; Cole 2003; Charlesworth 2004; Elden 2006). The historical events here recalled are well known and have been thoroughly discussed elsewhere in great detail. However, the purpose of revisiting them through the lens provided by the Agambenian concepts of *selva* and *città*, intended as geographical metaphors, is to gain new insight on the spatial and philosophical dimensions of Nazi geopolitics and biopolitics. Specifically, ours is an attempt to highlight how powerful spatial metaphors have contributed to the merging of the ideal and the factual realms of the Nazi geopolitical project; that is, how the anthropogenesis at the basis of Nazi biopolical ideology was translated into a set of spatial imaginations and spatial practices. Our chapter moves from the convinction that there is lot to learn from the ways in which thought and (imagined and material) space transversed the genocidal plans and practices of the Nazi perpetrators, by producing a specific geography of genocide, where (spatial) theory and the implementation of extermination came together.

In the original version of *Homo Sacer*, Giorgio Agamben presents the concepts of *selva* and *città* as the pillars of his theory of the sovereign ban (1995, 116–23). *Selva* is used in the dual sense of original "state of nature" *and* of actual forest; the Italian term *selva* implicitly recalls Servius's distinction between *lucus*, *nemus*, and *silva*: "*Lucus est arborum multitudo cum religione, nemus composita multitudo arborum, silva diffusa et inculta*" (King 1990, 227). *Città*, instead, is intended here, as for Agamben, as the locus of the *civitas* and of the *urbs* (Chignola 2007, 238). The *città* is not only the urbs, the city in literal terms. For Agamben (who, in that specific passage written in Italian, deliberately refers to the *città* in these broader terms), it is rather a principle of *communitas* and civilization. In this sense, the rural planning policies of the Nazis or even the wide-reaching plans for the reforestation of the occupied territories must be considered part of the production of the "*città*," of the spatialized attempt to build a racially purified *communitas*. The *selva*, then, is not *just* the forest, the woods; it is the spatialized principle of untamed nature against which the principle of a new German *communitas* must be imagined and, again, spatialized.

The *uomo-lupo*, the "werewolf," is the figure that Agamben adopts to explain the *soglia* ("threshold" in Italian) between this reading of *selva* and *città*: "a threshold of indistinction and of passage between animal and man, *physis* and *nomos*, exclusion and inclusion" (Agamben 1998, 105). The state of nature that the werewolf embodies is thus not an epoch that precedes the foundation of the *città*, but a principle internal to the *città* itself. Hence, Agamben's werewolf is a subject located between *selva* and *città*, excluded by but also constitutive of both. If *selva* and *città* are separated and defined by a true spatial threshold, then this space of indistinction is located in/on the body of the *ban*dit, of the werewolf "who dwells paradoxically within both while belonging to neither" (ibid.). It is here that Agamben situates the "survival of the state of nature at the very heart of the State" (ibid., 106), but also where the threshold between bare life and a-life-worth-living becomes visible.

This chapter intends to engage with these two spatial concepts in order to investigate how the broader geographies of the Holocaust and Nazi spatial racialized thinking were linked to the biopolitical project of crafting a German New Man (Fritzsche 2008, 150). Agamben's reflections on the relationship between the "human" and the "animal" are particularly useful here, since they allow us to conceive the "*selva*" as the real and metaphorical space/place where the "animal" was made to reside by modern political thought. This use of the concept helps to explain, we claim, why forests played such an important role in the production of the spaces of Nazi extermination. Drawing from this specific understanding of the *selva* we offer a geographical perspective on the links between the broader geopolitical project of realizing a greater purified and *judenfrei* German *città* and the spatial thinking behind the attempts to find a "territorial solution" to the Jewish question, then translated into the thanatopolitical spatialities of the Final Solution.

We begin by analyzing Agamben's reflections on the modern production of the "non-human in the human" and link them to his conceptualization of the *selva* as a fundamental space of the sovereign *ban*, but also as the place that the "non-humans" are supposed to inhabit and even embody. We then look at the spatialization of the *selva* within the grand territorial planning of the Third Reich, and in particular at the production of a "spatial rest" where the operation of separating, in the victims, the animal from the human took place. The third section concerns the perpetrators and their relationship with the *selva*; in particular, we reflect on the Nazis' need to momentarily a*ban*don their imagined *città*, in order to penetrate the *selva* together with their victims and perpetrate their murderous plans. A crucial question emerges here: if the *selva* is where the animal resides and where the labor of separating the non-human in the human must take place, how could the Nazi New Man penetrate this realm of primitive violence and, at the same time, remain fully "human" and a full citizen of the purified *Volk* in the name of which he committed murder? We will claim that this is where the attempted geographical separation of the *città* from the *selva* collapsed; not only did the perpetrators have to penetrate the uncertain threshold between the two in order to annihilate what remained of the "man-beast," to use Agamben's characterization, and put at risk their own "humanity," but these travels in the dark lands of extermination ended up reproducing the mutual penetration of *selva* and *città* and gave life to a vast space of exception where the separation between the two became literally impossible.

The Animal-in-Us

In *The Open*, Giorgio Agamben, inspired by Kojève's reading of Hegel, while questioning the existence of an original threshold between "man" and "animal" sets out what he describes as an "anthropophorous" animality; in the modern historicization of the human he instead identifies the negation of this very animality, that is, the attempt to separate, in a stable and possible permanent way, man from animal:

> Man is not a biologically defined species, nor is he a substance given once and for all; he is, rather, a field of dialectical tensions always already cut by internal caesurae that every time separate—at least virtually—"anthropophorous" animality and the humanity which takes bodily form in it. Man exists historically only in this tension; he can be human only to the degree that he transcends and transforms the anthropophorous animal which supports him, and only because, through the action of negation, he is capable of mastering and, eventually, destroying his own animality. (Agamben 2004, 12)

This process of "extraction of man" from the original anthropophorous animal translates into classifications and enumerations that represent attempts to speak the unspeakable, to capture what can never be captured by any language, never clearly defined (Agamben 2004, 13). Agamben thus suggests we focus our attention on the final product of this process, that is, its *resto* (rest), what remains of the modern biopolitical experiment of separating the human from the animal in order to create a sort of "total man," the citizen of a new paradise on Earth: "Perhaps the body of the anthropophorous animal . . . is the unresolved remnant that idealism leaves as an inheritance to thought, and the aporias of the philosophy of our time coincides with the aporias of this body that is irreducibly drawn and divided between animality and humanity" (ibid., 12). "The Jew," that is, "the non-man produced within the man" (ibid., 37), is a genuine product of what Agamben refers to as the "anthropological machine," a dispositive that establishes at its center:

> a zone of indifference . . . within which—like a "missing link" which is always lacking because it is already virtually present—the articulation between human and animal, man and non-man, speaking being and living being, must take place. Like every space of exception, this zone is, in truth, perfectly empty, and the truly human being who should occur there is only the place of a ceaselessly updated decision in which the caesurae and their rearticulation are always dislocated and displaced anew. What would thus be obtained, however, is neither an animal life nor a human life, but only a life that is separated and excluded from itself—only a bare life. (Agamben 2004, 37–38)

What is key to the argument of this chapter is the fact that this endless process of *dis-location* and *dis-placement* is the result of the *spatialization* of the work of the anthropological machine. The actual experiment of separating the human from the animal takes place on the human body, which is the very "territory" where the sovereign exception is translated into biopolitics.

> The division of life into vegetal and relational, organic and animal, animal and human, therefore passes first of all as a mobile border within living man, and without this intimate caesura the very decision of what is human and what is not would probably not be possible. It is possible to oppose man to other living things, and at the same time to organize the complex—and not always edifying—economy of relations between men and animals, only because something like an animal life has been separated within man, only because his distance and proximity to the animal have been measured and recognized first of all in the closest and more intimate place.
>
> But if this is true, if the caesura between the human and the animal passes first of all within man, then it is the very question of man . . . that must be posed in a new way . . . We must learn . . . to think of man as what results from the incongruity of these two elements, and investigate not the metaphysical mystery of conjunction, but rather the practical and political mystery of separation. (Agamben, 15–16)

If "the Jew" was the "rest" of the separation produced by the Third Reich's anthropological machine, and the camp an "extreme and monstrous attempt to decide between the human and the inhuman" (ibid., 22), then Nazi biopolitics must always be linked to a process of "localization," to a geography of real and imagined spaces *where* the virtual extraction of the animal from the body of the New Man and the parallel suppression of the human in the "man-beast" actually took place.

Nazi Spatial Politics and the Geographies of the *Selva*

In its extreme attempts to realize "a good life," a new, happy and perfect man, together with a new harmonization of "nature," the Third Reich tried to create a real topography of the separation between *selva* and *città* through the grand planning of an expansive German *Lebensraum*. This planning translated into an explicit geographical hierarchy of the occupied and/or annexed territories and produced a sort of "spatial rest," a real and imagined borderless *selva* separated from the Nazi Eden (Lower 2005, 186), from the reconciled and domesticated nature inhabited by the Aryan New Man. This *selva* was supposed to host and make invisible the "human rests" produced by the Nazis' endless categorizations and classifications of the "human," of all humans: the "Jew" *in primis*, a Jew dehumanized by the categories produced by a fictional racial continuum endlessly excised, articulated and rearticulated by the biopolitical/anthropological machine.

Nazi spatial politics of extermination were thus the expression, on the one hand, of an extreme rationalization of the concept of the "human" and of its cultural, biological, and, therefore, political definitions; on the other, of the (feared) reemergence of the "animal," both in the body of "the Jew" *and* in "the German," who, terrified by what this reemergence might mean—that is, the confirmation of the "non-existence" of a pure, civilized German—put an enormous effort into the progressive dehumanization and, eventually, destruction of the "Jew-turned-into-animal." The whole biopolitical machine at the core of the Nazi project seemed to be directed toward one main objective: the reduction of "the Jew" (and of other lesser humans) to the status of the animal, their return to primitive Nature. This was paralleled by the attempted biopolitical redefinition of the German people implemented by the programs of euthanasia; a major and very concrete experiment in the perfecting of Nature and, while getting rid of the monsters that its imperfection produced, in realizing a German Eden on Earth.

However, what the spatial deployment of this biopolitical project made immediately clear was that any attempt to translate an ideal *città* into a purified

Lebensraum could not live without the existence (somewhere) of a *selva*, of an "imperfected" (state of) nature. The *selva*-turned-into-real-space of the Nazis was a sort of "geographical-garbage-can," the other/outer space of a violent biopolitical geography. The *selva*-garbage-can was a "spatial rest," what remained of the *scarto* (gap, residual) (Agamben 1996, 25) between the "people" and the "population," that is, the people translated into a biopolitical body. In the ultimate German *civitas*, in the *Volk*, people and population had to coincide; the *selva*, instead, was where the "rest" of this operation could be figuratively *and* materially located and a*ban*doned or eliminated. Precisely for this reason, the *selva* could never disappear from the Nazi universe. The *selva* was indeed a crucial element of the operational strategy that aimed at realizing this hypothetical coincidence between the German people and population; it was the space-threshold in which the rest of this very operation ought to be *dis-located*, again, a real and metaphorical home for the biopolitical "human rest" – where the intensive labor of its extraction and elimination should take place. The (re)production of the "non-human," of Agamben's "*non-uomo*," required also a real location: again, the "geographical-garbage-can" produced by the spatial purification that aimed at translating the territories of the Third Reich into a *judenfrei* Eden.

Nazi spatial planning and spatial demographics—with their calculations of an orderly and geometric *Lebensraum*, with their endless racial categorizations and the related massive displacement of people/populations—played a fundamental role in the production of a Nazi *selva* on a grand geopolitical scale, a novel geography to be traced by the military occupation of a progressively larger part of Eastern Europe. In the first phase of the war, the *Generalgouvernement* seemed to represent a sort of ideal territorial garbage can, the "spatial rest" of the expansion of the Reich, an extended *selva* where the Jews formerly resident in Germany and in the annexed territories of Austria and Poland could be dumped. However, this role of the *Generalgouvernement* in the new geographies of the Reich was opposed by its governor Hans Frank, who wanted to convert it into a territory fully integrated within the new grand spatial order of the Nazi Empire. The new plans for the *Generalgouvernement*, submitted to Hitler in October 1940, helped to set the conditions for the identification, in the following months, of a new, even vaster, *selva*/garbage territory where the increasingly large "human rest" produced by the Nazi geopolitical/biopolitical machine, that is, the expelled Jewish population, could be deported.

Operation Barbarossa did not provide a "territorial solution" to this problem. Although initially the idea was to deport the Jews to the "Eastern lands" of the Asian continent and, literally, "a*ban*don" them there, in practice the expansion of the Nazi Empire to the vast spaces of the Soviet Union soon translated into a sort of growth by metastasis of the German *Lebensraum*. The expansion of the spaces to be purified and colonized made it necessary to link the conquered/occupied territories to the heart of the Reich through a colonial project founded on the creation of a new generation of Germans (or, lacking those, of "Germanized") settlers/soldiers for a radical reterritorialization of the "Eastern lands" (Burleigh 2000, 530–32). According to Hitler and the Nazi elite, Europe was supposed to become a vast civilized and Germanized space dominated by a new master race. This very project opened the ground for new calculative spatial rationalities; as noted by Götz Aly and Susanne Heim, "just one day before the German attack on the Soviet Union was launched, Himmler officially charged Meyer's planning department at the RKF with the task of drawing up a 'General Plan for the East.' This was intended as a blueprint for colonization and restructuring—not only for the whole of occupied Poland, but also for large expanses of the Soviet Union" (Aly and Heim 2002, 255). This project had a double effect: on the one hand, it was at the origin of a renewed effort in the attempt to fully "racialize" the new domains, as expressed by the biopolitical delirium that brought the Nazis to categorize each individual of the subjected populations in order to identify potential German or Aryan characteristics that would allow them to provide the *völkisch città* with fresh "human material of a good quality." On the other hand, in a Nazi Europe, putatively conceived as a comprehensive and definitive *città* in the making, there was to be no space left where to localize the *selva;* where to re-locate the "human rest" produced by the process of extraction and refinement of the New Man. In the long run, however, the Jews, but also all the other "leftovers" of the process of Germanization (among them, "Gypsies," "homosexuals," the disabled), were progressively to be deprived of any possible location/destination in the new Nazi world. Nobody wanted to host a growing mass of displaced Jews, no permanent *selva* seemed to be available anymore—neither in an expanded German Reich aiming at becoming a sort of total space, nor outside of it. There was literally no place to dump the "human rest" of the production of the new *città*, no manageable *selva* in sight anymore. This was the turning point at which the so-called "territorial solution" to the Jewish question began to be progressively translated into the dark horizon of an explicit politics of extermination (Browning 2004; for a more nuanced approach to the relation between the "territorial" and "final" solutions, see Longerich 2010). In geographical terms, this was the moment in which the spatial politics aimed at identifying a permanent localization of the *selva* external to the Reich—that is, the *Generalgouvernement*, Madagascar, Palestine, Siberia, etc.—was progressively replaced by the *punctual localization of the selva* in the real and imagined "forests" of Central and Eastern Europe *within* the newly formed empire.

This radical rescaling of the concept of *selva* in Nazi geopolitics is confirmed by the language that was used to describe the passage from a territorial solution to mass extermination. The secret protocols of the Wannsee Conference, held on January 20, 1942, show how high-ranking Nazi officials agreed that "emigration has now been replaced by evacuation of the Jews to the East. . . . The evacuated Jews will first be taken, group by group, to so-called transit ghettos, in order to be transported farther East from there" (Arad et al. 1999, 253, 256). In the Nazis' euphemistic jargon, the expression "farther East" denoted a space that canceled once and forever all ambiguity regarding the reterritorialization of Eastern Europe and that materialized into the emergence of the gas chamber. During the long phase of overlap between the "territorial" and the "genocidal" solutions, a new space of semiotic indeterminacy was created within which the project of extermination—the "surgical" removal of the Jewish presence in Europe—was expressed through the jargon of spatial planning. The massacres perpetuated during Operation Barbarossa were thus systematically described with the spatial language of the movement/relocation of populations (not only the Jews) and terms like *Umsiedlung* (resettlement), *Aussiedlung* (transportation) (Kogon 2000, 20–23), or *Entfernung* (distance, but also "expulsion" and "removal"—as, for a surgeon, the removal of a tumor),[1] reflecting a significant degree of continuity between the language of the territorial solution and the one adopted to describe the different stages toward the Final Solution, between regional and urban planning and biopolitical violence. In this process of *trans-lation*, the *selva* was no longer a potential permanent location of the "human rest" produced by the Nazi geopolitics, but, rather, a real and imagined space in which the violent tension implicit in the attempted realization of a purified German *Volk* could find full expression. The greater *città* envisaged by Himmler's General Plan for the East (*Generalplan Ost*) and planned by Meyer and geographer Walter Christaller (see, among others, Rössler 1989) required endless new "empty spaces" to fill with purified biopolitical substance—a new German *civitas* living in newly planned (mainly rural) *urbes*. The future spatial order of the Reich was to be the expression of a renewed (and utopian) equilibrium between the urban and the rural (see, among others, Schenk and Bromley 2003), opposed to the (spatial) degeneration produced by the uncontrolled expansion of modern cities and their suburbs, corrupted, as they were, by rootless cosmopolitan Jewry. In doing this, however, it also required a new *selva*-garbage-can where the remnants of this very spatial operation could be dumped. The real and metaphorical threshold between *selva* and *città* thus became a crucial battlefield for German biopolitical geopolitics.

The long-standing problem of "demographic density" (Aly and Heim 2002, 58–72) was then approached with an operative selection aimed at pushing the "lesser-human" toward the *selva* and their supposedly original animal condition, in order to inscribe on their bodies the battle for the survival of the fittest race. The proper place of the "Jew" became the *selva*, or better, its "border," where they could be reduced to the "rest" of this operation of "total humanization." The *selva* was therefore conceived as a space of exception where the human and the non-human were (re)produced, negotiated, and (re)localized. The interplay between *selva* and *città* became nothing more than a dialectic that produced ever new non-humans, in the perspective of the ultimate realization of a total, absolute New Man, the offspring of a new, dominant master race. The politics of extermination identified in the *selva*—a sort of necessary black hole in the apocalyptic geographies of the Nazis—was the fundamental principle of de-localization that the extraction of the animal in the human required. The no-longer-humans ought to be brought somewhere in order to translate their death into a "non-death," into the mere "production of corpses." The *selva*, in this geography of death, was therefore never really an-other space, never opposed to the Agambenian *città*: *selva* and *città* were, in fact, co-implicated in the production of the new *Volk*. The very existence of a real and imagined *selva* was what allowed the biopolitical machine to continue to function, precisely because the realization of a purified German *città* was an ideal geographical horizon, an imaginative space subject to endless purification. At the same time, the *selva*, by virtue of its spurious nature as a space devoid of humanity, could be identified for actions and functions related to this new total geography and constituted as an actual localization/site; that is, a crude spatialization of a monstrous state of exception.

It is not by chance, then, that extermination began in the forests of eastern Poland, Belarus, Ukraine, and the Baltic countries where the Einsatzgruppen accompanying the Wehrmacht performed the first vast systematic massacres of Jews and the local elites (Rhodes 2002). Also, it should not come as a surprise that the genealogy of the extermination camps was somehow linked to the forest. The Eastern European woods that witnessed mass murder and the Central European archipelago of camps appeared as the two "dislocating localization" (Agamben 1998, 175) of the *selva*—citing Agamben—right at the moment in which the initial aspiration of "dumping" the human rests in the open spaces of the "Far East" was rapidly replaced by the ambition of realizing a racialized colonial empire in the heart of Europe. The proximity and, in a sense, the continuity between the woods and the camp, played a crucial role in the production of the geographies of the Final Solution. From December 1941, in fact, the use of gas for mass murder began progressively to overlap and often replace the practice of shooting on the part of the Einsatzgruppen. The *Gaswagen*, the infamous gas chambers on wheels, became mobile thresholds between the forest and the camp (Kogon 2000, 72–97) and were soon transformed into key instruments in the operations of extermination implemented by the Einsatzgruppen and, eventually, in Chełmno, into the first extermination camp proper. The *Gaswagen* represented a true technical and symbolic passage in the shift from mass shooting to the industrial organization of murder; they were a *selva* determined by an uncertain topography, with no borders or location, which allowed for a new spatial mimetics of murder. The gaswagen was a true space-threshold onto which the victims were "loaded" (alive) in the *urbes* and "unloaded" (dead) in the periurban forests (ibid., 115).

Many camps maintained a sort of intimate (and functional) relationship with the forest. First, the forests surrounding the camps were often a sort of *tote Zonen*, a no man's land with the double function of hiding what was happening "inside" and making any attempt to escape extremely difficult; secondly, the original localization of many extermination camps was closely related to the existence/presence of a forest:

> two small wooden houses or peasants huts at the end of the footpath in the middle of a dense deciduous forest. . . . He [Wirth] first contemplated converting peasant huts into gas chambers by sealing them hermetically . . . "a primitive installation, consisting of a hermetically sealed shack hidden deep in the forest across from Galicia near Belzec" (Bodgan Musial as cited in Browning 2003, 205–6) in which gassing was tested. (Browning 2003, 28)

Also in Birkenau the first gas chambers were two converted peasants huts (Bunker 1 and 2, known as the white and the red cottages) in the forest located on the northern outskirts of Auschwitz's main camp (Pelt 2002, 267, 288). In addition, Birkenau literally meant "birch groves" and owed its name to the forest next to the extermination camp (Charlesworth and Addis 2002, 244).

Our claim is that the forest and the camp were closely related and that, in the geographies of genocide, the camp was a sort of punctual spatialization of the *selva*. Auschwitz-Birkenau was not only the pivot/fulcrum of the political topographies of German biopolitics, but it also represented, in our reading, the extreme spatial rationalization of the *selva*: the extermination camp was, in this sense, an extreme attempt to reproduce and, at the same time, "tame" the *selva* by rationalizing mass murder in a spatial laboratory where the human was constantly reconceived and separated from the animal.

It was the site where the ultimate non-human was produced as a remnant, as a human rest of the greater process of the realization of a new German *Volk*, purified by the very labor of performing and managing a "*selva*" marked by violence and the incumbent ghost of the return of the animal-in-us.

Travels in Darkness

In the biopolitical operation of extraction of the animal from the human, the Nazis not only had to closely approach "the Jews" to deprive them of their humanity, but also had to "penetrate" the anthropophorous nature of their victims and, quite literally, walk (with) them into the darkness of the *selva*. This New Man, in order to build an imaginary *völkisch civitas* devoid of any trace or presence of animality, had to transverse, through an endless series of "cuts/caesurae," the body of lesser beings whose lives were classified and manipulated to extrapolate the non-human. The *selva*, again, was the "natural" site for the final stage of this operation. However, once the *selva*, as part of the Final Solution, was translated into the extermination camp by the calculative rationalities of German geopolitics and biopolitics, approaching "the Jews" and entering with them into the *selva* lost its extemporaneous connotation and *became a permanent condition and location* (the camp) where the Nazi "integral human" was constantly exposed to the animal(ity) and the brutality of the *selva*. In violently performing the extraction of the-animal-in-the-Jew the perpetrators constantly risked revealing the animal-in-them and losing the very "total humanity" in the name of which they traveled to the darkness of the *selva* to commit murder. They were supposed to return, one day, to the safety of their purified *Volk*, finally rescued from the contamination of the *selva* that they created. The problem was that those who had traveled to the *selva* with the Jews were clearly exposed to the possibility of "remaining" in the *selva* forever, by becoming the beasts that they seemed to fear so much. It is as though "the-Jew-translated-into-animal" retained their perpetrators precisely in the same dimension in which both, albeit in dramatically different roles, found themselves once they penetrated together the primitive animal realm of the *selva*. This explains, perhaps, the concern often expressed by the higher ranks of the SS about the potential "animal forces" that the experience of mass murder could liberate in their own men: the *selva* could literally swallow up the New Man, who might never be able to return to the qualified citizenship/membership of their *völkisch civitas*.

One episode reported by Raul Hilberg bears witness to the attempts of the Nazi leadership to manage this dangerous threshold between the animal and the human in their own men, and its potential consequences in the accomplishment of the Final Solution:

> Once Himmler himself visited Minsk. He asked Einsatzgruppen B Commander Nebe to shoot a batch of a hundred people, so that he could see what one of these "liquidations" really looked like. . . . As the firing started, Himmler was even more nervous. During every volley he looked to the ground. . . . Himmler was visibly moved and decided to make a speech to all who were assembled there. He pointed out that the Einsatzgruppen were called upon to fulfil a repulsive (*widerliche*) duty. He would not like [it] if Germans did such a thing gladly. . . . Himmler told the men to look at nature. There was combat everywhere, not only among men but also in the world of animals and plants. . . . Didn't bedbugs and rats have a life purpose also? Yes, but this has never meant that man could not defend against vermin. . . . At the same time Himmler asked Nebe "to turn over in his mind" various other killing methods more human than shooting. . . . At the last, however, the RSHA technical unit (II-D) went to work in order to devise a different killing method, and the result of that experimentation was the gas van. (Hilberg 1961, 218–19)

It was as if, forced to de-localize the "rest" of this process of production of non-humans, the perpetrators themselves desperately tried to avoid being entirely (and irreversibly) absorbed by the tenebrous embrace of the *selva*, in order to retain some fundamental link to the dream of the *völkisch civitas* whose new foundation was the very reason why they penetrated so deeply into the dark labyrinths of the *selva*. The grotesque decorations of the gas van or of the gas chambers—together with the endless production of imaginative technocratic and bureaucratic jargon—certainly had the aim of deceiving the victims, but also possibly the more subtle objective of exorcising the terror/horror of the beast that was liberated by those very actions. The perpetrators had to be reassured that they were still human; they had to be reminded that, at the end, the horror of the *selva* would be left behind to return "home"; that these travels into the deep darkness of the non-human were necessary precisely to annihilate the *selva* that was among them and that threatened the ultimate realization of the German Eden on Earth.

This perception of the *selva* as a dangerous place for the German New Man, as a place of darkness where he could be retained forever, also played a central role in the structuring of the functions of the extermination camp. This explains, at least in part, the attempts to get the members of the SS away as far as possible from the direct management of the everyday practices of degradation and extermination of the victims, by delegating most of the tasks and the rituals related to the de-humanization and de-subjectivation of the inmates to other inmates, as was the case with the *Sonderkommando* in charge of the cleaning up of the gas chambers and the elimination of the corpses (see, among others, Venezia 2009). However, despite these attempts to keep the *selva* separate from the *città*—both through a specific spatial ordering of the camp and the "moral" code that sanctioned the SS for any form of sadistic behavior or familiarity with the victims—the fear of the reemergence of the "animal" in humans that were supposed to be racially purified was never entirely eliminated by the industrial management of the Final Solution. This zone of indistinction that the perpetrators created, and in which they found themselves, reflects the suicidal nature of the most ambitious biopolitical operation ever conceived: the attempted creation of a New Man who, to be "fully human" (that is, German), ought to (temporarily) return to his animal condition and enter into that very same *selva* he feared so much.

One key episode of strategic "return" of the forest-as-*selva* into the Nazi universe is worth mentioning. The Belarussian woods are where, since the end of 1941, Soviet and Polish partisans, together with many Jews, escaped from persecution and found refuge, and where they began to organize an active resistance against the occupation. The Nazi elite felt that, in order to take care of this unfinished job, they had to "return" to the forest: the biopolitical selection/cut operated on the body of the occupied populations had in fact produced these untamed "human rests," banned werewolves whose existence could potentially threaten the edification of a brave Nazi world. A key question was who should *re*-enter the *selva* in order to finally "*de*-forest" the new German lands. Tellingly, in February 1942, Himmler assigned this task also to the infamous Sonderkommando Dirlewanger (see MacLean 1998 and Ingrao 2006), named after its leader, and created in 1940 by Himmler himself. While this decision was part of a broader (and numerically much more significant) strategy to combat the partisans, the symbolic relevance of the Sonderkommando Dirlewanger in the attempted realization of a purified German *città* in the occupied Eastern territories is not to be underestimated. The overall idea was to give life to a special elite corps composed of poachers (and criminals) capable of operating in extreme and adverse conditions. The practice of enrolling hunters and poachers in special corps was not new to European military history; however, the deployment of this Sonderkommando went beyond the need for local "territorial" expertise, as was the case in past instances. By sending this group of "forest men" to fight the partisans in the woods, the high ranks of the SS tried to radically face the problem of the persistence of the *selva* in their universe; that is, of how to exercise the "bestial" violence required without indefinitely losing their "humanity."

The emphasis placed by both Hitler and Göring, on several occasions, on the passion that animates the hunter reveals how what was really at stake there, again, was the threshold between the human and the animal in the construction of the Nazi New Man (Ingrao 2006, 116–17). According to Ingrao, in the European *imaginaire cynégétique*, an excess of *sang noir* is what drives the hunter in seeking, in the violent experience of hunting, a sort of link between the human and the animal. However, if the hunter's violence and passion moved beyond the limits of what was normally accepted by civil society, then he became a poacher, a man of the *selva*, a sort of werewolf, somehow closer to his prey than to the *civitas* whose rules he transgressed. Here is a crucial point: extreme violence—that for the Einsatzgruppen ought to be practiced with detachment, as a sort of duty in the name of their civilized community (Stone 1999)—for the Sonderkommando Dirlewanger was an expression of their passion and of the animality they embodied.

> From the Nazi standpoint, evoking the *cynégétique* passion and cruelty was meant to bring back the *Sauvage* into the city at war. . . . What was allowed in the forest-like margins, liminal and disputed, of the millenary empire to be erected, it was not in the Nazi city. . . . The former [the *cynégétique* violence] would freely take place in sectors/spaces whose marginality was threefold. They were marginal in that topographically localized in the extreme East of the empire; they were also marginal as they were the settings of a war escaping the norms of the classical confrontation (*affrontement*). They were marginal, finally, because they acted in undomesticated spaces, where the German "mission civilisatrice" had not accomplished its tasks yet. (Ingrao 2006, 123–24)

As part of the attempted domestication of the "animality" exposed by the unruly actions of this cohort of werewolves, the brigade was progressively integrated with members coming from other marginal social groups: criminals, "a-socials," SS and Wehrmacht members expelled for disciplinary reasons, Russian deserters, and even political prisoners "reeducated" in concentration camps (Ingrao 2006, 21–61). This Sonderkommando became a unique consortium of "remnants" of the Nazi project, a strange regiment of marginalized people. The strategic "inclusion of the excluded" somehow represented a sort of "return of the animal" among the "humans" of the Sonderkommando, which liberated an escalation of extreme violence toward partisans and common citizens, but also within the brigade itself, with a sharp increase in episodes of coercion, desertion, and even execution from the summer of 1943 onward (ibid., 101–13). In this process of "retained animality," the Sonderkommando was progressively banned from the spaces of the Germanized *città* and confined to the *selva*, as shown by the documented clashes between Oskar Dirlewanger and the civil authorities, and by the attempts to expel the brigade from Lublin and the *Generalgouvernement* (MacLean 1998, 60–63). Nonetheless, in August 1944, the Sonderkommando Dirlewanger entered Warsaw with the task of repressing the insurrection organized by the Polish resistance (ibid., 175–98). This "return to the *urbs*," however, did not imply that the poachers and their unruly comrades were admitted to the *civitas* again; rather, it was the sign that, in Warsaw, *selva* and *città* were collapsing into a space-threshold of indistinction. The "mission" and the modus operandi of this special corps transformed "Varsovie insurgée en forêt urbane, en terrain de chasse pour la Sonderbrigade" (Ingrao 2006, 178). Not only the *un-localizable location* of the *selva* forced the Nazi New Man to return to the forest, but, in the last phases of the war, the *selva* (in the form of this unruly brigade) had, quite literally, to reenter and penetrate the purified space of the *città*.

Selva in *Città* / *Città* in *Selva*: The Collapse of Nazi Topographic Imaginations

In the previous sections, we reflected on the tension between the perfect world envisaged by Himmler and his peers—a *città* devoid of the *selva*, a geometric Eden filled with urbanized/civilized German humanity—and the fact that its realization required the permanence of the *selva*, of a (real and metaphorical) "outer space," a sort of counterworld compared to the geographies of permanent peace imagined by Nazi ideologues. This tension materialized on the border between these two worlds, and it is here that the future destiny of a Germanized Europe was negotiated. The Nazi territorialized Eden was supposed to be a garden carved out by harmonic geometries, inhabited by content and blond Aryans, where a de-forested *città* was finally at peace with a tamed nature. In this millennial picture, the *selva* was first imagined to persist somewhere far away, but when, with the evolution of the war, the spatial metaphor of the "Farther East" materialized in the extermination camp, the *selva* became, literally, a no man's land: only the *città* should have survived, inhabited by humans devoid of animality.

As Agamben would contend, any attempt permanently to isolate the animal from human life is doomed to failure. The greater Nazi geopolitical and biopolitical project was theoretically flawed from its outset: in the attempt to realize a Europe dominated by a German master race, the Jews became the "human rest" of a biopolitical "de-forestation" that could never be completed. The events around the Sonderkommando Dirlewanger bear witness, among many others, that the *selva*, the animal-in-us, fully survived the bio-geopolitical cuts operated by the Nazis in the European body politic. The "primitive" animality attributed to "lesser people" by Nazi demographic categories was, in fact, alimented by the bestial violence inherent in the very operations that were at the base of these endless caesurae of the population's body. The "rest" of these operations never left the real and imagined Nazi *città*; it was a constitutive element of its fabric, its hidden sewage system, its deep root ready to reemerge in the personae of the perpetrators and in the spaces of extermination.

The ultimate failure of the Nazi project was already present *in nuce* in the practice of deporting foreign workers (mainly Poles and Russians, but also French, Italian, and other Western nationals) to the very heart of the Reich (see Herbert 1997). The final result of this apocalyptic biopolitics was that *selva* and *città* ended up cohabiting in the very heart of the "Aryanized" geographies of the Third Reich. This is confirmed by the fact that the copenetration between *selva* and *città* was deeper and more permanent in the regions that, since the beginning, played a fundamental role in the geographical imaginations of the Nazi *Lebensraum*, that is, in Poland's western territories annexed to the Reich. The cities of Oświęcim and Łódź, in particular, became at once pivots of Nazi regional and urban planning *and* key sites (in the camp of Auschwitz-Birkenau and Łódź's infamous ghetto) for the production of biopolitical "human rests." They were thus assigned German names—Auschwitz and Litzmannstadt—and became literally "central locations" (in a Christallerian sense) in the spatial reorganization of that greater region. Their progressive "Germanization" aimed at realizing a new kind of German *città* in two locations that, while traditionally inhabited also by Germans, were identified above all as important clusters of Jewish presence (see, among others, Steinbacher 2005; Horwitz 2008).

Auschwitz/Oświęcim and Litzmannstadt/Łódź thus became the localization of both *selva* and *città* in the heart of the Nazi Eden shaped by the Final Solution. What is more important, this dual localization was neither temporary nor topographically separated, as it is often believed. The coexistence of "German" civilized space and primitive (human) nature (the Jews, the camp, the ghetto) lasted up to the very end of the war, giving life to a sort of gigantic permanent state—and space—of exception. Indeed, the camp and the ghetto were never radically separated from the *urbs*—despite the presence of endless gates, fences, and barbed wires in the Nazi concentrational universe (see Sofsky 1997, 55–64). Quite the contrary: they were kept distinct and, at the same time, integrated by a mobile spatial threshold where *selva* and *città* often overlapped and became indistinguishable. The construction of the *selva*-translated-into-an-extermination-camp in Auschwitz-Birkenau, for example, merged with the project for the realization of a greater *urbs* (see Dwork and Pelt 2002, 236–75), and produced a sort of hybrid space marked by uncertain borders and mobile geographies—this was the secret cipher of the relationship between *selva* and *città* as conceived by the Nazi geographical imagination. It was a relationship marked, at once, by fear and fascination. Otherwise, how can we explain the "exotic" tours of the ghettos organized by Germans eager to gaze at the "monkey men," the monsters produced by the biopolitical purification associated with Nazi spatial and demographic planning? The ghetto and its "banned" inhabitants were, in other words, rendered an exotic and fearful mirror of the human condition that the new *civitas* wanted to exorcise, expel, and isolate—the terrifying *selva*-in-us.

If we accept, with Agamben, that the *ban* is an intimate and violent *exclusive inclusion* (see Agamben 1998, 107), and that the sovereign decision is based on the mobile caesura that determines the threshold between *selva* and *città*, then their mutual "co-penetration" in the spaces here discussed shows how the spatial theory behind the project of a greater and purified German *Lebensraum* was flawed and untenable. The attempted selective animalization of the victims, in fact, ended up producing new forms of violent animality on the part of the perpetrators. This may also help explain the many episodes of irrational anger toward the victims, even in the last months of the war; it was as if the perpetrators knew that there was no possible return to a purified *città*, since the brutal violence of the *selva* that they feared so much had irreversibly penetrated their bodies and minds.

During the infamous Death Marches, in the final moments of the Reich, this fear took perhaps its most dramatic expression: wolves turned into cruel shepherds leading masses of "non-humans" in a meaningless march through a devastated land of violence and disorder, through the decomposing Nazi Eden, an immense *selva* with no rule and no *città*. Tellingly, the guards, despite the lack of clear instructions, kept on torturing and murdering the marching inmates during this endless, pointless performance of sovereign power (see Blatman 2003, 239). It is difficult to think of a more illustrative scene of the vanishing dream of a perfected German *civitas* devastated by the violent reemergence of the bestiality of primitive nature in the German body politic. During the march, many guards acted as real beasts: the only thing keeping them together was the possibility of killing and operating with cruelty—the animal that reemerged in them after having explored the darkest spaces of the *selva* did not a*ban*don them. After having experienced the end-of-man (of the human), they were traveling through a final indistinct land with no meaning and no direction, only pure violence. The safe shores of the *città* were no more.

What the ideologues of the Nazi Eden overlooked was that their envisaged radical separation between *selva* and *città* was pure (and poor) academic fiction. There is no Paradise on Earth; the human-in-us and animal-in-us are *always co-implicated*, with no "rest." That explains why the "human rest" produced by their lethal biopolitical machine was what they feared the most: that "rest" was their own mirror, the embarrassing residual of an impossible project, the witness of the animal-in-them, and the end of the millennial horizon for a self-appointed master race. This also explains many expressions of self-destruction; for the true believers, the German people could either be one thing with the *völkisch città*—or would not be. The predatory actions of gangs of Nazis and criminals—in some cases self-described as *werewolves* (Biddiscombe 2000, 20)—in the final phases of the war are possibly the most vivid example of the final sanction of the dark side of nature mobilized by Nazi biopolitics. In the arcane and tenebrous landscapes of an idealized *città* that was not to be, these gangs operated a sort of "self-banning" and a return to an "animal" condition through parasitical forms of survival, typical of an apocalyptic view of life and society. It is precisely here that the entanglement of the *selva* and the *città* appears to be, at the same time, the driving force of the genocidal project and its ultimate point of breakdown, something that can perhaps be found also at the core of other genocidal practices. The primitive nature reawakened by the Nazis' de-humanization of others seemed to have definitely colonized their own bodies and minds, shocked by the vanishing image of a *città* lost forever. All that remained was the cruel (animal?) immediacy of the *selva* and its precarious political economy of theft and rape: "The transformation into a werewolf corresponds perfectly to the state of exception, during which . . . time the *città* is dissolved and men enter into a zone in which they are no longer distinct from beasts" (Agamben 1998, 107).

Note

1. The same ambiguity is found in *Ausrotten*, that translates as "to destroy," but also "to stamp out," "to eradicate" (see Shermer and Grobman 2009, 205–8).

References

Agamben, Giorgio. 1995. *Homo Sacer*. Turin: Einaudi.
———. 1996. *Mezzi senza fine*. Turin: Bollati Boringhieri.
———. 1998. *Homo Sacer*. Stanford, CA: Stanford University Press.
———. 2004. *The Open*. Stanford, CA: Stanford University Press.
Aly, Götz, and Susanne Heim. 2002. *Architects of Annihilation*. London: Phoenix.
Arad, Yitzhak, Yisrael Gutman, Abraham Margaliot, Lea Ben Dor, and Steven T. Katz. 1999. *Documents on the Holocaust: Selected Sources on the Destruction of the Jews of Germany and Austria, Poland, and the Soviet Union*. Lincoln: University of Nebraska Press.
Biddiscombe, Perry. 2000. *The Last Nazis*. Stroud: Tempus.
Blatman, Daniel. 2003. *Le marce della morte*. Milan: Rizzoli.

Browning, Christopher. 2003. *Collected Memories*. Madison: University of Wisconsin Press.
———. 2004. *The Origins of the Final Solution*. Lincoln: University of Nevada Press.
Burleigh, Michael. 2000. *The Third Reich*. New York: Hill and Wang.
Charlesworth, Andrew. 2004. "The Topography of Genocide." In *The Historiography of the Holocaust*, edited by Dan Stone, 216–51. Basingstoke: Palgrave Macmillan.
Charlesworth, Andrew, and Michael Addis. 2002. "Memorialization and the Ecological Landscapes of Holocaust Sites." *Landscape Research* 27 (3): 229–51.
Chignola, Sandro. 2007. "Civis, Civitas, Civilitas: Translations in Modern Italian and Conceptual Change." *Contributions to the History of Concepts* 3 (2): 234–53.
Clarke, David B., Marcus A. Doel, and Francis X. McDonough. 1996. "Holocaust Topologies: Singularity, Politics, and Space." *Political Geography* 15 (6–7): 457–89.
Cole, Tim. 2003. *Holocaust City: The Making of a Jewish Ghetto*. London: Routledge.
Dwork, Debórah, and Robert Jan van Pelt. 1996. *Auschwitz, 1270 to the Present*. New York: Norton.
Elden, Stuart. 2006. "National Socialism and the Politics of Calculation." *Social and Cultural Geography* 7 (5): 753–69.
Fritzsche, Peter. 2008. *Life and Death in the Third Reich*. Cambridge, MA: Belknap Press of Harvard University Press.
Herbert, Ulrich. 1997. *Hitler's Foreign Workers*. Cambridge: Cambridge University Press.
Hilberg, Raul. 1961. *The Destruction of the European Jews*. Teaneck, NJ: Holmes and Meier.
Horwitz, Gordon J. 2008. *Ghettostadt: Łódź and the Making of a Nazi City*. Cambridge, MA: Belknap Press of Harvard University Press.
Ingrao, Christian. 2006. *Les chasseurs noir: La brigade Dirlewanger*. Paris: Éditions Perrin.
Kay, Alex J. 2006. *Exploitation, Resettlement, Mass Murder*. New York: Berghahn.
King, Richard. 1990. "Creative Landscaping: Inspiration and Artifice in Propertius 4.4." *Classical Journal*: 225–46.
Kogon, Eugen. 2000. *Les chambres à gaz secret d'État*. Paris: Éditions de Minuit.
LaCapra, Dominick. 2004. *History in Transit*. Ithaca, NY: Cornell University Press.
Longerich, Peter. 2010. *Holocaust*. Oxford: Oxford University Press.
Lower, Wendy. 2005. *Nazi Empire-Building and the Holocaust in Ukraine*. Chapel Hill: University of North Carolina Press.
MacLean, French L. 1998. *The Cruel Hunters: SS-Sonderkommando Dirlewanger, Hitler's Most Notorious Anti-Partisan Unit*. Atglen, PA: Schiffer Publishing.
Mazower, Mark. 2008. "Foucault, Agamben: Theory and the Nazi." *Boundary 2* 35 (1): 23–34.
Minca, Claudio. 2007. "Agamben's Geographies of Modernity." *Political Geography* 26: 78–97.
Moyn, Samuel. 2010. "In the Aftermath of Camps." In *Histories of the Aftermath*, edited by Frank Biess and Robert G. Moeller, 49–64. New York: Berghahn.
Pelt, Robert Jan van. 2002. *The Case for Auschwitz: Evidence from the Irving Trial*. Bloomington: Indiana University Press.
Rhodes, Richard. 2002. *Masters of Death*. New York: Knopf.
Rössler, Mechtild. 1989. "Applied Geography and Area Research in Nazi Society: Central Place Theory and Planning, 1933 to 1945." *Environment and Planning D: Space and Society* 7 (4): 419–31.
Schenk, Tilman, and Ray Bromley. 2003. "Mass-Producing Traditional Small Cities: Gottfried Feder's Vision for a Greater Nazi Germany." *Journal of Planning History* 2 (2): 107–39.
Shermer, Michael, and Alex Grobman. 2009. *Denying History*. Berkeley: University of California Press.

Sofsky. Wolfgang. 1997. *The Order of Terror: The Concentration Camp*. Princeton, NJ: Princeton University Press.

Steinbacher, Sybille. 2005. *Auschwitz: A History*. New York: Ecco.

Stone, Dan. 1999. "Modernity and Violence: Theoretical Reflections on the Einsatzgruppen." *Journal of Genocide Research* 1 (3): 367–78.

Venezia, Shlomo. 2009. *Inside the Gas Chambers*. Cambridge: Polity Press.

Witte, Peter, Michael Wildt, Martina Voigt, Dieter Pohl, Peter Klein, Christian Gerlach, Christoph Dieckmann, and Andrej Angrick. 1999. *Der Terminkalender Heinrich Himmlers 1941/42*. Hamburg: Christians.

12

Geographies of Ghettoization: Absences, Presences, and Boundaries

TIM COLE

Where do you begin an essay on geographies of ghettoization? On which side of the ghetto wall do you stand—the outside looking in, or the inside looking out? Those questions of our own positionality as we approach these landscapes are more than merely self-reflective posturing. Where we position ourselves—outside or inside the ghetto wall—draws on, and contributes to, different historiographical traditions on the ghettos established in occupied Europe as one key landscape of the implementation of the so-called "final solution of the Jewish question." Reflecting broader divisions within academic writing on the Holocaust, in pioneering work scholars tended to approach ghettoization and the ghetto from either the perspective of the motivations of Nazi planners or the experiences of segregated Jewish communities (Gringauz 1949; Hilberg 1961). Given the focus of this book, which foregrounds the geographical imagination within Nazi ideology and practice, my intention in this essay is to hone in primarily on perpetrator perspectives, although ghettoization was, as I hope to show, something that impacted both Jews and non-Jews.

However, even given that starting point of thinking about ghettos as aspects of Nazi policy, I am still left with the question of where to start—on the outside looking in, or the inside looking out—framed around a further set of historiographical divisions. Ghettos as demarcated, concentrated, and segregated spaces reflected an attempt to rid an area (variously other parts of the city, other towns and villages, rural areas, the Reich) of Jews, *and* to concentrate Jews within a single controlled space. In short, ghettoization was both an act of creating spaces of Jewish absence—literally making parts of the urban and rural landscapes *judenfrei*—and an act of creating spaces of Jewish presence within the walls of the ghetto. The relationship between these two (sides of the wall) was not straightforward. Of course ghettoization involved both. The act of moving Jews to a chosen spot meant that they no longer lived in other places. The act of clearing Jews from a variety of places meant that they had to be relocated somewhere. However, the priority of one over the other was something that shifted and changed on the ground from place to place as ghettoization was implemented in the locality (Cole 2003, 2004).

But, and here is the nub of my question of whether to start outside the ghetto wall (conceptualizing the ghetto as creating *judenfrei* space) or inside the ghetto wall (conceptualizing the ghetto as creating Jewish space), positioning ourselves by that wall is to place ourselves within far wider debates about the nature of Nazism and the motivations behind the attack on European Jews (and other groups). At the danger of reducing a complex literature to simplistic binaries, an approach which situates the Holocaust within a broader context of Nazi imperial expansion tends to start outside the ghetto wall, seeing ghettoization as one element of ethnic cleansing and the creation of German space within the East (Bloxham 2009). In contrast, focusing on ideological anti-Semitism tends to hone in on the world within the ghetto wall, seeing ghettoization as part of a making concrete of anti-Semitic perceptions of the *Ostjuden* that the Nazis encountered as they occupied the East (Michman 2011).

In this chapter, I move between the two sides of the ghetto wall, starting with the absences created in the world outside the ghetto before focusing on the presences created within the ghettos and exploring the interplay of absences and presences in two cities. But I also interrogate the ghetto wall itself, suggesting that we do well to examine the nature of this boundary in past time / past place. Here I want to explore ghettoization as the exercise of power through space and yet also to examine the ways that things are complicated when we shift from the spatial abstract to historical ghetto places where segregation and concentration played out in complex ways.

Absences (or Starting Outside the Ghetto Walls)

Ghettoization, in part at least, meant making space *judenfrei*. Ghettos were not only sites of gathering together Jews but also were a product of a wider reimagining and cleansing of space. This clearing of Jews took place at a variety of scales from the national, through the regional, to the urban. At the national scale, German Jews deported eastward as the Reich was made *judenfrei*, were dumped into ghettos in the East. At the regional scale, ghettoization often meant a clearing of rural areas, with Jews placed into cities. Certainly in the Hungarian case, the key criteria for choosing ghetto sites in

1944 was whether a town or city had a population of more than 10,000. In Hungary ghettoization amounted to the forced urbanization of the Jewish population (Cole 2011, 57–59). Moving to a local scale—and it was ultimately at a local scale that the precise siting of ghettos was discussed, contested, and determined—specific districts and streets in the city were cleared of Jews. In Łódź, ghettoization was part of a much broader reimagining (and renaming) of this city not as a Polish (and Jewish) city called Łódź, but a German city called Litzmannstadt (Horwitz 2008). Here ghettoization was part of a broader reshaping of the world into "Jewish," "Polish," and "German" space within a context not only of German occupation but also German imperialist expansion that sought to reshape the physical and imaginative rural and urban landscape (Dwork and Pelt 1994, 127–59).

But there was more to that reshaping of the city into "Jewish," "Polish," and "German" space than merely carving up Łódź into arbitrary zones. As Raul Hilberg was aware, the decision to site the ghetto in "a slum quarter" was fueled by concerns with appropriation (Hilberg 2003, 224; Smith 1994, 258–61). As Hilberg explained,

> As the Jews moved into the ghetto, they left most of their property behind. This "abandoned" property was confiscated. It can readily be understood now that the choice of the ghetto location was of utmost importance to the success of the operation. As a rule, the preferred ghetto site was a slum, for in that way the better houses, apartments and furniture were left behind. But this solution also had its difficulties, because the slums were often filled with warehouses and factories. (Hilberg 1961, 241–42)

Not only were officials concerned with where the ghetto was sited, but also where it was not. They were keen to ensure that those parts of the city with "the better houses, apartments, and furniture" became available for ethnic Germans resettled within the expanding Reich.

Situating ghettoization as act of imperialist clearing characterized the early so-called "functionalist" historiography of the Holocaust that saw the creation of ghettos not as forerunner to a policy of extermination as characterized the early "intentionalist" scholarship of Philip Friedman, but the freeing up of real estate (Friedman 1954). For Hans Mommsen, "the primary motive" behind ghettoization, "was revealed in the executory provisions of Himmler's decree for the *Reichsgau Wartheland*: 'The purging and protection of the new German areas' [which] was designed to provide housing and employment prospects for the ethnic German settlers" (Mommsen 1991, 243). Once established, these ghettos presented increasingly intolerable conditions and became a place where "functionalist" historians saw the radicalization of anti-Jewish policy taking place. In Martin Broszat's words, "epidemics and a high mortality rate [in the ghettos] suggested the possibility of 'helping nature along' in a systematic fashion" (Broszat 1970, 97). As is clear, the starting point for this historiography was very much outside the ghetto walls, focusing on the priority of clearing and emphasizing the significance of the cleared spaces left behind by ghettoization in explaining Nazi motivation.

These interpretations were challenged in an important intervention by Christopher Browning, who signaled the shifting interplay of concerns on the ground in Polish cities and the view from Berlin. While advocating a need to think local about why—and where—Polish ghettos were set up, Browning argued for a need to move from the periphery to the center in explaining the radicalization of policy on the ground. Taking the important ghettos in Łódź and Warsaw as examples, Browning suggested that while both "were created at different times for different reasons" (as "a strictly temporary device for extracting Jewish wealth," and to limit the spread of epidemics respectively), once in existence the German authorities faced "identical problems": the "imminent starvation" of those living in these closed ghettos (Browning 1986, 347–48). Local Nazi officials differed on their responses to this problem ghettoization threw up, with those Browning dubbed "productionists" arguing for the rational exploitation of the productive potential of the ghetto population and those he dubbed "attritionists" arguing for deliberate attrition (ibid., 355). While "productionist" arguments prevailed in both places in the fall of 1940 (Łódź) and spring of 1941 (Warsaw), these were replaced during late 1941 and into 1942 by policies from Berlin that spelled the liquidation rather than maintenance of ghetto populations following the decision to kill Europe's Jews. However, within this chronology of shifting meanings being given to Polish ghettos across 1940–42, there remains a place to imagine ghettoization—especially in its earliest stages—as not simply an act of gathering together Jews, but also a policy of redistributing urban space between Germans, Poles, and Jews. In short, ghettoization was not only about making ghettos but also about remaking the remainder of the city along racialized lines.

Central to Christopher Browning's intervention was an emphasis on working at the local scale of individual cities in understanding the motivations behind the setting up and siting of individual ghettos, given that, "ghettoization was in fact carried out at different times in different ways for different reasons on the initiative of local authorities" (1986, 344–45). Here he differed from an earlier—"intentionalist"—historiography that laid greater stress on Heydrich's order to the chiefs of Einsatzgruppen units in Poland in September 1939 to concentrate Jews in those cities with a Jewish population of more than 500 close to railway junctions. Taking this order as a centralized plan for ghettoization as forerunner to deportations, Philip Friedman argued that ghettoization was a self-conscious "step towards genocide" (Friedman 1954, 61). However, while such arguments for central planning at this early stage have been replaced with greater focus on initial localism, there is a literature that focuses on ghettoization as an act of gathering Jews together and not simply an act of clearing living space for ethnic Germans or other non-Jewish population groups.

Presences (or Starting Inside the Ghetto Walls)

This work brings us inside the ghetto walls to consider the ways in which ghettos not only created *judenfrei* space beyond those walls but also Jewish space within those walls. Here, the recent work of Dan Michman is significant in positioning ghettoization as an act of making concrete preexisting anti-Semitic perceptions of the *Ostjuden* that the Nazis encountered in Poland. Arguing for the influence of Peter-Heinz Seraphim's 1938 book *Jewry in the Territory of Eastern Europe* on Nazi policy makers, Michman suggests that Jewish quarters within Polish cities were seen as threatening spaces. Thus, ghettoization, he suggests, was not an act of "*ex nihilo* creation, because . . . ghettos, as the Germans understood them, already existed . . . the Germans merely demarcated their boundaries" (Michman 2011, 74). While Seraphim's book may have been "on the desks of German administrators," I remain less convinced than Michman that despite Seraphim being "not of high enough rank to dictate the extent to which the idea of concentrating Jews in ghettos would be implemented; nor was he ever actively involved in these efforts at ghettoization . . . I believe, nevertheless, that his logic, as well as his understanding of the concept of ghettoization . . . lay at the basis of the actions taken on the ground" (ibid., 62, 94–95). Michman's "belief" in Seraphim's influence in the nitty-gritty of where to site ghettos on the ground remains unsubstantiated.

However, where this turn to look from the inside out is important is in highlighting the connections—and disconnections—between prewar Jewish demographic patterns in the city and the wartime location of ghettos. In one sense ghettoization was relatively simple. It did not involve building a camp from scratch, but simply drawing lines on a map of the city and then the building of walls around that demarcated space. However, in another sense, ghettoization was extremely complex. Those walls were not built on *terra nulla* but rather in towns and cities with their own complex histories and geographies of Jewish and non-Jewish residential and business patterns. While Jews lived in larger or smaller proportions in different parts of towns or cities, they nowhere made up the entire population of a street, let alone a district, as ghettoization dictated.

Placing ghettos within longer-running histories and geographies of Jewish residence within urban centers raises a key question to the fore. In particular, it points to a historiographical clash that is at least implicit in the literature. For Michman, ghettos were a hardening and concretizing of anti-Semitic notions of the Jew's place within the city or nation. By contrast, Hilberg's positioning of ghettoization as process within a wider model of the evolution of a machinery of destruction suggests that ghettoization can be seen as one important aspect of the broader process of constructing the "Jew" in and through space. Given his European-wide analysis, Hilberg sought a catchall category that covered both those places in the occupied East where physical ghettos were constructed, and those places in the West where they were not. Thus ghettoization, for Hilberg, was about more than simply constructing a segregated place, but involved the five-step process of social isolation, crowding into "special houses," movement restrictions, identification measures, and "the institution of a Jewish administrative apparatus through which the Germans exercised a stranglehold on the Jewish population" (Hilberg 2003, 180). Ghettoization was about identification and isolation through imaginative, institutional, and spatial separation. Here it fits within a broader context of what David Sibley has described as,

> a history of imaginary geographies which cast minorities, "imperfect" people, and a list of others who are seen to pose a threat to the dominant group in society as polluting bodies or folk devils who are then located "elsewhere": This "elsewhere" might be nowhere . . . or it might be some spatial periphery, like the edge of the world or the edge of the city. (Sibley 1995, 49)

Such approaches suggest that we read ghettoization as act of imaginative as well as physical removal framed less around the material impulses envisioned within a focus on the creation of *judenfrei* real estate than a more ideologically driven concern to rid the nation (and so the city) of Jews, and to quite literally, put the Jews in their place.

Of course it is artificial to somehow seek to keep the two sides of the ghetto wall and these different aspects of ghettoization entirely separate. Ghettos enclosed and separated. Constructing ghettos was both an act of creating spaces of Jewish absence and an act of creating spaces of Jewish presence within towns and cities that varied from place to place. Here I need to shift briefly from the abstract to the concrete to explore some of the ways these themes of absence and presence played out in two of the largest cities in occupied Europe where ghettos were established: Warsaw and Budapest.

Absences and Presences on the Ground: The Ghetto and City in Warsaw (1940) and Budapest (1944)

Reading the pages of the diary of Adam Czerniakow, leader of the Jewish Council in Warsaw in the aftermath of the German occupation of the city, the earliest traces of a reimagining of the city along segregated ethnic lines emerge. Two days after noting that signs reading *Achtung! Seuchengefahr. Eintritt verboten* were to be placed at the outer edges of the largely Jewish quarter in the city, Czerniakow explained that "the 'ghetto' will be marked with the sign *Achtung! Seuchengefahr. Eintritt verboten*. The military authorities will add signs placing the area off limits to the troops" (Hilberg, Staron, and Kermisz 1979, 91). What for Czerniakow was the first stage of the construction of a "ghetto," was focused on creating an absence of sorts, but not—at this stage at least—absences in terms of freeing up real estate for Germans or Poles. This was not about creating a *judenfrei* Warsaw, but a Nazi-German-*frei* Jewish quarter in the city, on the grounds of fears of infection.

The idea of a "quarantine area" off limits to German troops persisted throughout 1940, with the Jewish Council instructed to begin building sections of walls to mark this area out (Hilberg, Staron, and Kermisz 1979, 134–41, 145; Gutman 1982, 50). However, while Jews were increasingly concentrated within this "quarantine area" in the traditional Jewish quarter during 1940, there was no immediate systematic attempt to clear this area of its non-Jewish population aside from not permitting non-Jews to take up new tenancies in the area (Hilberg, Staron, and Kermisz 1979, 183, 186–88). Although the Jewish diarist Chaim Kaplan acknowledged that "a formal ghetto complete in every detail has not yet been established," he suggested that "a German quarter, an Aryan quarter, and a Jewish quarter" were becoming increasingly fixed as Jews were being required to leave the German quarter and "in the Aryan quarter, mainly inhabited by Poles, the only Jews permitted are those who already live there" (Katsh 1966, 169–70). In October 1940, the issuing of the ghetto order restructured Warsaw into separate and distinct "Jewish," "German," and "Polish" zones (Hilberg, Staron, and Kermisz 1979, 206; Sloan 2006, 72). Now all Jews had to move into the ghetto area, and all non-Jews had to move out into properties outside the "German quarter." It seemed to Czerniakow that in late October and early November 1940, "everybody was preoccupied with moving" (Hilberg, Staron, and Kermisz 1979, 214). While ghettoization may not have forced "everybody" to relocate, the German authorities did estimate that around one-fifth of the city's population were on the move as the city was restructured along ethnic and racial lines (ibid., 40).

Ghetto planners were well aware of the impact of ghettoization on the city's population and on the infrastructure of the city itself. It seems that officials initially considered the suburbs of Warsaw as the potential site for the ghetto rather than the heart of the city (Hilberg, Staron, and Kermisz 1979, 396). In Czerniakow's diary, "rumours about a ghetto in Praga" were clearly circulating in late 1939 and early 1940 (ibid., 101, 117, 122). A little later, two further sites on the edge of the city appeared to be under serious consideration (ibid., 19–56, 396). The attraction of suburban ghettos was, as Waldemar Schön, who was responsible for supervising population relocation in the Warsaw district, argued, because "ghettos of this kind, located at the periphery of the city, would be a relatively minor obstacle and would be least harmful to the economy, industry, and transportation in Warsaw" (Gutman 1982, 52). Here was a desire—albeit short-lived—for clearing the central business district of Jews. However, ultimately plans for locating the ghetto in the city's suburbs were abandoned and the ghetto was sited in the traditional Jewish quarter of the city, although the ghetto was created in such a way as to allow one of the major tram lines in the city to continue running, hence limiting the impact of closing off part of the city on the transport infrastructure. In effect, the decision was made to create a closed ghetto where a *de facto* ghetto of sorts was already being created and thus to take the ghetto to the Jews. However, this appears to have been driven less by an ideological making concrete of the Jew's place in the city and more by a concern with the pragmatics of implementing ghettoization rapidly and with the least impact on non-Jews. Put simply, it was easier to site the ghetto where Jews already made up the majority of the population for all sorts of reasons, not the least that this meant that the numbers of non-Jews forced to move would be kept to a minimum.

This picture was repeated in Budapest where a series of ghettos was planned and implemented in 1944, a year after the Warsaw ghetto had been emptied through mass deportations and razed to the ground following the uprising. Early ghetto plans in Budapest appear to have been equally concerned with ghettoization as opportunity for cleansing parts of the city as much as the strategic placing of Jews in a series of ghetto areas. Plans drawn up on May 9 to create seven ghetto areas on both the Buda and Pest sides of the city explicitly identified a series of major streets and squares across Budapest that were to be "completely cleansed" of Jews. Removed from this prime real estate, Jews were to be housed in "Jewish quarters" that made either demographic or geostrategic sense. The two largest ghetto areas envisioned on

May 9 fitted with the central districts of Pest where the majority of the city's Jews lived, while the other five seemed to have been more concerned with creating a human shield of sorts in a city that had recently come under Allied air attack (Cole 2003, 84–92).

However, as was the case in Warsaw, these plans were ultimately shelved and replaced with more pragmatic plans that sought to find a speedy and pragmatic solution to the problem of where to locate the ghetto. After the completion of a citywide survey in late May, 2,639 houses that stretched across the entire city were designated for Jewish use on June 16. These were apartment buildings where a majority of Jews lived. As in Warsaw, the ghetto came to the Jew. Yet it was still estimated in the press that 40,000 Jewish families and 12,000 non-Jewish families would have to move, which was offered up as evidence that ghettoization was "not merely a Jewish affair, but closely affects the Christian population also," who were shown to be reluctant to leave their homes. This reluctance comes through in the hundreds of petitions delivered to the city authorities in the days after the ghetto list was published. The largest number was from non-Jews asking that their apartment building be excluded from the list (Cole 2003, 101–49).

However, there were some non-Jews who saw ghettoization as an opportunity for upward mobility and a chance to exchange their poor-quality present housing with better apartments soon to be vacated by Jews (Cole 2003, 142–45). These sentiments certainly meshed with the concerns of the authorities. They clearly had an eye—as Hilberg suggested was also the case in Łódź—on the relative quality of the real estate included in the ghetto and that excluded from the ghetto. The former was to be of poor quality. The latter was to be of high quality. Here were socioeconomic concerns with the real estate left behind in the process of ghettoization. Making parts of the city *judenfrei* was not simply about creating imaginative absences, but also material absences in the city. Jews were not simply to have less space to live, but worse housing to live in as a result of ghettoization, with the intentional corollary that non-Jews would have more and better housing. This is evident in the second, definitive list of ghetto houses issued on June 22, 1944, after a week of frantic petitioning and thorough investigation. With 840 properties canceled from the list and 149 properties added to the list, the concern was not only to draw up a smaller list of 1,948 apartment buildings where Jews were in the majority but also to intentionally exclude higher-quality and highly desirable villas in the city's suburbs (ibid., 156–59).

It seemed that non-Jewish petitioners had won their week of battling with the authorities over the precise siting of the ghetto in the city. Not only had the authorities removed smaller, sought-after villa properties from the ghetto list, but also, more significantly, had shifted the scale at which ghettoization was implemented. A few days after the list of ghetto houses was issued, the authorities followed up with a list of regulations managing ghetto space. One point in a long list forbade non-Jews from hiding Jews in their homes or even letting Jews "for no matter how brief a period into either Christian houses *or the Christian-tenanted portions of Jewish houses*" (Braham 1994, 856). As this regulation made clear, in reality non-Jews were permitted to remain living in their apartments within those buildings that made up the ghetto. The end result was that the ghetto wall in Budapest was, in many cases, the apartment wall throughout the summer and autumn of 1944 before a closed ghetto was established in the traditional Jewish quarter in Pest in the winter of 1944 alongside a more dispersed, so-called "International Ghetto" housing Jews with protective paperwork issued by the neutral authorities (Cole 2003, 191–220). When the Pest ghetto was created, 144 of the 162 apartment buildings within this area, designated for Jews back in June, still had non-Jews living alongside Jews (ibid., 155). If this was typical of the rest of the city then it is clear that Budapest in the summer and fall of 1944 was a place where total segregation of Jews and non-Jews was never fully achieved despite ghettoization, which brings me to focus on the ghetto wall.

Boundaries and the Ghetto as Territorial Solution

"Fundamental to the very idea of the ghetto," wrote Hilberg of the first ghettos established in Poland, "was the sheer segregation of its residents. Personal contacts across the boundary were sharply curtailed or severed altogether, leaving in the main only mechanical channels of communication: some telephone lines, banking connections, and post offices for the dispatch and receipt of letters and parcels. Physically the ghetto inhabitant was henceforth incarcerated" (Hilberg 2003, 237). Focusing on the ghetto fence or wall, these landscapes can be seen within a broader history of sites of incarceration. Within the Nazi context, ghettos can be positioned as part of a wider and longer-running project of physical removal of those deemed "undesirable" to incarceratory spaces that ranged from the early concentration camps, through the ghettos of occupied Eastern and Central Europe, to the most notorious of such sites: the purpose-built death camps that aimed at a final removal. Ghettos were not purpose-built sites of concentration in the way that many of the camps were. Rather, they were more ad hoc reworkings of the ordinary streets and buildings of towns and cities into segregated and concentrated space. Critical in this reworking of urban space was the creation of some kind of boundary, which like any boundary, served to separate and confine.

However, what ghettos did share with the camp system was that both fitted within a broader policy and practice of confinement. Ghettoization involved an attempt to demarcate and then control Jewish space and so demarcate and control Jews. Such territorial solutions had their advantages to the perpetrators, in particular in terms of allowing for greater degrees of control. As Robert Sack has pointed out, "it is easier to supervise convicts by placing them behind bars than by allowing them to roam about with guards following them. Controlling things territorially may save effort" (Sack 1986, 22). Behind guarded walls, the ghetto functioned, Hilberg suggested, as "a closed-off society, its gates permanently shut to free traffic," thereby ensuring "the segregation of its inhabitants from the surrounding population" (Hilberg 1981, 157).

While Hilberg saw the ghetto fence or wall as an effective block to continuing Jewish/non-Jewish interactions, he was careful "not to imply the total absence of contacts with Germans or Poles. There were telephone, electrical, gas, and water connections; removal of human waste; exports of manufactured goods; imports of coal, food, or raw materials; mail and parcel shipments through ghetto post offices; loans from banks, payments of rents, and so forth." Yet these continuing links were simply "institutional transactions that had to be maintained if the ghetto was to function," rather than "private bonds, across the boundaries, that could no longer be tolerated because they were incompatible with the function of the ghetto" (Hilberg 1980, 100–101). The end result was, Isaiah Trunk argued, that the "Nazi-created Jewish ghetto necessarily became a Jewish city sui generis, linked to the main city only by slender links which had to be maintained for purely technical reasons, e.g. lighting, water-supply, sewage, and limited telephone connection" (Trunk 1972, 44).

This separation was achieved by boundary making—first sketched out on a map, and then enacted on the ground through the constructing of fences and walls (paid for by the Jewish community themselves) to separate out Jewish and non-Jewish space (Cole 2011, 56–70). For survivor Erwin Baum, the ghetto wall "was like a border" between two countries (USHMM 1994). Jean Améry went further, positioning the ghetto wall as nothing less than "the demarcation line that separated the Jews from the human being" (1984, 23). These acts of boundary marking and making were critical to the effective exercise of power through control over space. The ghetto wall was intended as a barrier to both the flow of people and products. Wall and fence building around the ghetto not only sought to separate out Jews from non-Jews and limit possibilities for escape from the ghetto over to the so-called "Aryan side" but also to ensure careful control over the entry of such essentials as fuel and foodstuffs into the ghetto. Policing the boundaries of the ghetto, it was possible to control nutritional intake in the ghetto and implement a policy of effective starvation rations. The carving out of three broad spaces within the city—Jewish, Polish, and German—was about carving out three states of sustenance varying from starvation through plenty (Hilberg, Staron, and Kermisz 1979, 55).

In order for those in the ghetto to be fed, supplementary food had to enter through nonofficial channels—not in official supplies through one of the ghetto gates but over or through the walls in the pockets of smugglers, or through the gates in the pockets of returning work groups who had been on the "Aryan side." In practice then, ghettos never functioned as entirely hermetically sealed containers despite all appearances to the contrary with the erection of fences and walls and the policing of boundaries. The length of boundary walls could stretch—as they did in Warsaw—for literally miles. Listening to oral history interviews and reading memoir accounts, the Warsaw ghetto wall always seemed to have holes in it that could be wriggled through. Typical here are the diary entries of Abraham Lewin, who frequently reported on the appearance of holes in stretches of the ghetto wall (Lewin 1990, 66, 77, 89, 118, 124). He was particularly interested in "the gap in the wall outside my window" that separated the ghetto from Przejazd Street over on the "Aryan side" (ibid., 118). Known as a "target," this was, he reckoned, "wide enough for a sack with 100 kg of potatoes or corn or other foodstuffs," and he watched constant smuggling "without a break from dawn at half past five until nine in the evening" (ibid., 77, 124). Although "blocked up countless times," gaps in the wall reappeared, Lewin noted, as, "each time the Polish and Jewish policemen on both sides of the wall are bought off and before the lime has a chance to dry the bricks are taken down and the smuggling continues" (ibid., 102). The ongoing difficulties of policing the boundary led to the relocation of the wall in the center of streets to prevent openings being made in the backs of houses along the boundary and thus allow for more effective policing (Browning 1988, 26).

Continuing contact between these two worlds—inside and outside the ghetto wall—extended not only to foodstuffs but also to people. It was not only goods that were being smuggled into the ghetto but also people who were being smuggled out. Escapes took place over, under, and through the wall itself, as well as through the ghetto gates, which could be breached in vehicles, with passes, and through a combination of bribing and distracting the guards. Not only were the Jewish and Polish policemen stationed there open to bribery, but a number of (ordinary) German gendarmes were also. In an important book, Steve Paulsson has pointed to the porosity of the ghetto wall in a reimagining of wartime Warsaw not simply as a city divided into

Jewish and non-Jewish space by the ghetto wall, but a more complex place where what he dubs a "secret city" existed underneath the official geography of distinct zones of residence. This "secret city" outside the ghetto walls was home, Paulsson (2000) estimates, to 28,000 Jews being variously helped by networks of 70,000 to 90,000 Poles and pursued by 3,000 to 4,000 extortionists and blackmailers.

These networks of relationships between Jews and non-Jews that persisted despite the segregatory logic of ghettoization is also something seen in Budapest, in particular during the summer and fall of 1944 when, as I have already noted, the "Aryan side" could be as close as the neighboring apartment. On paper, non-Jewish apartments in ghetto houses were out of bounds to Jews, however, in practice much depended on the degree of control apartment building caretakers exercised. From oral history accounts, it is clear that non-Jewish neighbors helped in a variety of ways, from arranging hiding to getting hold of protective paperwork. These opportunities were particularly possible where long-standing prewar social networks could be drawn on in apartment buildings designated as part of the ghetto and therefore, while involving concentration, ghettoization did not mean relocation (Giordano and Cole 2011).

Although in many ways ghettos were a radical breach with prewar residential patterns, and in particular the *co*-location of Jews and non-Jews within the city, they tended—to varying extents—to build upon (and harden) prewar residential patterns, as can be seen in the case of Warsaw and Budapest. The end result was that ghettos were evolutionary rather than revolutionary spaces that still had—at least initially—some sense of familiarity about them. In a telling passage, Debórah Dwork and Robert Jan van Pelt write that, "the inhabitants of east European ghettos were connected to the history of the place in which they were now compelled to live. These streets, synagogues, and markets had grown over centuries to meet the Jewish community's needs; they now suggested that life could go on" (2002, 217).

Of course everything was now so very different. Ghettoization involved an act of concentration that led to terrible overcrowding. According to German estimates, the population density of what they preferred to call the "Jewish residential district" in Warsaw was reckoned to be roughly three times that of residential space in the remainder of the city. Adding such overcrowding to the starvation rations being officially permitted to enter the ghetto, it is no surprise that disease and death rates rocketed within the ghetto. However, ghettos were not—at least in most cases—built, from scratch, seemingly in the middle of nowhere, but were constructed out of the existing infrastructure of the city, reshaped for the purposes of the concentration and segregation of Jews (and the clearing of other areas of the city). In this way there was something familiar, and embedded, about ghettos that contrasted with the alienating experience of transportation, in some cases over hundreds of miles, to the literally foreign places of camps such as Auschwitz (Gigliotti 2009).

I have always been struck by Elie Wiesel's reflections of his own progressive experience of a "shrinking universe" as he moved from his family home into the ghetto, and then onto the train, and into the camp. "The universe began shrinking," wrote Wiesel, "first we were supposed to leave our towns and concentrate in the larger cities. Then the towns shrank to the ghetto, and the ghetto to a house, the house to a room, the room to a cattle car" (Wiesel 1985, xv). His experience was not simply one of spatial concentration from crowded apartment, through cattle car, to the ultimate spatial reduction of gassing and cremation. It was also an experience of movement away from home. Of course the route from ghetto to killing was not always one that was so distant (and distanced). In the waves of killings that took place along the Polish-Soviet border in 1941–43, ghetto inhabitants were generally taken from the ghetto and shot into ditches dug in nearby woods. However, for those Jews like Wiesel living in ghettos in Hungarian territory in the spring and summer of 1944, the liquidation of the ghettos meant the long train journey north to Auschwitz. It was very much an experience of leaving home.

That ghettos were places close to home is something that emerges in the work of the pioneer of Holocaust geographies, Andrew Charlesworth. In his important essay on "Holocaust Topographies," Charlesworth—drawing on years of site visits and landscape study—takes readers on a virtual tour of a handful of Polish landscapes. As he shows, "ghettos were connected to the existing urban system" rather than existing in a conceptual, institutional, or material vacuum (Charlesworth 2004, 242). The result is that Charlesworth brings the ghettos much closer to home materially and also imaginatively, and places them not over there—distant and distanced—but here—cheek by jowl with the urban world on the other side of the fence that continued to operate, seemingly as normal, during the war. In a sense he forces us to look closely at European towns and cities again as landscapes scarred by the coexistence of death and life on the two sides of a thin dividing wall (which was, as I have suggested, more porous than perhaps we imagine). He forces us to confront the fact that Nazi geographies not only constructed camps on the periphery but also ghettos within urban centers.

References

Améry, Jean. 1984. *Radical Humanism: Selected Essays*. Translated and edited by Jean Améry, Sidney Rosenfeld, and Stella Rosenfeld. Bloomington: Indiana University Press.

Bloxham, Donald. 2009. *The Final Solution: A Genocide*. Oxford: Oxford University Press.

Braham, Randolph L. 1994. *The Politics of Genocide: The Holocaust in Hungary*. New York: Columbia University Press.

Broszat, Martin. 1970. "Hitler and the Genesis of the 'Final Solution': An Assessment of David Irving's Theses." *Yad Vashem Studies* 13: 73–125.

Browning, Christopher R. 1986. "Nazi Ghettoization Policy in Poland, 1939–1941." *Central European History* 19: 343–63.

———. 1988. "Genocide and Public Health: German Doctors and Polish Jews, 1939–41." *Holocaust and Genocide Studies* 3: 21–36.

Charlesworth, Andrew. 2004. "The Topography of Genocide." In *The Historiography of the Holocaust*, edited by Dan Stone, 216–52. Houndmills: Palgrave Macmillan.

Cole, Tim. 2003. *Holocaust City: The Making of a Jewish Ghetto*. New York: Routledge.

———. 2004. "Ghettoization." In *The Historiography of the Holocaust*, edited by Dan Stone, 65–87. Houndmills: Palgrave Macmillan.

———. 2011. *Traces of the Holocaust: Journeying in and out of the Ghettos*. London: Continuum.

Dwork, Debórah, and Robert Jan van Pelt. 1994. *Auschwitz, 1270 to the Present*. New Haven, CT: Yale University Press.

———. 2002. *Holocaust: A History*. London: John Murray.

Friedman, Philip. 1954. "The Jewish Ghettos of the Nazi Era." *Jewish Social Studies* 16: 61–88.

Gigliotti, Simone. 2009. *The Train Journey: Transit, Captivity, and Witnessing in the Holocaust*. Oxford: Berghahn.

Giordano, Alberto, and Tim Cole. 2011. "On Place and Space: Calculating Social and Spatial Networks in the Budapest Ghetto." *Transactions in GIS* 15: 143–70.

Gringauz, Samuel. 1949. "The Ghetto as an Experiment of Jewish Social Organization." *Jewish Social Studies* 11: 3–20.

Gutman, Israel. 1982. *The Jews of Warsaw, 1939–1943: Ghetto, Underground, Revolt*. Brighton: Harvester.

Hilberg, Raul. 1961. *The Destruction of the European Jews*. Chicago: Quadrangle Books.

———. 1980. "The Ghetto as a Form of Government." *Annals of the American Academy of Political and Social Science* 450: 98–112.

———. 1981. "The Ghetto as a Form of Government: An Analysis of Isaiah Trunk's Judenrat." In *The Holocaust as Historical Experience: Essays and a Discussion*, edited by Yehuda Bauer and Nathan Rotenstreich, 155–71. New York: Holmes and Meier.

———. 2003. *The Destruction of the European Jews*. 3rd ed. New Haven, CT: Yale University Press.

Hilberg, Raul, Stanislaw Staron, and Josef Kermisz, eds. 1979. *The Warsaw Diary of Adam Czerniakow: Prelude to Doom*. New York: Stein and Day.

Horwitz, Gordon J. 2008. *Ghettostadt: Łódź and the Making of a Nazi City*. Cambridge, MA: Belknap Press of Harvard University Press.

Katsh, Abraham I., trans. and ed. 1966. *Scroll of Agony: The Warsaw Diary of Chaim A. Kaplan*. London: Hamish Hamilton.

Lewin, Abraham. 1990. *A Cup of Tears: A Diary of the Warsaw Ghetto*. London: Fontana.

Michman, Dan. 2011. *The Emergence of Jewish Ghettos during the Holocaust*. New York: Cambridge University Press.

Mommsen, Hans. 1991. *From Weimar to Auschwitz: Essays in German History*. Cambridge: Polity Press.

Paulsson, Gunnar S. 2002. *Secret City: The Hidden Jews of Warsaw, 1940–1945*. New Haven, CT: Yale University Press.

Sack, Robert, D. 1986. *Human Territoriality: Its Theory and History*. Cambridge: Cambridge University Press.

Sibley, David. 1995. *Geographies of Exclusion: Society and Difference in the West*. London: Routledge.

Sloan, Jacob, ed. 2006. *Notes from the Warsaw Ghetto: From the Journal of Emanuel Ringelblum*. New York: ibooks.

Smith, David M. 1994. *Geography and Social Justice*. Oxford: Blackwell.

Trunk, Isaiah. 1972. *Judenrat: The Jewish Councils in Eastern Europe under Nazi Occupation*. New York: Macmillan.

USHMM. 1994. Oral History Archive RG-50.030*0016, interview with Erwin Baum.

Wiesel, Elie. 1985. Introduction to *The Holocaust in Hungary: Forty Years Later*, edited by Randolph L. Braham and Béla Vágo. New York: Columbia University Press.

13

Spaces of Engagement and the Geographies of Obligation: Responses to the Holocaust

MICHAEL FLEMING

Introduction

The spatialities of responses to the Holocaust remain relatively under-researched and under-theorized. In this chapter I make use of two key concepts—spaces of engagement and geographies of obligation—in order to highlight how different spatial imaginaries influenced responses to the Holocaust. The notion of spaces of engagement draws on strands of geographical research, including David Livingstone's (2005) ideas relating to the geography of reading, as well as Edward Said's (1991) traveling theory, among others. It refers to how encounters, whether with texts, ideas, problematics, or events, are influenced by the places and communities in which they take place. For Livingstone, analysis of the geographies of reading not only highlights "the collective character of interpretation and the way any individual reading is located in the reader's membership of a community sharing some foundational assumptions and interpretive strategies" (Livingstone 2005, 395) but also demonstrates that the "generation of knowledge involves interpretation as well as invention" (ibid.). This dual aspect characterizes spaces of engagement as well. On the one hand, particular places, including institutions, can encourage people to share assumptions and interpretative strategies to understand phenomena, and on the other, help to generate knowledge through the dialectic of interpretation and observation. Understanding the interpretative communities embodied at various sites where news of the unfolding Holocaust was received is crucial to comprehend communities' varied responses.

In addition and related to the notion of spaces of engagement is the concept of geographies of obligation, which describes the hierarchies of bonding between different groups.[1] Social distance between different groups is a key factor underpinning geographies of obligation, but so too are physical distances and the varied potentialities to engage with specific others. The notion therefore augments the concept of spaces of engagement by highlighting the strength/weakness of bonds between different groups, allowing a critical assessment of their actions/inactions in response to the Holocaust.

Livingstone's notion of geographies of reading is informed by Said's contention that theories "travel." Traveling theory is especially helpful in understanding the imaginative geographies constructed and negotiated by various "witnesses" to the Holocaust. According to Said, theories (processed, considered propositions, and/or explanations) are reshaped within new environments. They must satisfy certain conditions of acceptance within these new contexts and are reformed under the impact of resistances. Theories are transformed by the uses to which they are put, and, importantly, by their position in a new, different time and place. In the context of the Holocaust, the fostering of an imaginative geography in Britain that sometimes equated physical distance with social, ethnic, and cultural distance played a key role in determining how different actors responded to news of the mass murder of European Jews by Nazi Germany.

The overdetermining of "distance" in Britain not only helped shape British Foreign Office responses to the Holocaust, but also impacted on how other Allies responded. For, while anti-Semitism was widespread throughout Europe both before and during the war, the way in which Allied governments whose Jewish citizens were being murdered by the Nazis responded was, at least in part, influenced by the hegemonic conception of Jews' appropriate place in society promoted in Britain. Allied governments exiled in London could not ignore British policy toward (and theories about) Jews and had to reach some form of accommodation with British ideas while reevaluating their own "imported" conceptions. The co-construction of similar geographies of obligation that marginalized Jews allowed negative synergies toward the Jewish tragedy to develop among the Allies, but such synergies were not inevitable.

In this chapter I provide an indicative overview of three different spaces of engagement: the British Foreign Office, the Polish government-in-exile—first based in France, and then, from 1940 onward, in London—and the Polish Underground in Poland itself. Through these three case studies I indicate how these spaces of engagement related to and shaped (changing) geographies of obligation. In the conclusion I draw attention to the privileged position of the British Foreign Office to influence both the dissemination of news about the Holocaust in Britain and the responses of Allied governments to this terrible news. I maintain that the Foreign Office played an important role in shaping geographies of obligation in Britain and beyond. British "theories" relating to Jews traveled.

The British Foreign Office

Writers such as Tony Kushner (1994) and David Cesarani (1998) have illustrated the way in which British policy toward Jews prior to the Holocaust was marked by liberal anti-Semitism. For Cesarani (1989, 7), "the state, society and culture in Britain operated a discourse about Jews that was exclusive and oppressive that eventuated in and legitimated discrimination." The British government concern about Jewish migration to Britain during the first half of the twentieth century is relevant to discussion on the Foreign Office's responses to the Holocaust, as the assumptions made about Jews during the period following the passing of the 1905 Aliens Act, which sought to limit the flow of Eastern European Jews to Britain, continued into the postwar period throughout the state apparatus. Indeed, as late as May 1945 Herbert Morrison argued for Jewish refugees in Britain to be displaced from the country, as he was "seriously alarmed regarding the possibility of antisemitism in this country" if the economic situation deteriorated.[2] The logic behind such arguments—whether the 1905 Aliens Act, the 1919 Aliens Act, or the subsequent Alien Orders that granted the Home Secretary a great deal of discretion in relation to immigration policy (a regime that survived until 1938, when visas were introduced for immigrants from Germany and Austria)—was that cultural and social differences posed a challenge to social harmony. Minorities threatened the majority.

The worldview, in which homogeneity was favored and some groups—in particular Eastern European Jews—were seen to be problematic, had a broad constituency within the Foreign Office during the period of the Holocaust. Bloxham and Kushner (2005, 194) have argued that "between 1939 and 1950, the idea that Jews bring anti-Semitism with them was taken to its logical extreme, becoming in itself ingrained into bureaucratic mentality and practice."

Information about the Holocaust, for officials in the Foreign Office, was colored by two powerful filters: liberal anti-Semitism and institutional instrumentalism. Institutional instrumentalism describes the practice of negatively apprehending information and/or events that brought into question or could hinder, in this case, the prosecution of Foreign Office policy. Such information was to be effectively controlled to limit its impact on Foreign Office policy and practice. Since minorities, and the Jews especially, were seen as problematic for a variety of reasons (discussed below), the Foreign Office response was to manage information about them not in terms of that news, but to ensure that such information did not compromise established Foreign Office policy.[3]

The view that the Jews were problematic and brought anti-Semitism with them was widespread throughout the British state apparatus. The Ministry of Information, which liaised closely with the Foreign Office over (overseas) propaganda through the Political Warfare Executive, outlined British policy toward news about Jews. As early as July 1939, before the war started, the Ministry of Information decreed that "no special propaganda addressed to Jews is necessary outside Palestine."[4] The policy of only addressing Jews as citizens of the states in which they resided could not adequately respond to the particular anti-Jewish nature of the Nazi genocidal program. News of atrocities against the Jews was also more tightly controlled than against other groups. In July 1941, a planning document from the Ministry of Information stated which information could be distributed. It called for the suppression of news about the Jews.[5] Nevertheless, by December 1942 information about the fate of European Jewry was widely known. On July 9, 1942, the Ministry of Information hosted a conference that revealed data brought to Britain in the Bund report of May 1942. This conference interrupted British censorship policy.

However, through the summer of 1942, the flow of information stalled. Firstly, British censorship policy was tightened. Information about the death camps at Treblinka, Sobibór, and Bełżec revealed in July 1942 could have been repeated and requests for additional information from the Polish government could have been made. No significant repetition took place or requests were made. Secondly, during July 1942, Poles and Swedes involved in couriering reports out of Poland to Britain via Stockholm were arrested. As a security measure, the distribution of new reports from Poland was embargoed through the summer by the Polish deputy defense minister (Stola 1997, 10). Thirdly, the Polish government was concerned that reports were not being believed and adjusted how information was presented to British colleagues. Dariusz Stola (ibid., 8) argues that Polish foreign minister Edward Raczyński's "opinions were influenced by the attitudes of his British colleagues in the Foreign Office," and in these circumstances the Polish government did not always emphasize Nazi genocidal programs when it had the opportunity to do so.[6]

The US State Department confirmation of the veracity of the "Riegner Telegram" in November 1942 opened up the possibility for various groups and governments to speak clearly about the Holocaust. The Polish government, subjected to strong pressure from the Polish National Council, especially from its two Jewish members, asked for an Allied declaration and issued a diplomatic note on the extermination of the Jews on December 10, 1942. German atrocities in Poland were discussed on December 15, 1942, in the House of Lords, and the Nazi killing of Jews was debated in the House of Commons on December 17. The British foreign secretary delivered an important speech

(UN Declaration) at the House of Commons the same day, and MPs stood in silence for the murdered Jews; however, action was not forthcoming.[7]

British responses to news of atrocities against the Jews were influenced by three further factors. Firstly, they sought to narrate the war as a British fight against tyranny in order to sustain the British people's morale and commitment to the war effort. Focus on the Jewish tragedy, it was thought, could undermine British unity and actually increase anti-Semitism in Britain. Secondly, they wished to avoid the excesses of World War I propaganda, in part, because it was felt some information would not be believed both at home, leading to mistrust in information propagated by the state, and abroad, thereby alienating allies and potential allies in enemy countries. Thirdly, the Foreign Office also had concerns about how attention on what was happening to Jews in Europe would be understood in the Middle East.

News of atrocities against Jews had real policy implications for the Foreign Office. The choreographed announcements of July 9, 1942, and December 17, 1942, stimulated civil society for action, for retaliatory strikes, and for a more relaxed refugee policy. Immediately after the foreign secretary had issued the declaration to the House of Commons on December 17, the Independent Labour MP for Glasgow Shettleston, John McGowan, asked Eden whether those who escaped occupied territories would be given sanctuary. Eden's response was noncommittal and highlighted "security formalities" and "immense geographical and other difficulties."[8]

For the Foreign Office, demands to initiate retaliatory measures or to accept Jewish refugees into Britain were distinctly unwelcome, and it worked hard to regain the policy initiative. The Bermuda Conference of April 1943 achieved this, in part, by providing a mechanism for civil society demands to be dissipated, without providing any commitments to retaliate or relax British refugee policy. In addition, information about the mass killing of Jews during 1943 and 1944 was tightly controlled. None of the significant information passed to British officials about Auschwitz during 1943, for example, was published in the mainstream press or broadcast on the BBC's Home Service. By withholding confirmation of Polish intelligence reports that reached London and were circulated, the Foreign Office, together with the Ministry of Information, effectively inhibited their publication in the national press.

However, the inaction of Britain and her allies to the unfolding Jewish tragedy was contested. Through the summer of 1942, Shmuel Zygielbojm—a member of the Bund and a member of the National Council of the Polish government-in-exile—petitioned the British government for action, spoke to an audience from the British Labour Party, delivered a broadcast on the

BBC outlining the fate of Polish Jews and, in December, telegrammed prime minister Winston Churchill. With no action forthcoming, and following the crushing of the Warsaw Ghetto Uprising in May 1943 (and probably with awareness of the latest news from Auschwitz),[9] Zygielbojm committed suicide on May 12, 1943. He left letters for the president of Poland, Władysław Raczkiewicz, and the prime minister of the Polish government-in-exile, Władysław Sikorski, in which he argued that the passivity of the Allied nations to the Nazi murder of Jews made them "partners to the responsibility."[10]

The suicide of one of the most determined advocates of Polish Jewry did not alter the policy of the British Foreign Office. The destruction of the Jews of Warsaw in 1942 and 1943 prefaced the liquidation of Hungarian Jewry in the spring and summer of 1944. The fate of Hungarian Jews was well understood by British officials in the Political Warfare Executive. An *aide memoire* dated March 24, 1944 (the same date as Roosevelt's warning to the Germans) titled "Special annexe on the persecution of the Jews" provided details of the incremental process ending in the death camps.[11] It also discussed Auschwitz, drawing on Polish reports from March 30, 1943, and January 18, 1944—indicating the Polish sourced information was seen as credible.[12] Again there were numerous voices calling for the German extermination policy to be named and highlighted, and for action to be taken, such as warning Hungarian Jews of Nazi extermination plans. This was mooted in the April 20, 1944, Political Warfare Executive directive to the BBC's Hungarian Service, but there was no mention of the death camps until June 8, 1944, by which time thousands of Hungarian Jews had been gassed at Auschwitz.[13]

The Polish Government-in-Exile

The Polish government-in-exile throughout World War II struggled to maintain influence with her erstwhile Allies. Beginning in September 1939, Polish/British interests diverged in relation to the Soviet occupation of eastern Poland. The British sought not to step "on Russian toes," for example.[14] In addition, the Poles were sensitive to the damage that British perceptions of Polish anti-Semitism could have not only on Polish/British relations, but also on Polish relations with the Jewish diaspora both in the United Kingdom and, more importantly, in the United States. The Polish government, once it was established in London, following the fall of France in 1940, also had to confront British "anti-alienism." In 1941 Alexander Cadogan, the permanent undersecretary for foreign affairs, was moved to write to other British ministries advising them it was

> of the highest importance from the point of view of the foreign policy of His Majesty's Government that the status of these Governments [i.e., the various governments-in-exile] should be fully recognised and protected, and their representatives should be treated not merely as the representatives of Foreign Governments, but as Allies.[15]

The Polish government also faced a less than welcoming British press. The pro-Soviet *The Daily Worker* maintained an uncompromisingly hostile stance, while right-leaning papers caricatured the Poles through cartoons.[16] Emblematic of the difficulties the Polish government faced was the Pritt affair, which rumbled on through 1940.

Denis Pritt, a left-wing Labour MP, wrote to Foreign Office minister R. A. Butler on April 2, 1940, complaining about anti-Semitism in the Polish army, at that point based in France. An investigation conducted by Roger Makins of the Foreign Office concluded that "it seems quite on the cards that anti-Semitism is in fact quite rife in the Polish Army, but in this particular case [an anti-Semitic speech by a Polish army captain] some allowance must be made for the mess-room exuberance of a Polish officer." Butler wrote to Pritt on July 27, 1940, stating that the British government had no reports of active persecution and referred Pritt to the official Polish government position. On August 5, 1940, General Sikorski tried to end the scandal and declared that any soldier who took up arms for Poland was a Pole, regardless of race or religion, and that the military authorities would take active care that these orders would be observed.[17] Repeatedly (on October 6, 1939, November 3, 1940, February 23, 1942), the Polish government attempted to distance itself from the accusation of being anti-Semitic by proclaiming its pluralist principles based on the equality of Polish citizens regardless of race or religion.

The Pritt affair placed the British and Polish governments in an awkward position. Both the British and Poles incorporated Jews within their respective polities with a degree of ambivalence. The fact that Sikorski was forced to make a specific declaration regarding Jewish soldiers serves to highlight this point and illustrates that the government-in-exile was cognizant of how accusations of anti-Semitism could erode its political capital with its allies—notwithstanding the fact that those same allies exhibited anti-Semitic tendencies themselves.[18] It should also be noted that Pritt was seen in Westminster as rather too close to the Soviet Union (he was expelled from the Labour Party for defending the Soviet invasion of Finland), and during 1940 the USSR sought to portray the Poles as having been unfair toward national minorities in order to legitimate its occupation of eastern Poland. Pritt's intervention, like that of *The Daily Worker*, was certainly aligned with the USSR's effort to portray the Polish government as a reactionary clique.

Recognizing the wider diplomatic impact of accusations of anti-Semitism, the Polish government-in-exile suppressed the sections of Jan Karski's 1940 report on the situation in Poland describing Polish attitudes to Jews (see Engel 1987, 1993) and in 1944 vigorously denied knowledge of Roman Knoll's memorandum that suggested an alternative homeland for Polish Jews after the war—when it was leaked to prominent Polish Jewish figures in the United States.[19]

The Polish government-in-exile played an active role in highlighting the fate of Polish Jewry to her allies and received reports from government couriers such as Jan Karski, Jan Nowak-Jeziorański, Jerzy Salski, and Tadeusz Chciuk-Celt, as well as via radio messages. However, the Polish government-in-exile frequently framed Nazi atrocities against Jews within the wider context of atrocities in Poland, including those against Poles. This has been understood as demonstrating that the Polish government sought to marginalize the particularity of the Nazi genocide of Jews.[20] However, such a contention assumes that the Polish government was free to narrate the war in any manner it chose. This was not the case. The ability of the Polish government to act independently was heavily constrained by the British context. Polish publications in Britain depended on British paper supplies (administered by the Paper Control Office) and largely conformed to policy laid down by the British (PWE, Ministry of Information, Foreign Office). The Polish government's official publications (*Dziennik Polski* and *The Polish Fortnightly Review*) were carefully vetted and subscribed to the British "voluntary" censorship regime.[21] News about Jews *was* narrated, in the British press, as part of a wider national story, but this was not due to the Poles alone. It is true that some members of the Polish Cabinet needed no persuading that this was the right course—for example, Marian Seyda and those on the political right. In Britain, where anti-Semitism was repeatedly reported in Home Intelligence reports, the propaganda policies of the Ministry of Information and Foreign Office heavily impacted on what could be published, how, and when. There was limited scope for the Nazi genocide of Jews to be granted attention outside the carefully choreographed revelations of July and December 1942.

The British environment in which the Polish government-in-exile operated worked in several ways to undermine that government. Through British censorship, the British public was not advised of the fate of people who endured the Soviet occupation of 1939–41, or of the Katyń massacre.[22] The British public therefore misunderstood Polish hostility toward the USSR, and this allowed the pro-Soviet *The Daily Worker* to successfully paint the Polish government as reactionary (aided no doubt by the anti-Semitic statements issued by some right-wing politicians and military personnel). The limited publicity given to the Holocaust and the manner in which material published was presented was also seen as evidence of Polish ethnocentrism, whereas in reality British sensitivity over news of Jews and atrocities played an important and indeed controlling role.

The British context is often implied as moderating Polish ethnocentrism and encouraging the Polish government to respond to Jewish demands for equality (Engel 1987, 1993). This is only part of the story. The British influence on the Polish government reporting the Holocaust was far from benign. Information received by the Polish government, including information about Auschwitz, was passed on, but was marginalized by the British. Only the obscure newspaper *The Polish Jewish Observer*—a subscription-only supplement to *The City and East London Observer*—was able to publish news received from the Polish government on the Holocaust regularly. There were also infrequent broadcasts on the BBC's Polish Service (in Polish). As late as March 1944, a Polish government press release which highlighted that more than half a million people, most of whom were Jews, had been killed at Auschwitz, was ignored by the mainstream press in Britain.[23] In the United States, with the creation of a more welcoming discursive environment following the creation of the War Refugee Board and the expectation of the issuance of a warning to the Nazis by President Roosevelt, three major newspapers published the information.[24]

The fact that the Polish government did not energetically protest at the British restriction of the circulation of news about Jews reflects the government's awareness of hardening sentiment in Poland, the influence of right-wing sections of the Polish Cabinet, and the strategic (and financial) need to maintain decent relations with the British and American governments. Despite this, the Foreign Office's Frank Roberts in August 1942 noted that "several of our allies are susceptible to pressure. General Sikorski in particular has in the past given us reason to suspect that he was not averse from embarrassing His Majesty's Government if he could at the same time appease influential Zionists."[25] Polish government responses to the Holocaust can only be comprehended if due weight is given to the British context that affected them.

The Polish Underground

In an important book from 2002, Gunnar Paulsson discusses the significance of evasion as a tactic adopted by numerous Jews to avoid the fate the Nazis planned for them. Focusing on Warsaw, Paulsson (2002, 230) concludes that "but for the Hotel Polski and the Warsaw Uprising, the survival rate among Jews hiding in Warsaw would have been about the same as that in Western Europe, contrary to all expectations and contemporary perceptions."[26] One reason for the surprising survival rates was the fact that by 1939 Polish was spoken by many Warsaw Jews and "around 50–60,000 were acculturated native Polish speakers, and of these, 25,000 or so mainly middle-class Jews were assimilated" (ibid., 30). Those who had only recently arrived in the city or were poor Yiddish speakers were more likely to have perished. For members of the Polish Underground, reaching out to Jews with whom they had some prewar contact was easier—practically and relationally—than seeking to aid those less assimilated to the dominant Polish culture. For as Fein (1979) has argued and Paulsson (2002, 42) has reiterated:

> Even the most progressive elements of Polish society regarded the Jews as outside their "universe of obligation," an alien element whose fate was of little concern to Polish society as whole. . . . [This was not] the result of some Polish policy of apartheid, but had existed by mutual consent for centuries. . . . [This] protected Jews from assimilation. The disadvantage became apparent when they desperately needed their neighbours' help and found them, even when friendly, preoccupied with their own affairs.

In short, as the Nazi genocidal program unfolded, Poles and the Polish Underground extended some aid, but this aid was not equally available to all Jews. Those more acculturated to dominant Polish norms, who spoke Polish without an accent, and had "good looks," had better chances of maintaining contacts with non-Jewish Poles and hence receiving assistance. Even the reach of Żegota, the Underground organization that aimed to assist Jews that existed between 1942 and 1945, was limited despite the relaying of Polish aid through Jewish organizations.[27]

Through the war, German anti-Semitic propaganda was commonplace in Poland. In addition, the Germans instituted a system of punishments and incentives to hand over Jews in hiding. Reports received in London, such as a report written by Roman Knoll in 1943, highlighted that the Polish peasantry expected a postwar Poland to be more ethnically homogeneous. A new homeland for Polish Jews was also suggested. Recent work by Barbara Engelking (2011) has demonstrated that the murder of Jews by Poles was fairly frequent in the countryside, and was committed by individuals not part of the criminal underclass. During the Warsaw Uprising of 1944, Paulsson (2002, 177) highlights the case of soldiers under the command of Captain Wacław Stykowski who were responsible for the murder of tens of Jews, noting that "the moderate right defined the Jews as enemies of Poland" (ibid., 181).

Despite the antipathy toward the Jews that was expressed by sections of Polish society, the Polish Underground systematically collated, organized, and sent information about the Holocaust to London. Details of atrocities, including those the Nazis committed against Jews, were incorporated in the "Aneks" series of reports. News of the Holocaust was also occasionally included in other reports the Underground sent West by courier and/or radioed to London.[28]

Awareness of growing intolerance among Poles toward minorities through the course of the war was noted by Underground leaders, but these leaders—Stefan Korboński, Stefan Rowecki, Stanisław Jankowski—did send information about the Jews to the Allies. This includes information from inside Auschwitz, sent out by Witold Pilecki and by Jozef Cyrankiewicz among others. Adam Puławski (2009) records that news about German actions against Jews was often marginalized or delayed in its transmission/dispatch to London. However, the contention that this was solely due to social distance between Poles and Jews, due to Polish ethnocentrism, or victim competition must be supplemented by the acknowledgment that news about the Jews was not deemed to be an urgent priority by Western Allies.[29] After the war, Stefan Korboński, the head of the Civilian Struggle Directorate (Kierownictwo Walki Cywilnej), wrote on the Polish Underground State and noted that the reports he sent to London on the deportations of Jews from Warsaw to death camps were ignored. Korboński recorded that he finally received information from the Polish government (it is not clear from whom exactly) explaining that "not all your telegrams are fit for publication."[30] News about German actions against Jews sent West did not generally accord with the "voluntary" censorship regime the British operated.

The mass murder of Jews was well understood by the Polish Underground and the Polish government-in-exile in London by mid-1942, following the receipt of the May 1942 Bund report. Stefan Korboński sent radio messages about the killing of Jews throughout the war. Radio messages were also sent by the head of the Home Army (Stefan Rowecki until the summer of 1943) and the government delegate. The Polish Ministry of the Interior received a steady flow of information about the death camps and *did* disseminate this data to the Polish Cabinet and members of the Polish National Council. Verbatim Underground reports were published with some delay in *The Polish Jewish Observer.* Any delays in Warsaw in sending information about the Holocaust to London were generally of secondary significance when compared with the profound difficulty in disseminating the information in the West outside the sanctioned and choreographed announcements of July and December 1942. News of the Holocaust reached British and American officials, but its dissemination to the wider public was managed by the Western propaganda and information agencies.

Conclusion

The notions of spaces of engagement and geographies of obligation can help account for the various responses of the different organizations discussed in this chapter to the Holocaust as it was unfolding. Drawing on Livingstone's ideas about the geography of reading and Said's traveling theory enriches our comprehension of how information about the Holocaust was understood and responded to at different sites. Said's work is significant here insofar as it helps to highlight how British ideas about Jews "traveled" and were reworked at other sites—notably within the Polish government-in-exile. The spaces of engagement discussed in this chapter were not entirely discrete, and ideas from one site overflowed to other sites unequally: the influence of Foreign Office views on Jews was felt by the Polish government-in-exile. This is clearly illustrated through an assessment of the geographies of obligation. The "traveling" of British conceptions of Jews impacted, at least in part, how the Polish government reacted to the Holocaust. Negative synergy between Foreign Office policy and the stereotypes the Polish right promoted played an important role in ensuring that the Polish government did not always highlight information about Nazi atrocities against Jews in public announcements and in official publications, though information *was* passed to the British.

The particular ideas about the Jews dominant in each organization largely militated against an effective response. In the British Foreign Office liberal anti-Semitism and institutional instrumentality together inhibited action even after December 1942, when the systematic extermination of Polish Jewry was widely recognized. Indeed, the Foreign Office, together with partner ministries and institutions, such as the Ministry of Information and the Political Warfare Executive, worked to minimize the reporting of news about the Jews in Britain. In this, they were largely successful, and this malign policy not only prevented civil society organizations pressuring the government to aid, inform, or provide refuge to European Jewry being annihilated by the Nazis, it has also blinded scholarship to the scope of information that reached the West about the Holocaust through 1943–44. The continual traction of the contention that the genocide taking place at Auschwitz was unknown in the West until June 1944 with the dissemination of the Vrba/Wetzler report is powerful evidence of the success of British censorship policy.

Also the Polish Underground, a broad coalition of political interests, demonstrated some ambivalence toward the Jews, but individual and organizational aid was also extended, through Żegota, for example. The ambivalence of certain sections of the Underground toward the Jewish tragedy was linked to the social distance between Poles and Jews, in which the notion of being a singular community of fate struggled to find a constituency in the face of the Nazi onslaught, to the traditional anti-Jewish stance of the Catholic Church, to the belief that there was a limited amount that could be done in the face of the Nazi determination to murder European Jewry, and later, to the support the Allies gave to the idea of nationally homogeneous states in East Central Europe. The failure of the British to seek or to highlight news about Jews outside the two choreographed news releases of 1942 (in July and December) certainly sent a (negative) signal to the Polish Underground.

The specific situation the Jews faced was not universally seen as a demand for action, and prewar *idées fixes* of the Jews were repackaged by some in the Underground to inhibit the provision of aid—if only to cause individuals to remain bystanders. Recent work by Engelking (2011) and Grabowski (2011), among others, has highlighted that Polish violence against Jews, including murder, was far more common during the war than has been previously thought.

The Polish government-in-exile's policy toward the Jews was shaped both by prewar Polish political orientations and by the situation of exile. Explicit anti-Semitism could seriously undermine the government's political capital, and it worked hard, at least publicly, to challenge accusations of anti-Semitism. The Polish government also sought to engage the Jewish diaspora by mentioning the Nazi policy toward Jews in Poland—as far as this was possible in the British context. This is often seen as a strategy to focus attention on what was happening in Poland and to Poles. Such an assessment is partial. British censorship policy also demanded that news be framed as being about nationals of particular countries. Even when news of atrocities focused on non-Jewish Poles, it remained difficult for the Polish government to persuade Allies to respond. It took eight months of continuous lobbying in 1943 for the Western Allies to issue a statement on the deportation (including to Auschwitz) of Poles from the Lublin/Zamość region, for example. Publicizing any news of atrocities committed in Poland was difficult, publicizing news about Jews was even more problematic.

Throughout the war, the Polish government-in-exile delivered reports to its allies and the press of what was happening to Polish Jewry. Its inability to encourage interventions to aid Jews specifically was mirrored in its inability to influence Allied policy on other important issues such as the future Polish borders and sovereignty. And it should be recognized that the British Foreign Office's professed skepticism of specific reports on Nazi atrocities was aligned with *the policy not to act.* The potentiality for the Polish government-in-exile to persuade the British and other allies to look beyond their own particular interests was limited.

The British state apparatus, in particular the Foreign Office, largely failed to move beyond its own specific interests and idées fixes—about the Jews, about the war effort, about Jewish influence, and about the stimulants of anti-Semitism. The recognition of the Nazi murder of Polish Jews, which commanded a great deal of interest in December 1942 in the press, in the British Parliament, and on the airwaves was not repeated. Even when Hungarian Jewry was threatened in early spring 1944, no strenuous action was forthcoming from the British Foreign Office. British foreign secretary Anthony Eden was even reluctant to issue a statement, but was pressured into delivering one following Roosevelt's warning to the Nazis in late March. The wide circulation by the Polish government of another report about the mass murder of Jews at Auschwitz did not change the Foreign Office's reticence to speak out.

The Foreign Office, in liaison with associated institutions such as the Political Warfare Executive and the Ministry of Information, had the capacity to shape the discursive environment in which news about the Holocaust presented by the Polish government was received in Britain. By choosing to restrict the flow of information in order to inhibit anti-Semitism, defuse pressure to change refugee policy, and restrict calls for retaliation, the Foreign Office controlled the geography of obligation in Britain. Civil society calls to aid the Jews of Europe remained weak. The chief rabbi of British Jewry, Dr. Joseph Herman Hertz, in a speech at Central Hall, London, on February 29, 1944, alluded to the malign influence of British censorship:

> Our great foe is the ignorance of the rank and file of the people concerning the heartbreaking and utterly unprecedented horror of the situation. If the press throughout the country would but unveil to them the scientific fiendishness displayed by the Nazis in the foul massacring of an entire people, we should have a large and determined body of public opinion behind us in our representations to Governments.[31]

In this controlled context, the efforts of the Polish Underground, intelligence agents, and representatives on the Polish National Council, including its Jewish members, in gathering, checking, preparing, encrypting, de-encrypting, rewriting, printing, and circulating news of the Holocaust to British officials and journalists were poorly rewarded. Information about the Holocaust was marginalized by the Foreign Office, the PWE, the Ministry of Information, the British mainstream press, and the BBC's Home and European Services, especially through 1943 and the first half of 1944. The construction of the particular geographies of obligation within the Allied camp was not inevitable, but the consequence of specific entanglements of ideas about Jews in different places under the defined priorities of the Allied war effort. Shmuel Zygielbojm's efforts to change those priorities and reshape the geographies of obligation came to nothing, leaving his accusation of partial responsibility to haunt us.

Notes

1. Helen Fein (1979) and David Engel (1987, 1993) use the notion of "universe of obligation" in order to explain the relationship between Poles and Jews during the Holocaust. However, this concept offers two preclusive options—inclusion or exclusion. I refer to "geographies of obligation" in order to overcome some of the problems associated with the notion of "universe of obligation" and to highlight the spatiality of feelings of obligation toward various others.

2. National Archives (Kew). Hereafter, NA, CAB 95/15: May 15, 1945.

3. For a fuller discussion of the importance of these two "filters," see Fleming (2014). Milland (1998, 56) notes that "there can be no doubt that anti-Semitism, or at least a deep unwillingness to challenge it, were present at many levels in PWE."

4. NA, FO INF 1/770 Memo 314 document 20 (p. 7) Publicity Division, Planning section meeting 10/7/1939.

5. NA, INF 1/251: July 25, 1941, p. 2.

6. It is worth noting that Polish monthly/fortnightly reports specifically dealing with the German terror campaign in Poland that were created by the Military Historical Bureau of the Information and Press office of the Home Army Main Command—the so-called Aneks series—provided regular details of the Holocaust and they were routinely handed to the British.

7. It was the Polish government, not the British government, the origin of the declaration. NA, FO 371/34361 CM 255 (26).

8. Hansard, House of Commons Debates, December 17, 1942, vol. 385 cc 2082-7.

9. Walter Laqueur (1982, 112) notes that Zygielbojm was not the most trusting of men and that he had no complaints that the Polish government withheld information from him. A great deal of information arrived in London from Auschwitz during March and April 1943. An important report was intercepted by British Postal and Telegram Censorship and circulated to the Foreign Office and the Political Warfare Executive in May 1943. NA, FO 371/34522.

10. http://yad-vashem.org.il/about_holocaust/documents/part2/doc154.html (accessed August 3, 2011).

11. NA, FO 371/39014 (439–49).

12. Hoover Institute Archives, Mikołajczyk Box 10, File "Radio Station ŚWIT Correspondence January–March 1943"; Polish Underground Movement Study Trust MSW 1944 *Sprawozdania Krajowe*, vol. 4, London (*Sprawozdanie* Nr 1/44 p. 85).

13. NA, FO 371/39272 PWE Directive 20/4/1944 (Hungarian Service).

14. NA, FO 371/23097/ 203–9. Following the Soviet invasion of eastern Poland on September 17, 1939, Ivone Kirkpatrick in a document that was circulated among officials at the Foreign Office, titled *The German Peace Offensive*, argued that it was "possible for [the British] to stand for an ethnographical and cultural Poland without standing on Russian toes." Throughout the war, the British maintained flexibility on the integrity of Polish borders.

15. NA, FO 371/26463, February 19, 1941.

16. David Low's cartoons in Lord Beaverbrook's *Evening Standard* were particularly damaging.

17. NA, FO 371/24481 (89/92/104/109/110).

18. Anti-Semitism in Britain and the United States was pervasive. Wyman (1985, 9) notes that "in the spring of 1942, sociologist David Reisman described anti-Semitism in the US as 'slightly below the boiling point.'"

19. Roman Knoll was the director of the Commission for Foreign Affairs of the government delegation in Poland. Knoll's document can be found at AAN 202/XIV-9, 135: "Uwagi o naszej polityce zagranicznej," no. 1 (1943).

20. This is a point made by David Engel (1987, 1993) and Joanna Michlic (2007).

21. The *Polish Telegraphic Agency*, which distributed news from Poland to the press, disseminated news about the Holocaust, including about the mass murder of Jews at Auschwitz during 1943 and 1944. The British mainstream press did not publish much of this information.

22. On Katyń see NA, FO 371/34577 C166. See Sanford (2005) for discussion of the O'Malley report, Foreign Office responses, and the British and American suppression of news of Katyń. For British advice to the Polish government on publishing details of the Soviet occupation of eastern Poland in 1941, see NA, FO 371/26725 C7040 (June 24, 1941).

23. Hoover Institute Archive, Polish Information Center Papers, Cable from London Box 3.9, march 21, 1944.

24. See *Los Angeles Times*, the *Washington Post*, the *New York Herald Tribune* (March 22, 1944).

25. NA, FO 371/30917 Roberts to Lias (August 21, 1942).

26. In summer 1943, the Germans set a trap to entice Jews out of hiding with the promise of transferring Jews for Germans interned abroad. "Volunteers" for the scheme were permitted to live legally at Hotel Polski, Warsaw. In total some 3,500 Jews were captured in this way. Only one to two hundred survived. See Paulsson (2002, 139).

27. Żegota—The Polish Council to Aid Jews—was founded under the auspices of the Polish government in exile through the government delegation in the country in October 1942, and included members from Jewish and non-Jewish organizations alike.

28. For a discussion of the different types of reports that the Polish Underground sent West, see Ważniewski and Marczewska (1968, v–xvi).

29. The British assessed intelligence that the Poles passed to them and outlined the kind of information they wanted. Information about Jews was not requested. Indicatively, see Michał Protasewicz's (head of the VI Bureau, Polish Intelligence, London) messages to General Stefan Rowecki (head of the Home Army, Warsaw) AAN 1326 203/1-27.

30. Cited in Laqueur (1998, 113). Information about the deportation of Jews from Warsaw was initially met with incredulity across the Polish government and Polish National Council.

31. *Polish Jewish Observer*, March 3, 1944.

References

Bloxham, Donald, and Tony Kushner. 2005. *The Holocaust: Critical Historical Approaches*. Manchester: Manchester University Press.

Cesarani, David, ed. 1989. *The Making of Modern Anglo-Jewry*. Oxford: Blackwell.

———. 1998. *Britain and the Holocaust*. London: Holocaust Education Trust.

Engel, David. 1987. *In the Shadow of Auschwitz: The Polish Government-in-Exile and the Jews, 1939–1942*. Chapel Hill: University of North Carolina Press.

———. 1993. *Facing a Holocaust: The Polish Government-in-Exile and the Jews, 1943–1945*. Chapel Hill: University of North Carolina Press.

Engelking, Barbara. 2011. *Jest taki piękny słoneczny dzień: Losy Żydów szukających ratunku na wsi polskiej, 1942–1945*. Warsaw: Stowarzyszenie Centrum Badań nad Zagładą Żydów.

Fein, Helen. 1979. *Accounting for Genocide: National Responses and Jewish Victimisation during the Holocaust*. London: Macmillan.

Fleming, Michael. 2014. *Auschwitz, the Allies, and Censorship of the Holocaust*. Cambridge: Cambridge University Press.

Grabowski, Jan. 2011. *Judenjagd: Polowanie na Żydów, 1942–1945*. Studium dziejów pewnego powiatu. Warsaw: Stowarzyszenie Centrum Badań nad Zagładą Żydów.

Kushner, Tony. 1994. *The Holocaust and the Liberal Imagination: A Social and Cultural History*. Oxford: Oxford University Press.

Laqueur, Walter. 1998. *The Terrible Secret: The Suppression of the Final Truth about Hitler's "Final Solution."* New York: Henry Holt.

Livingstone, David. 2005. "Science, Text, and Space: Thoughts on the Geography of Reading." *Transactions of the Institute of British Geographers* 30: 391–401.

Michlic, Joanna. 2007. "The Soviet Occupation of Poland, 1939–1941, and the Stereotype of the anti-Polish and pro-Soviet Jew." *Jewish Social Studies: History, Culture, Society* 13 (3): 135–76.

Milland, Gabriel C. 1998. "Some Faint Hope and Courage: The BBC and the Final Solution, 1942–1945." PhD diss., University of Leicester.

Paulsson, Gunnar. 2002. *Secret City: The Hidden Jews of Warsaw, 1940–1945*. New Haven, CT: Yale University Press.

Polish Ministry of Information. 1941. *The German New Order in Poland*. London: Hutchinson.

Puławski, Adam. 2009. *Wobliczu Zagłady: Rząd RP na Uchodźstwie, Delegatura Rządu RP na Kraj, ZWZ-AK wobec deportacji Żydów do obozów zagłady (1941–1942)*. Lublin: Instytut Pamięci Narodowej.

Said, Edward. 1983. "Travelling Theory." In *The World, the Text, and the Critic*. Cambridge, MA: Harvard University Press.

Sanford, George. 2005. *Katyn and the Soviet Massacre of 1940: Truth, Justice, and Memory*. London: Routledge.

Stola, Dariusz. 1997. "Early News of the Holocaust from Poland." *Holocaust and Genocide Studies* 11: 1–27.

Ward, Stephen. 1993. "Why the BBC Ignored the Holocaust: Anti-Semitism in the Top Ranks of Broadcasting and Foreign Office Staff Led to the News Being Suppressed." *The Independent*, August 22.

Ważniewski, Władysław, and Krystyna Marczewska. 1968. "W świetle akt delegatury rządu RP na Kraj." In *Zeszyty Oświęcimskie Numer specjalny (I)*, v–xvi. Oświęcim: Wydawnictwo Państwowego Muzeum w Oświęcimiu.

Wyman, David. 1985. *The Abandonment of the Jews: American and the Holocaust, 1941–1945*. New York: New Press.

14

Hello Darkness: Envoi and Caveat

ANDREW CHARLESWORTH

Though I have only now, in 2003, been asked to close this hearty, hopeful set of articles with a dose of realism about humanity—realism about how much humans may do to humanize themselves—I chose my title in Lithuania in July 1995. I was relaxing after a wonderful alfresco meal, prepared by my Polish traveling companions and guides in our railway compartment. But train journeys in Europe can never be enjoyed completely: darkness always finds its way to consciousness. I spent a part of March 1996 in Oświęcim. After three days in the Auschwitz complex with my students and two colleagues, I was quite happy to start all over again. I was shocked by one of my colleagues, a researcher on the landscapes of Italian fascism and hence no stranger to the atrocities of World War II, when he said that he simply wanted to get out of Auschwitz and not come back for a very long time. He had seemed to be, as I was, absorbed in the ambiguities and enigmas of the camp landscapes—so absorbed, in my case, that I had been unaware of the shock to his psyche. I have been fascinated by Birkenau since my first visit in 1989. When I walked into Birkenau, I had a feeling of coming home, but not as a victim.

How are we to connect with the Shoah?—the question is not infrequently asked and the answers have all become conventional. I find that cheap music can be especially potent as a reference point, a sign of connection, linking the ordinary and ridiculous to the pit. Those in the cattle cars had popular tunes, lines from songs sunk deep in their memories, and these would surface from time to time on their journey to Auschwitz. Sitting in the opera house in Lvov watching a production of *Carmen*, Ben Helfgott remarked that the tunes held such memories for him because he had been taught them by a fellow prisoner in the slave labor factory in his home town of Piotrków Trybunalski. The words, the prisoner had changed, but the tunes were the same. To Ben, each time a big tune came up, the opera faded and images of the glass factory occupied the stage. So my subheadings here, all musical, all weirdly potent, will attempt to bridge worlds that we strive to separate by a chasm and yet, in vain, we are forced to acknowledge are one world.

The Name of the Game

Place names are the tracer of nationalism's deadly game across tracts of Central, Eastern, and Southern Europe. Whichever place name I choose to use, someone will be disoriented, someone else, offended: Oświęcim/Auschwitz, Brzezinka/Birkenau, Vilno/Vilna/Vilnius, Kovno/Kaunas, Lvov/Lemberg/Lviv. "Supermarket to be built at Auschwitz" reads differently than "A supermarket is to be constructed in Oświęcim by the entrance of the site of the former death camp of Auschwitz I." The former appears to mark Auschwitz as a place apart, a site occupying a flat, uninhabited landscape; hence, the sense of violation that the construction of a supermarket would bring. The latter sentence does at least make the reader reflect that Auschwitz is situated in a town, in a Polish urban context, and that the supermarket may thus not be so out of place. Jonathan Webber has pointed out some of the difficulties that arise from the false imagery associated with the name of Auschwitz. Given the power of names, it is surprising how little reflection there has been on names that have become so embedded in the Shoah. Take Auschwitz, Treblinka, Sobibór, Chełmno, Majdanek, and Bełżec. No one seems to comment on the anomaly that, of the six death camps, only Auschwitz was given a German name. The others were given Polish names, place names that would be dispensed with, just as the Poles could be, once their purpose had been served. Only German names would be spoken in succeeding centuries; the rest would eventually be silenced for good.

Inside the camps themselves, there are place names that are referred to all the time and not much questioned. Kanada is perhaps the most famous—the subcamp at Birkenau where all the personal effects of those brought to die were sorted and graded before being dispatched into the war economy of the Third Reich and its collaborators. Mexico was the name given to B III, the unfinished third subcamp at Birkenau constructed for the massive influx of Hungarian Jews. Why these continental American references? The topology of the camp has Kanada and Mexico linked by an unnamed territory. Was this gallows humor? Was the United States—land of the free—the unnamed territory linking subcamps B II and B III? But we must not forget that there is a Polish dimension to the camp slang. *Kanada* and *Mexico* were slang terms in a Polish vocabulary referring to the social geography of any Polish town—

Kanada, the wealthy part of town; Mexico, the most dangerous part. Peking was the poorest area of town, but there was no need to invoke that name in a place where all were, in Primo Levi's words, at the bottom. *Texas* in Polish slang is not so much a place as a state of mind, meaning freedom—and there was no need to invoke that quality or quantity in a place where the slogan "Work Bestows Freedom" mocked you to death. "What's in a name?" is a question that well-read sophisticates—we unconventional conventionalists—like to ask. Names that bear the mark of Cain, and mark all of us now living, meant precisely nothing to the Jews who arrived in Kanada or Mexico and lost their lives.

What is Auschwitz? Where is Auschwitz? When will the Poles recover Oświęcim? When they choose to have their dead cremated? Crematories are not likely to be built in Polish towns this side of a hundred years. But why, one wonders, has "To the gas chambers!" as equivalent to "Fuck off!" become so rapidly a part of Polish slang? And why has a local supermarket in Pruszków (near Warsaw), whose design resembles the gas chambers/crematoria at Auschwitz, been nicknamed (according to Colin Richmond) the "Crem?" Is this the blackest humor or something darker than blackest? When I mentioned this example in a lecture in Kraków, I was stopped in mid-sentence by howls of protest from Polish students. Had I touched a raw nerve?

Climb Every Mountain

The *topography* of death we tend to ignore when we read accounts of the destruction of European Jewry. How often do publishers of survivors' memoirs fail to provide a map? Even when they do, as in the case of *Schindler's List* by Thomas Keneally, the contours are not shown. One assumes the Nazis chose flat open spaces for their concentration and death camps. The shock of Płaszów, not captured in Spielberg's film, is its basin-like undulating form.

It was like that in Lithuania, where we seemed to climb and nothing but climb. We were often climbing out of the valley of the Neris, one of the fluvial lifelines of Lithuania, to the places where the shadow of death hung throughout the Nazi occupation. Those climbs were unexpected, despite the fact that on the leaflet from the Jewish Museum in Vilna—our first stop there—is an engraved picture showing a group of Jews climbing as they leave the ghetto on their way to the killing site at Ponary. But there were more surprises to come. The man who asked for money by the railway track at Ponary to direct us to the killing site and when refused pointed in the wrong direction—he was not unexpected. Nor was the fact that the signs to Ponary are regularly destroyed or misaligned. Some Lithuanians wish to make it a hard road to get to, not out of shame, but out of a feeling that getting to the killing site should be hard and uncomfortable. The main railway line at Ponary *was* a surprise and should not have been. The site had been under construction by the Russians for bulk oil storage when the German invasion occurred. Oil could only be brought economically to the site by rail. The vast execution pits at Ponary have nearly all been manicured and monumentalized and so have lost their potency. But one has been saved from impotence, and its clay outlines give some feel of the rawness of the site, raw like an exposed wound that will not heal. The Latin cross placed at the entrance to the site did not evoke forgiveness in me.

In Kaunas, the street that starts the terrible climb to the IXth Fort from the ghetto is named Panerai Street—again the Neris, again the river to the mountaintop. In *Das Rheingold*, the terrible deed, the forsaking of love, is done at the bottom of the river and then Wagner has us climb to the mountaintop for more failings of humanity to be revealed. So I was thinking in the silence as our car climbed from the ghetto to the hill fort. Others had climbed in Lithuania from the river to the hilltop in a jubilant mood—killers and their spectators. IXth Fort is but one of a series of hilltop forts. They became places of spectacle and sport in 1941; the road to IXth and VIIth Forts became a road of celebration, the victors' procession. Before I went to Kaunas, I thought of basketball as a game of American college students and African Americans—clean-cut, noncontact, youthful, enthusiastic, Harlem Globetrotting wizardry. But basketball is a national sport of Lithuania. As a reward, in the summer of 1941, the Kaunas basketball team, after achieving a famous victory, was taken to the VIIth Fort and allowed to shoot into the crowd of Jews imprisoned there. Some of the prisoners would have been fans a few months earlier.

On the ascent to IXth Fort, the Jews would have seen a sign that read "Road to Heaven." *Himmelstrasse* appears to be a name that the Germans assigned only where topography allowed. The rise that the Jews had to walk up to the gas chambers at Treblinka was so named. Irony had its limits—a sign or a word might disturb the Jews but it was not supposed to panic them. On the other hand, what are we to make of Jews, like those in the Kovno Ghetto, naming the local road to death "Via Dolorosa," the name given to the road ascended by another Jew to his martyrdom? Did some Jews believe that there was still a way out, the route the church had offered them endlessly in the past—conversion? In the church at Chełmno, a holding station before they were driven into the gas vans, Jews cried out to Our Lady.

In Vilnius, you climb from the ghetto to the shrine of the Virgin in the Ostra Brama. The devout climb, often on their knees, to pray to the Miraculous Madonna, an icon looking out, now as then, from the Ostra Brama to the site of the ghetto. We met Czesław Miłosz, Nobel laureate, a Pole born in Lithuania, at the foot of this climb—Miłosz, the author of "A Poor Christian Looks at the Ghetto," a poem that challenges anyone who professes Christ as savior but challenges especially the Catholics of Eastern Europe. The Mother of God looked on the Vilna Ghetto, her impassive stare a daily sign to the Lithuanians to bystand. She performed no miracles for Jews. I knelt at the foot of the icon in the Ostra Brama and felt my own faith eroding.

As we began our daily climb from our hotel in Vilnius, as often as not we passed the cathedral. It was only on the last night that we examined the building in detail. How prophetic that the frieze on the gable of the cathedral at Vilnius has represented on it an image of the burnt offering made by Noah in response to being saved from the Flood. For Catholic Lithuanians, the redemption to remember is the rescue of the cathedral from Soviet hands. Yet there the frieze remains, an allusion to the great holocaust of Hebrew Scripture, for those who have eyes to see. As we watched Miłosz walking across Cathedral Square, my companions and I talked about why the frieze had not inspired him to compose a poem.

There Is a Green Hill Far Away

At IXth Fort in Kaunas, now a museum and memorial site, we were shown around with a care and patience that should be the norm at all such places of infamy. How fortunate Lithuania and the world were that the curator of this site was then Zita Nekrosyte, a person of intelligence and energy. Here it was the site itself, not the monument to the dead, that exerted power over visitors. The monument was impressive enough as it towered over us, but it was the awesome size of the meadowland surrounding the fort—thirty-five hectares—that made the expression "killing fields" so suddenly significant. Our belief in the events was tested by the meadow, the sun, the birds. The Russian fortification returned us to normality with its dungeons and cells, dank, dark, and damp—just a curator's nightmare for exhibits and artifacts.

That the huge green hill is under threat was the shock of visiting the IXth Fort. A new housing estate laps up against the killing fields. Forget the supermarket complex at Auschwitz I—that area has long since been developed by light engineering, car repair shops, wholesalers: it was in fact the SS industrial zone. But this site at Kaunas had been untouched by any buildings and monuments not designed to be commensurate with history, and now the sense of place is being hacked away. Until now, roads out of Kaunas have had to deviate around the site, but there are proposals for the housing development to be linked by a new road across the mass gravesite to the main road into town, reducing commuters' costs in terms of time and gasoline, both precious commodities in free market Lithuania. Museum authorities feel powerless against the new capitalists and developers.

In Kaunas proper, there is also a green hill, named Green Hill, on which the depths and heights of the human condition were once reached. At its bottom, on June 24, 1941, nationalist Lithuanians were cheered on by a crowd of bystanders as they beat a group of Jews to death. The event occurred within sight of a place called Peaceful Park or something like that—the Lithuanian adjective does not quite permit a one-to-one translation, and the linguistic ambiguity reinforces the ambiguous nature of a place where children have played, lovers have walked, ceremonies of church and state have been staged, and Jews have been murdered to roars of approval. Still, you can climb Green Hill, as thousands of Jews did in 1941, and at the top you come to the house of the Japanese consul, Sugihara, who signed visas to allow Jews to escape to neutral territories, despite the opposition of the German, Soviet, and Japanese governments. How they and their families must have rejoiced as they descended the hill into Peaceful Park, probably not noticing the Tatar mosque on their descent. The Muslim community of Kaunas was left undisturbed by the Germans and their Lithuanian neighbors, and the mosque is still open today. Did Muslims join the killing crowd in the garage complex?

It was in any case not Sugihara who most moved me on the day of my visit in 1995, but Miss Ulvydaite, a neighbor of Sugihara's in 1941 and still living above the former Japanese consulate. She asks nothing for herself; she dismisses what she and her brother offered the Jews lining up to see Sugihara—use of her toilet, drinks of water, small amounts of food—as negligible. All she wants now is for this foreigner, Consul Sugihara, to be recognized in her home country and for her house, on her death, to be turned into a memorial museum to him. At the end of a day when I had found my anger toward Lithuanians rising step by step, tale by tale, atrocity by atrocity, Miss Ulvydaite, a devout Catholic, a Lithuanian lady in her seventies, gave me the kind of relief that rethinking brings. Can one person redeem a nation? Sugihara was punished on his return to Japan for the aid he gave to Jews.

Looking out from Miss Ulvydaite's window on to the tree-lined slope of Green Hill and beyond to Peaceful Park, I thought of David Kahane's description in *The Lvov Ghetto Diary* of yet another window looking down from a green hill. I remember standing in the children's playground at the foot of that hill, looking up the grassed and lightly wooded slope to the Archbishop's Palace and its highest windows. Kahane's narrative amounts to a layered memory of the many times he had looked out of that window—looked out, not impassively like the Ostra Brama Madonna, but helplessly. Kahane, a rabbi, was hidden in the palace by the archbishop and his staff. Compared to the Ukrainians in the park below, Rabbi Kahane was unfree. Compared with his fellow Jews, on the other side of the city, in the Lvov ghetto, Rabbi Kahane was a free man. Still, his freedom to act was of course severely circumscribed. Like God, he had to watch events unfold. Below was a landscape of pleasure and relaxation. Daily, he must have watched Ukrainians arrive to recreate—then one day a Jewish boy came to the park. The boy kept his distance from the Ukrainians, but the Ukrainian children became suspicious; they questioned him, they sensed something was not right. Parents came to see what was going on, the questioning continued. Finally, a policeman was sent for, and the boy was led away. A place of pleasure had become through the agency of innocents a place of terror. The boy was found out, found guilty, and condemned by people at play. In Lithuania, the Jews were led to Green Hill to die. In Ukraine, one was led away from a green hill and put to death.

The Flowers That Bloom in the Spring

Before the journey to Lithuania, I spent time in Płaszów and Birkenau setting up a research project on the ecological landscapes of these two camps with a colleague, Mike Addis. The flowers of Birkenau, like the hares, the deer, the birds and frogs, are a shock for many visitors. Perhaps that experience explains why the proposed management plan for Birkenau would replace the rich variety of flora and fauna with manicured lawns of grass, one of the most ecologically sterile habitats there is. The rationale is that this kind of sterility comes closest to how the camp looked in the years 1943 to 1945. But in 1943–45, the camp was a sea of mud or dry dirt, depending on the season. As Kitty Hart remembers, if there was a blade of grass, it was immediately eaten. Watching from Kanada, Kitty Hart saw a child in the waiting crowd outside Gas Chamber IV, picking flowers. This act so incensed an SS guard that he snatched up the child and smashed its head against the side of the gas chamber.

Conceivably, the SS guard was incensed because he was himself studying the flora of Birkenau. One SS officer wrote a book on the birds of the area. In that remarkable museum in Podgórze, the site of the Kraków Ghetto, the Apteka, there is a picture of SS guards on the hillock above Amon Goeth's house (Spielberg inverted the topography of the camp to create his Holocaust morality play), a picture that shows them relaxing in the meadowland of flowers and one almost imagines that one guard has been picking the flowers that he appears to hold in his hand. The SS were modern men as regards nature, viewing it as an object of study. It was also a place for Aryans to recreate before returning to the job at hand.

In Lithuania, the job at hand involved the deflowering of Jewish girls, some as young as eight. In the ghetto prison at Vilna, Lithuanians and Germans performed a variety of rapes on Jewish females. Raping daughters in their mothers' presence, performing oral rape on girls as young as eight, forcing them to say how like milk and honey the semen tasted, and then in front of the defiled daughters pulling out every hair on their mothers' bodies. As this information was related to us in the stuffy Jewish Museum at Vilna, I nearly vomited. I wanted to run out of the place, out of the subject. I had the same suffocating feeling, gasping for breath, when Colin Richmond related the details of the oh-so-English pogrom against the Jews of York in Clifford's Tower. Clifford's Tower: my boyhood playground on our annual day trips to York in a coach from Garside's sweet shop. I wonder sometimes if that hilltop fortress, like the mountain in *Close Encounters of the Third Kind*, drew me to uncover its awful secret. How happy was I to throw people off those battlements in my childhood games. I was completely ignorant, at age eleven, of the past of Clifford's Tower. Should children in Vilnius remain ignorant of the inhumanity of their great-grandfathers? The wall of the ghetto prison, still extant, should bear testimony of what happened to those Jewish girls and their mothers at the hands of Lithuanians. Lithuanians have a knack for laying the blame for all that happened to the Jews at the door of the German invader/liberator. The wall of the ghetto prison today bears a small, uninformative memorial tablet and a freshly painted swastika. Both should for decency's sake be removed.

At Birkenau, where the ash is particularly deep, the lovely plant, betony, flowers with its glowing pink. Before coming to Birkenau, Mike Addis had associated this plant with a dear friend of his whose garden was in summer awash in betony; and his home was named Betony Cottage. Mike tells me he will never be able to visit Betony Cottage again without thinking of Birkenau. The Garden of Eden was, I suppose, not so innocent a place either.

Sounds of Silence

Sitting, at Birkenau, on the stones of the harsh, sterile monument to the dead of Auschwitz, my thoughts went back to the fiftieth anniversary of the camp's liberation. I remembered watching on television a rabbi chant the Kaddish while, behind him, two Polish flag bearers, men in their fifties, cracked a joke. (Likewise, Redshirt in Lanzmann's *Shoah* smiles as he mocks and imitates the Yiddish spoken in the cattle cars at night.) Polish anti-Semitism went unremarked by the television commentators; I should perhaps have expected no better. But later that same night, the Channel 4 reporter said that the rabbi had intoned the "Christian names" of the dead; when the error was pointed out, Jon Snow compounded the damage by pleading that on a day of such emotion mistakes of that kind were inevitable. On a day of that kind and in a place of such significance, I would have thought the reporting would naturally be scrupulous. I remembered Richard Dimbleby's famous "Panorama" report on his return to Belsen and the final shot with a cross seen behind one of the mass-grave sites. Have we learned nothing since then? Do we *want* to learn nothing? My letter to the national newspapers in Britain, asking those questions, was met with silence. The Shoah had had its allotted time.

Perhaps I am wrong to complain. The ordinary is what makes the extraordinary extraordinary. Consider the murder of Zalman Osovsky, chief rabbi of Slobodka, in Kaunas. His head was hacked off with an ax or spade by Lithuanians as he learned Talmud. His head was displayed in the front window of his house, overlooking the main road into Kaunas. Today the window looks onto a bus stop. As I was staring, wondering, a bus drew up. What if, I thought, the passengers looked into the rabbi's window and saw his severed head, his opened holy book. What would *they* have seen? The head of a Soviet agent disguised as a religious man? A book in which the ritual of making *matzah* with the blood of Christian children was described? The bus moved off toward the center of Kaunas, a journey of probably ten minutes. I wished that that window could display a photograph of that head forever and a day but knew that we have all agreed to be silent—except in the extraordinary demarcated places of what Alan Bullock called the geography of hell. Yet, in Europe anyway, there are openings to hell at nearly every railway station, town center, bus stop. At the conclusion of that ten-minute bus journey in Kaunas stands the cathedral, outside of which there is a bust of Archbishop Skvireckas, the metropolitan of Kaunas at the time of the Jewish massacres. Skvireckas broadcast greetings to the German liberators and offered prayers for their well-being. When asked to send a message to his compatriots condemning the slaughter of Lithuanian Jews, he said nothing.

We should be careful about laying blame: the conspiracy of silence is fairly universal. At Birkenau, I watched a German in his fifties, with camera and tripod, arranging to photograph his three teenage daughters on the new ramp. He painstakingly positioned them, barking his orders, so they would be exactly framed by the infamous entrance gate. And this, on a day with the temperature at least 100 degrees Fahrenheit—weather from hell. Forgive him, for he knows not what he does? To hell with forgiveness, I thought; he knows just what he is doing. His age, his thoroughness, his barked orders, made his intent clear. But I did nothing, said nothing, stood by in silence. Did that make me a bystander or a participant?

Or take a different silence, again at Birkenau. Walking along the new ramp, you often meet groups carrying flowers to lay in the gas chambers or on the memorial at the end of the railway track. On the same day as I watched the barking father and his teenagers, I met a group of German youth bearing flowers, walking toward the memorial. When I arrived back at the gatehouse, I saw a group of Canadian Jewish youth alighting from their bus and assembling with flags. (Flag carrying should be banned from Auschwitz as behavior inappropriate to the place.) I knew that this group of Jewish youth would meet the returning group of German youth at the very point of selection. They would pass each other without a word, probably without a glance. What an opportunity missed—no reconciliation here, no hope there. The museum authorities meanwhile expend their creative energies turning the site into a sort of theme park.

Vollendet das Ewige Werk

So here I am, returned, as always, to the ramp at Birkenau—all railways seem to lead here, my journey's end, the place I journey to meet myself. But now, for the first time, the scale of the renovation and reconstruction hits me. After climbing to the top of the guard tower, all I can now hear as I look out from the main gate are words of Wotan:

Vollendet das ewige Werk
Mannes Ehre
ewige Macht
ragen zu endlosem Ruhm . . .
Wie im Traum ich ihn trug
Wie mein Wille ihn wies
Stark und schon
Steht er zur Schau
Hehrer, herrlicher Bau.

The eternal work completed
Man's honor
eternal might
stretch out to endless fame . . .
Just as I saw it in dreams
just as my wishes intended it
strong and beautiful
it stands on show
majestic, marvelous building.

And here is the camp the SS wanted—orderly, clean, and *judenrein*. Unwittingly, it is this image that the museum authorities are creating: the sterile, spotless ramp, the manicured lawns of grass alongside the ramp and between many of the huts, huts themselves under active reconstruction, the bright red tiles of the new half-roof over the entrance. At any moment you are expecting the next train to arrive so that this SS Camelot (remember Arthur's song), having made itself *judenrein*, can extend its achievement worldwide. In fact, a traffic sign outside the entrance warns you of an imminent train.

The vast expanse of Birkenau can still—but for how much longer?—take the breath away. I see a French teenager as she emerges from the claustrophobic stairs of the gatehouse; she tries to calm herself with a sharp intake of breath and exhales a *C'est énorme*. The boys I watch stand uneasily, twitch, or try to look macho through sly glances at each other. But they know, in that same split second, that before them lies the most masculine landscape the world has ever seen—right-angled, straight, obsessed with ranking. Here machismo was manifest at the height of its rationality, the full extent of its muscular logic. That French girl, just into puberty—into her capacity to give life—was confronted, as she lost her breath, by the power of undeterred masculinity to destroy lives, more than a million in this place alone. Wotan greets the completion of his great work Valhalla (created by males for males) with the most phallic of objects, his spear.

Which is why I say: nature should take over Birkenau. Brzezinka should, like Dunsinane, come again—return the sacred birch grove, close up the wound (for the scar will never go), smooth out this geometric maleness, lend us hope, give us protection and vigilance against all this happening again. A reconstructed, a recreated, male landscape will do nothing of the kind: it will mesmerize us with the power of logic, the right (angles) of rationality. The manicured landscape will give rise to comments like those overheard by my students—American tourists complaining that Birkenau should have waxwork dummies dressed in SS and camp uniforms to bring the place back to life. Theme Park Auschwitz is but a short step from its present tidy, ordered re-creation. Let us hope that Samuel Willenberg, imprisoned at Treblinka, was just a prisoner and not a prophet. He wrote:

> "If what you're saying is true," I said, "the bastards want to set up the illusion of a little heaven on earth. I understand why Sidow ordered us to hoe the ground around the huts and to plant some flowers. The next thing they'll do is set up a merry-go-round and turn this death factory of theirs into an amusement park."

I have reached my journey's end. But indulge me in one last visit to the neglected old ramp at Birkenau with the rail spur to the camp still intact. It now passes through people's gardens and, with the points frozen in an indeterminate position, waits for someone to open or close the way to the new spur forever. The locals, the museum authorities, the living marchers, have no inclination to be that someone; and the latter two never go there. Are you that someone? I am wondering. Or is there someone yet to come after you?

Acknowledgments

The author wishes to thank the University of Gloucestershire for a grant that made his travels possible. He extends gratitude as well to the late Andrzej and Bogu Nowodworski, Colin Richmond, Pamela Hope Levin, Irena Guzenberg of the Jewish Museum in Vilnius, and Zita Nekrosyte of the IXth Fort Museum at Kaunas. The last person who is clearly identified in this essay has been removed from her position at the museum.

SECTION 4

Microgeographies of Memory, Witnessing, and Representation

15

The Interruption of Witnessing: Relations of Distance and Proximity in Claude Lanzmann's *Shoah*

RICHARD CARTER-WHITE

Introduction

Claude Lanzmann's *Shoah* (1985) is one of the most celebrated, important, and controversial films to take the Holocaust as its subject matter (see Skloot 2012). A genre-defying meditation on the killing processes of the Nazi death camps told through the accounts of a wide range of witnesses, the film has been lauded for its unique emphasis on the material sites of the destruction, with the director himself describing *Shoah* as "a topographical film, a geographical film" (Chevrie and Le Roux 2007, 39). In this chapter I explore an alternative interpretation of this phrase, by analyzing a more sophisticated spatial logic at play in the film. Specifically, I argue that *Shoah*'s performance is based upon a paradoxical series of *distances* and *proximities* that can help inform our understanding of the historiographical force of witness testimony.

In recent decades Holocaust testimony has emerged as a significant force in the public sphere, across multiple literary and audio-visual genres (see Wieviorka 2006). The legitimation of testimony—particularly survivor testimony—as an authoritative mode of narrating the Holocaust has been reflected in changing perspectives on the role of witnessing within historiography. While some prominent historians of the Holocaust have shared a distrust of testimony due to the inevitability of factual errors in subjective personal accounts (for example Dawidowicz 1975), this position has been countered by recognition of the complexity and historiographical value of survivors' perspectives, beyond their "accuracy" as a source of historical data (Carter-White 2009). Reflecting on his own historical account of the Holocaust, Saul Friedländer observes that "an individual voice suddenly arising in the course of an ordinary historical narrative of events such as those presented here can tear through seamless interpretation and pierce the (mostly involuntary) smugness of scholarly detachment and 'objectivity'" (Friedländer 2008, xxvi). Similarly, while historians have generally limited the incorporation of witness testimony into their narratives for fear that the emotional content of individual memory might compromise the critical distance necessary for the production of historical knowledge (Wieviorka 2006, 90), Friedländer observes that such an intervention is in fact "essential to the historical representation of mass extermination and other sequences of mass suffering that 'business as usual historiography' necessarily domesticates and 'flattens'" (Friedländer 2008, xxvi). If the critical perspective of history has been conventionally associated with a sense of objective "distance," then, witness testimony has become valorized within interdisciplinary debates about the historiographical normalization of the Holocaust for precisely the way that it interrupts the distancing effects of historical representation (Hatley 2000). Implicit in this spatial imaginary is the possibility that witness testimony offers some kind of contrasting *proximity* to the Holocaust; the transmission of experiential knowledge as a more visceral trace of the event (see Felman and Laub 1992; Bernard-Donals and Glejzer 2003).

The recognition that testimony might offer a valuable sense of proximity that is otherwise lost in historical writing is complicated by work drawing attention to alternate types of "distances" inherent in the act of witnessing. Paul Harrison draws on the testimonies of Primo Levi, Elie Wiesel, Jean Améry, and Charlotte Delbo to demonstrate a common acknowledgment among Holocaust survivors of the inadequacy of language to describe their experiences; instants that "open gaps in the text, within the words of the text itself, between the referent and reality, between representation and experience and, perhaps first and foremost, a gap between the addressor and the addressee" (Harrison 2010, 162). These gaps can be attributed to a disjuncture within testimony itself, between the instant of witnessing and subsequent acts of recollection (Lyotard 1988; Derrida 2000). Testimony is necessarily dependent upon the delay between these moments of witnessing, during which the intervention of language and subjectivity leads to uncertainty not simply about the accuracy but more fundamentally about the intentions and nature of any given act of witnessing (Carter-White 2012). If Holocaust testimony promises to deliver some kind of proximity to the event, it is a proximity permeated with distance.

The complex spatiality of testimony has assumed a place in wider debates on Holocaust representation partly through increasing interest in the dynamics of trauma in witnessing. As Dominick LaCapra writes, the inability of those who live through traumatic events—who are most proximate to those events—to fully integrate their experiences into memory without gaps and absences in recollection "raises the issue of the way in which the historian or other analyst becomes a secondary witness, undergoes a transferential relation, and must work out an acceptable subject-position with respect to the witness and his or her testimony" (LaCapra 1998, 11). By highlighting the personal and ideological implication of scholars in their narratives of the Holocaust, the paradoxical spatiality of testimony—as expressed in trauma—opens up onto ongoing debates on the relationship between history and memory and the potential for a more supplementary relation than the traditional binary opposition that underpins much of Holocaust representation (ibid.). In the absence of direct access to traumatic experience, cultural representations are an important arena in which to trace the interaction of history and memory (see Rothberg 2000), and *Shoah* provides a particularly distinctive opportunity to explore the spatial imaginaries of proximity and distance through which witness testimony has been incorporated into Holocaust historiography.

With its nonchronological structure, exclusion of archival images, and staged elements, *Shoah* is in no straightforward sense a documentary film (Burns 2002). It is a film about witnesses, but moreover it is a film about the dynamics of witnessing itself—the complexity, immense labor, and spatial ambiguity of bearing witness. In the first part of this chapter I suggest that Lanzmann's choice of witnesses, depiction of material sites, and interview style are indicative of a filmmaking strategy aimed at gaining the greatest possible proximity to the historical reality of the Holocaust. However, this starting point opens up a series of proximities and distances in the grammar of the film and in the experience of viewing that considerably complicate the spatial logic of *Shoah*. I argue that the relations of distance and proximity expressed in this film demonstrate the necessity of a historiography of testimony based on the indeterminate spatiality of witnessing.

An Accumulation of Proximities

Over the course of its nine and a half hours *Shoah* unfolds a complex geography incorporating the sites of former concentration camps and their surrounding areas in Poland; West German urban landscapes; various domestic, countryside, and resort settings in Israel; government and administrative buildings in Washington, DC; New York highways; and endless lines of Polish railways. Interviews with witnesses are split between two types of locations: what Margaret Olin refers to as "places of refuge" mainly in the United States and Israel, and the "tainted ground" of the "sites of extermination" (Olin 1997, 2) in Poland.

The prominence of physical locations in *Shoah* can be partly attributed to the director's objection to a particular type of distance established between image and viewer in conventional Holocaust documentary. Lanzmann describes archival footage of Nazi violence as "images without imagination" (Chevrie and Le Roux 2007, 40) whose illusion of direct representation relieves viewers of the effort of understanding, while he similarly critiques the conventional use of voice-over to simulate the authority of institutionalized knowledge and thereby impose an external interpretation on events that suspends any imaginative interaction between film and audience (ibid.). For Lanzmann, such conventional documentary techniques produce the Holocaust as a spectacle (Liebman 2007, 14), kept at a safe distance from the present and from the imaginative involvement of viewers. This amounts to "the worst moral and artistic crime that can be committed in producing a work dedicated to the Holocaust" (Lanzmann 2007, 35), which is to frame the Holocaust as "past" and therefore amenable to comprehendible representation, rather than as an ongoing present that continues to haunt both people and places in its singular violence (Clarke, Doel, and McDonough 1996).

In place of "direct" representation and the comfortable distance of spectacle, Lanzmann substitutes an almost childlike literality in his efforts to gain proximity to the historical event of the Holocaust. This literality can be observed in his choice of survivor witnesses to feature in the film. Primo Levi observed that the "true witnesses" of the Nazi genocide are those who were killed (2004), but in the absence of this impossible testimony Lanzmann sought out witnesses who were as close as possible to the killing without falling victim to it: notably Filip Müller, a former member of the Auschwitz *Sonderkommando*, the division of prisoners responsible for the preparation, sorting, and burning of those killed in the gas chambers; Abraham Bomba, a barber who cut the hair of prisoners as they waited in the gas chambers of Treblinka; and Simon Srebnik, a thirteen-year-old boy at the time of his imprisonment at Chełmno who survived a gunshot wound to the head shortly before the camp's liberation. Apart from their past physical and organizational proximity to the killing, in their interviews these witnesses also display an ongoing emotional proximity to the traumatic event. LaCapra (1997) notes Lanzmann's directorial preference for witnesses who visibly act out their memories in the form of emotional breakdown over those who have mastered and incorporated the past into a coherent narrative, which is to say, a preference for witnesses who have not imposed a psychic distance between themselves and the past. "One must start with the naked violence" (Lanzmann 2007, 34).

The accumulation of proximities in *Shoah* reveals itself in Lanzmann's unique interviewing technique. Lanzmann's strategy was not simply to have each witness recount their experiences but rather to "revisit" the past (Didi-

Huberman 2007, 117) by enacting various types of *return* to the killing sites. Most obviously, he *physically* returns his witnesses to the material sites where extermination camps were situated. Thus, former assistant switchman Jan Piwonski acts as a guide through the forest of Sobibór, providing a geometric mapping of the former "camp site":

LANZMANN: Exactly where did the camp begin?
PIWONSKI: I'll show you exactly. Here . . . there was a fence that ran to those trees you see there. And another fence, that ran to those trees over there.
LANZMANN: So I'm standing inside the camp perimeter, right?
PIWONSKI: That's right.
LANZMANN: Where I am now is fifty feet from the station, and I'm already outside the camp.
PIWONSKI: Yes.
LANZMANN: So this is the Polish part, and over there was death.
PIWONSKI: Yes.
[. . .]
LANZMANN: [Pointing] The ramp began there. [Walking] So here we're outside the camp, and back here we enter it. Unlike Treblinka, the station here is part of the camp. And at this point, we are inside the camp.
PIWONSKI: This track was inside the camp.
LANZMANN: [Pointing] And it's exactly as it was?
PIWONSKI: Yes, the same track. It hasn't changed since then.
LANZMANN: [Walking] Where we are now is what was called the ramp?
PIWONSKI: Yes, those to be exterminated were unloaded here.
LANZMANN: So where we're standing is where 250,000 Jews were unloaded before being gassed?
PIWONSKI: Yes.

What is striking about this exchange is that, as viewers, we have little sense of *where* Lanzmann, Piwonski, and Barbara Janica, Lanzmann's translator, are standing, because we are provided with no wider spatial context. The scene begins with the three already sitting on a railway station platform, and as they wander around an area of grassland the camera remains at ground level and circles around the protagonists, occasionally deviating to follow one of Piwonski's gestures into the distance but never providing an orienting sense of location. We are instead left with an assertion of pure proximity. The same is true of Simon Srebnik's description of the former site of Chełmno, now a peaceful countryside landscape; even as the site elicits his testimony, the camera is fixed on his face and our attention on his words. "It's hard to recognize, but it was here. They burned people here. A lot of people were burned here. Yes, this is the place."

A second type of return enacted in the film is achieved through the *practices* of the witnesses. One controversial aspect of *Shoah* is that, despite its documentary appearance, several interviews contain staged elements: Lanzmann rented a barber shop for Abraham Bomba to trim a "customer's" hair while describing his experiences in Treblinka, and hired a locomotive for former train driver Henrik Gawkowski to complete the route that he used to take when bringing Jews to the same camp; in neither case is it revealed that these setups are staged. The film opens with Simon Srebnik repeating the same journey down the Ner River that he made to dispose of the ashes of victims, singing the same song that would entertain the SS and keep him alive for another day. These provide clear examples of the accumulation of proximities in *Shoah*: with Gawkowski and Srebnik, Lanzmann takes the witnesses *who were there* back to the place *where it happened* to repeat some of the things *that they did*. Although Bomba's act of witnessing takes place far away from the scene of his suffering, the act of describing cutting the hair of women before they were gassed while repeating the embodied practice of cutting hair brings about an emotional breakdown. In this case, it could be argued that the carefully staged repetition of practice causes a third type of return, the *emotional* return of traumatic recollection (see LaCapra 1997).

The staged elements of *Shoah* are for Lanzmann born of the necessity of turning witnesses into "characters" (Chevrie and Le Roux 2007, 44–45). Witnesses remember, and in Lanzmann's view memory is the institution of distance between past and present; by contrast, his aim for the film was "the abolition of all distance between past and present" (ibid., 45) by placing witnesses in conditions where they might relive the past in unexpected ways, even if this amounted to creating a situation that would provoke harrowing recollection for the witness.

In the process of turning witnesses into "characters," Lanzmann aims to highlight continuities between past and present, to collapse distance and generate proximity. But proximity to what? These disarmingly literal attempts at getting close to the event—going to the place, finding the nearest people, making them recreate their actions—are constantly interrupted by the abyss that separates the peaceful Sobibór of the 1980s from that of the extermination camp active from 1942 to 1943, and that distinguishes the middle-aged Simon Srebnik who sings along the Ner River at the behest of a film director from the thirteen-year-old Simon Srebnik who did so en route to emptying sacks of human remains into the water. As viewers we are forced to maintain the "contradictory essence" (Skloot 2012, 263) between the multiple proximities assembled by Lanzmann and the immeasurable distance that remains in every frame. With every push against the barrier separating past and present, the absolute density of that barrier is registered.

A Geography of Trauma

If the interplay of places, characters, and practices in Lanzmann's interviews provides a remarkably literal gesture of proximity toward the Nazi genocide, which is ultimately generative of a heightened sense of distance from the space-time of the killing, the rhythm of *Shoah* is characterized by a comparable dynamic of immediacy and belatedness. This is illustrated by a striking sequence two hours into the film. From the harrowing testimony of Treblinka survivor Richard Glazer the film abruptly cuts to an elderly couple happily dancing together in an empty but plush dance hall. Having become accustomed to pastoral scenes and agonized expressions over the previous two hours, the effect of this transition to a jolly scene is shocking and perplexing, and then with no explanation the visuals cut away to a Berlin highway at night. As we proceed along the busy highway the soundtrack from the dance hall is gradually replaced by the testimony of an unidentified witness describing the persecution of Jews in Berlin. Only after minutes of testimony do we learn the significance of the previous scene; the still-unidentified witness describes how the Jews were concentrated in a large dance restaurant before being separated into numerous transports. As the witness talks about the guilt she feels in having escaped this fate she is finally identified, first visually amid scenes of trains departing from a station and finally by name: "Inge Deutschkron. Born in Berlin. Lived there throughout the war. In hiding from the beginning of February 1943. Now lives in Israel." As this testimonial micronarrative settles into belated comprehension the film makes another unexpected jump, to a van parked in a quiet street. A slow zoom shot reveals that inside the van works a camera team secretly recording an interview between Lanzmann and "Franz Suchomel, SS Unterscharführer," and a new micronarrative for the viewer to unravel begins.

The belatedness of this sequence derives from the montage style used throughout the film, whereby the visuals frequently cut from one location to another without any obvious connection or explanation, while a voice that may be familiar from earlier in the film or may be entirely new offers testimony that relates in a more or less oblique way to the unfolding scenes. It is the viewer's responsibility to piece these elements together retrospectively, waiting for moments of convergence between sound and image that might connect with the emerging themes of the sequence, before being plunged back into confusion with the next unexpected cut. And it is here that immediacy strikes: not with "stock images of shock" but with "a shocking clash of images" (Doel and Clarke 2007, 890) that with each juxtaposition forces the viewer into direct acknowledgment of their incomprehension. At these moments the viewer perceives but does not understand the significance of the unfolding event; subsequently, with the delayed understanding of previous scenes, the shock of immediacy is forgotten and comprehension settles into place, ready to be interrupted again by the next cut.

This dynamic of immediacy, belatedness, and repetition is analogous to the structure of traumatic recall. As Cathy Caruth writes, traumatic experience is inassimilable to thought and comprehension at the moment of its occurrence; psychic trauma "registers the force of an experience that is not yet fully owned" (Caruth 1995, 151). Thus, the immediacy of a traumatic event is, by definition, *perceived* but not *understood*. Because it has not yet been "fully owned" it remains "largely inaccessible to conscious recall and control" (ibid.). Traumatic recall, in flashbacks and other embodied forms of reliving (see Bondi 2014), is instead fundamentally characterized by belatedness; in the absence of conscious integration into knowledge, it continually recurs, unbidden, after the event. Efforts to integrate traumatic reliving into a coherent narrative entail "the loss, precisely, of the event's essential incomprehensibility" (Caruth 1995, 154), the overwhelming immediacy of the event at the moment of its occurrence; analogously, the recomposition of *Shoah*'s fragments into a narrative whole entails a forgetting of its disorientating cuts and juxtapositions.

The rhythm of *Shoah* can thus be considered according to a geography of trauma: a distance between perception and understanding that can only be closed at the cost of effacing the shocking proximity to an event of incomprehension. For English-speaking audiences, this is most vividly illustrated by the use of translation in the film. Several languages are featured in *Shoah*, and while the accounts of witnesses speaking in German and French are subtitled in the conventional documentary style, those spoken in Polish, Hebrew, and Yiddish are not. Audiences who do not speak these languages—including Lanzmann—have to wait for a French translation provided by a translator in real time, and only this is subtitled. The inclusion of both the original account and the French translation of so many witness testimonies not only contributes significantly to the running time of the film, it also transforms the testimonial interview—a staple of Holocaust documentary (Kerner 2011)—into a disconcerting experience for viewers.

Inherent in any act of translation is the possibility that, in the transition between two languages, some change in the meaning of a statement has taken place. Translation therefore operates in an "economy of loss" (Stoicea 2006) that mirrors the structure of trauma. The belatedness of traumatic recall is predicated on an original distance separating the traumatic event from the comprehension of its victims, and in an analogous fashion translation proceeds according to the presupposition of a distance between languages that cannot ever be fully closed; instead, it has to be bridged, with the necessity of this mediator indicating the distance that remains. This is highlighted in exchanges where Lanzmann interrogates the veracity of his own translator:

TRANSLATOR: The Poles had to serve them and work.
LANZMANN: I heard her use the word "capital."
TRANSLATOR: The capital was in the hands of the Jews.
LANZMANN: Yes . . . You didn't translate that. Ask her again. So the capital was in the Jews' hands?

In translation and trauma alike there exists a pervasive sense of inaccessibility, and through repetition—another central component of traumatic recall (Caruth 1996, 2)—Lanzmann attempts to gain greater proximity to the truth of witnessing. But each repetition amounts to a new iteration of the same abyss that structures translation and reaches back to the original distance that propels each survivor's testimony: the struggle to translate trauma into an intelligible form for their interlocutor and for themselves (Stoicea 2006; Harrison 2007).

While the director has the option of questioning and prompting his translator, the viewer of *Shoah* has no opportunity to enter into such dialogue. The only option is to wait, and search expressions, gestures, and recognizable phrases for clues about the translation as yet undelivered. An infamous scene outside of Chełmno Church presents a group of local Poles, celebrating the birth of the Virgin Mary and steadily growing in number, happily recognizing and greeting Simon Srebnik. Due to the large number of speakers this scene presents a logistically difficult task for the translator, and most of the discussion remains untranslated. Lanzmann speaks directly to a couple of individuals in a translated exchange, and, as he asks the crowd why they believed the Jews were killed, one man takes center stage. In between intervals of translation non-Polish-speaking viewers are left to speculate on the meaning behind his gesticulations, the laughs and jokes shared among the crowd, and the polite expression on Srebnik's face. This scene must be important—why else would the director dedicate so much time to it?—but comprehension is suspended. We *see*, but we do not *understand*. Eventually, the translation informs us that the man had been explaining that the Holocaust was expiation for the

Jews' role in the death of Christ. This belated knowledge simultaneously casts a new light on the preceding scene while clearly falling short of a comprehensive transcription of the multiple interactions of the group discussion. It is *belated*, it is *inadequate*, and it nonetheless strikes with the immediate force of a revelation. Distance and proximity: a geography of trauma.

If the traumatic structure of *Shoah* estranges the viewer from unproblematic comprehension of the unfolding events of the film, this estrangement is generative of yet another form of proximity. The various ways in which the film incorporates the dynamics of trauma into the viewing experience might be characterized as "necessary to transmit to the viewer a muted trauma required for empathetic 'understanding'" (LaCapra 1997, 250). Through the film's paradoxical distances of belatedness and immediacy, it positions the audience as witness to the psychic geography of the survivors.

An Ethical Distance of Viewing

> Nobody meets anyone in *Shoah*. I already said this, but there is a corroboration in spite of this—I make them meet. They don't meet actually, but the film is a place of meeting.
> CLAUDE LANZMANN, RUTH LARSON, AND DAVID RODOWICK (1991, 84)

Lanzmann's words apply as much to the places as the people of *Shoah*. Throughout the film Lanzmann juxtaposes places typically separated by distance, time, and circumstance in a manner that exemplifies the multiplicity of space (Massey 2005) and acknowledges the wider historical context of the Holocaust within the Nazis' plans for *Lebensraum* (see Doel and Clarke 2002; Giaccaria and Minca 2011). Crucially, the film activates this geography of distance and proximity in the imagination of viewers by foregrounding the responsibility of the viewers in making sense of *Shoah*'s opaque structure.

Shoah is famous for its exclusion of archival images of the genocide, but such images are not entirely absent. LaCapra points out that the ironic quality of scenes shot in the peaceful sites of camps and ghettos depends on the audience's prior familiarity with archival documentary footage (1997, 243). Lanzmann makes explicit use of the overwhelming presence of these absent images through ambiguous conjunctions of image and voice-over. Early in the film Lanzmann visits a forest in Israel with two survivor-witnesses of Vilna, where they initially discuss similarities between the forestry of the two places. The witnesses proceed to describe the act of digging up bodies from mass graves, and as they do so the camera pans over the dug-up landscape, with smoke rising in the background. Exposed to this conjunction of sound and image, it is difficult not to recall those iconic archival images to the extent that this color, contemporary scene itself takes on the character of a piece of archival footage, transforming before our eyes. Through an editing style that uses the absence of anticipated imagery, and depending on the recognition and imagination of the viewer when confronted with this absence in the film, the *here and now* transforms into the *there and then*: Vilna is mapped onto Israel, and contemporary footage becomes a stage for the play of excluded archival imagery.

This play of contemporary and archival footage is made possible by an initial observation about the relative thickness of forestry in Israel and Vilna. In the absence of an obvious chronological narrative the structure of *Shoah* hinges on Lanzmann's obsession with such details, using them to weave a multitude of overlapping micronarratives. An example of this can be traced over the first hour of the second half of *Shoah*, beginning with Lanzmann's interview with Franz Suchomel. Lanzmann questions the former SS officer about the processes beginning with the arrival of prisoner transports at the railway ramp of Treblinka and ending with death. In the course of his descriptions Suchomel mentions "the funnel," a path lined with barbed wire leading into the gas chamber building. Lanzmann shows particular interest in this subject, pausing as he gazes at the map of Treblinka pinned to the wall before asking:

LANZMANN: Can you describe this "funnel" precisely? What was it like? How wide? How was it for the people in this "funnel"?

SUCHOMEL: It was about 13 feet wide.

[Lanzmann and Suchomel both gesture from one side of the room to the other.]

SUCHOMEL: As wide as this room. On each side were walls [pointing upward] this high or this high.

LANZMANN: Walls?

SUCHOMEL: No, barbed wire. Woven into the barbed wire were branches of pine trees. You understand? It was known as "camouflage." There was a "Camouflage Squad" of twenty Jews. They brought in new branches every day.

LANZMANN: From the woods?

SUCHOMEL: That's right. So everything was screened. People couldn't see anything to the left or right. Nothing. You couldn't see through it.

The interview continues on these terms, before moving on to discuss the dimensions of the camp accompanied by footage of the contemporary site of Treblinka. The conversation then returns to the funnel:

SUCHOMEL: You had to climb up to [the gas chamber].
LANZMANN: The "funnel" was called the "Road to Heaven," right?
SUCHOMEL: The Jews called it the "Ascension." Also "The Last Road." I only heard those two names for it. [. . .]
LANZMANN: The people got into the "funnel." Then what happens? They're totally naked?

Suchomel goes on to describe the process of organizing prisoners before and during their entrance into the gas chamber. The film then cuts to an interview with Treblinka survivor Abraham Bomba in what appears to be a barbershop—presumably Tel Aviv, based on the location of Bomba's previous interviews in the film. Here again conversation briefly turns to the funnel, which Bomba insists the Nazis called the "Road to Heaven." The next scene returns to contemporary Treblinka, with the Suchomel interview providing voice-over. The voice-over begins with Suchomel again referring to the funnel, and specifically the "death-panic" that overwhelmed prisoners waiting to die. As he describes this scene the camera moves forward through a narrow clearing between trees, with the site of the gas chambers in the background and centered on the screen, in a manner that is evocative of Suchomel's description of the funnel, before the visuals return to the covert footage of Lanzmann and Suchomel in conversation. Suchomel goes on to discuss how certain prisoners were led to the "infirmary" rather than the gas chamber, and as this description continues the visuals again return to the contemporary site of Treblinka. Suchomel's voice-over describes the "passage" that led to the infirmary while the camera proceeds through a narrow clearing between trees reminiscent of the "funnel" previously depicted. At this point the image of a funnel- or passage-shape has been so densely woven into the grammar of the film that it is difficult not to see it repeating itself everywhere. As Suchomel concludes his description of killing at the "infirmary," and while the visuals pan over the ground of the contemporary site, the film jump cuts to Treblinka survivor Richard Glazer sitting at a city riverside, and as he describes the "narrow site" of the "infirmary" the camera closely follows boats passing through narrow arches beneath a bridge—here, the funnels and passages of Treblinka are mapped in the unlikeliest terms onto a pleasant urban scene in Switzerland. Glazer directly compares the "narrow passage" of the "infirmary" to the "funnel," and shortly afterward the film jump cuts to a train following its own narrow passage in "Auschwitz today," with the voice-over of Auschwitz survivor Alfred Vrba. The visuals quickly cut to Vrba: his location is not disclosed, although viewers may remember from previous interviews that he is in New York, and in doing so recall the scenes of vehicles passing under bridges on busy New York highways from earlier in the film. His description of the ramp at Auschwitz is intercut with shots of trains passing through the contemporary sorting station. The visuals jump again to the site of Auschwitz, and then again to a plaster model recreation of terrified prisoners crammed through a long, narrow passageway leading to the gas chambers of crematoria II and III of Auschwitz, narrated in voice-over by Auschwitz survivor Filip Müller (although again he is not identified). As Müller's description of the process and spaces of gas chamber killings continue, the visuals cut to the contemporary site of the crematoria, and also to the heart of the film's focus on the killing process of the camps.

In the middle of this nine-hour film and over the course of one hour, Suchomel's Austrian hotel room, Treblinka, Tel Aviv, Switzerland, New York, and Auschwitz are juxtaposed upon one another by the abstract image of a funnel, one that emerges from Lanzmann's initial insistence that Suchomel's description should bring us *as close as possible* to the exact spatial dimensions of Treblinka, and the way in which this insistence expands within the viewer's imagination. This technique is used throughout the film: as Liebman (2007) notes, *Shoah* possesses a loose structure, but it is a structure revolving around the capacity of *specific details* (see Brinkley and Youra 1996) to intersect and tie together disparate people and heterogeneous places within the space-time of the Holocaust rather than the conventional chronological structure of historical documentary.

The examples cited here emerge from my subjective interpretation of the film, and this is a critical point. *Shoah* is loaded with resonant imagery and editing, but the experience of viewing the film is fundamentally propelled by the withdrawal of authorial intervention. As viewers we are given no clear guidelines about the meaning of *Shoah*'s narratives, jump cuts, and imagery; we are left to fill that distance ourselves. In the absence of an authoritative narration of the unfolding scenes, the responsibility of the viewer for making an *imaginative engagement* with the film emerges. It takes the active involvement of the viewer to piece the puzzle of *Shoah* together, to move closer to that world and take something comprehensible from it. "Nobody meets anyone in *Shoah*," but the film is nonetheless a "place of meeting." By establishing these distances in the experience of watching the film, by refusing to represent the Holocaust in a definitive manner or provide any other overt guidelines for viewing, *Shoah* creates the conditions for a more intimate, singular encounter between viewer and the event (Saxton 2007; Skloot 2012), an ethical proximity, beyond representation.

Conclusion

Given that the testimonial accounts featured in *Shoah* are not those of the director, the question of whether a film of such artistic vision "qualifies" as an act of witnessing remains open (see Apel 2002). Whether or not *Shoah* is better categorized as an act of secondary witnessing or even an outcome of uncontrolled transference and acting out (LaCapra 1997), the geographical logic it exhibits can be productively read alongside the indeterminate spatiality of testimony. The senses of proximity present in *Shoah* arrive as both the outcome of and foundation for a series of distances that interrupt and suspend coherent interpretation. In the same way, Holocaust testimony is too complex a genre to be associated definitively with proximity to the event whether or not this association is meant positively or negatively; its belatedness and imbrication in the drift of language (Agamben 1999) are constitutive features that ensure the experience of reading testimony is riven with uncertainty about the authorial intentions and generic categorizations of individual acts of witnessing, riven, in other words, with gaps, absences, and distance. The historiographical input of testimony should therefore not be limited to the personal idiosyncrasies that would either invalidate it as a source of historical data or valorize it as an injection of emotional proximity, depending on one's historiographical perspective, and it is here that Lanzmann's "geographical film" provides a model for developing a historiography of testimony.

Despite Lanzmann's efforts to frame *Shoah* as an effort toward the abolition of distance between past and present—a type of proximity with which survivor testimony is commonly associated—the length, structure, and performance of the film institute multiple distances in the experience of the viewer that threaten to confuse, frustrate, and even bore. Yet it is precisely the institution of these distances that demands from the viewer a "geography of individual responsibility in which distance is not allowed to disable engagement" (Hones 2010, 479). In the absence of narration, chronology, or any other standard conventions of the documentary film genre, it is up to the viewer *in the present act of viewing* to forge a meaningful proximity with the past. In *Shoah*, distance becomes the motor of proximity. The constitutive uncertainty of witnessing has an analogous capacity to interrupt the interpretive frameworks of its audience, and in addition to the historical data and emotional connections that can be derived from testimony, the case of *Shoah* demonstrates the value in exploring the historiographical potential of such acts of interruption. This amounts to a subtle shift in how we approach the historiography of witnessing not only in terms of the fragments that can be used to supplement historical discourse (Waxman 2012), but also in terms of how specific witness accounts address and engage their audiences in an indeterminate spatiality of proximity *and* distance. Such a shift could contribute toward a historiography of the Holocaust that does not preexist but that emerges from and with the ongoing process of witnessing, in the here and now of reading, listening, and viewing.

References

Agamben, Giorgio. 1999. *Remnants of Auschwitz: The Witness and the Archive*. New York: Zone Books.

Apel, Dora. 2002. *Memory Effects: The Holocaust and the Art of Secondary Witnessing*. New Brunswick, NJ: Rutgers University Press.

Bernard-Donals, Michael, and Richard Glejzer, eds. 2003. "Introduction: Representations of the Holocaust and the End of Memory." In *Witnessing the Disaster: Essays on Representation and the Holocaust*, 3–22. Madison: University of Wisconsin Press.

Bondi, Liz. 2014. "Understanding Feelings: Engaging with Unconscious Communication and Embodied Knowledge." *Emotion, Space and Society* 10: 44–54.

Brinkley, Robert, and Steven Youra. 1996. "Tracing Shoah." *PMLA* 111 (1): 108–27.

Burns, Bryan. 2002. "Fiction of the Real: Shoah and Documentary." *Immigrants and Minorities* 21 (1–2): 80–88.

Carter-White, Richard. 2009. "Auschwitz, Ethics, and Testimony: Exposure to the Disaster." *Environment and Planning D: Society and Space* 27 (4): 682–99.

———. 2012. "Primo Levi and the Genre of Testimony." *Transactions of the Institute of British Geographers* 37: 287–300.

Caruth, Cathy. 1995. *Introduction to Trauma: Explorations In Memory*, 151–57. Baltimore: Johns Hopkins University Press.

———. 1996. *Unclaimed Experience: Trauma, Narrative, and History*. Baltimore: Johns Hopkins University Press.

Chevrie, Marc, and Hervé le Roux, eds. 2007. "Site and Speech: An Interview with Claude Lanzmann about *Shoah*." In *Claude Lanzmann's Shoah: Key Essays*, edited by Stuart Liebman, 37–49. Oxford: Oxford University Press.

Clarke, David B., Marcus A. Doel, and Francis X. McDonough. 1996. "Holocaust Topologies: Singularity, Politics, and Space." *Political Geography* 15 (6–7): 457–89.

Dawidowicz, Lucy. 1975. *The War against the Jews, 1933–1945*. Middlesex: Penguin.

Derrida, Jacques. 2000. *Demeure: Fiction and Testimony*. Stanford, CA: Stanford University Press. Published in the same volume as *The Instant of My Death*, by Maurice Blanchot.

Didi-Huberman, Georges. 2007. "The Site, Despite Everything." In *Claude Lanzmann's Shoah: Key Essays*, edited by Stuart Liebman, 113–23. Oxford: Oxford University Press.

Doel, Marcus A., and David B. Clarke. 2002. "Figuring the Holocaust: Singularity and the Purification of Space." In *Rethinking Geopolitics*, edited by Gearóid Ó Tuathail and Simon Darby, 39–61. London: Routledge.

———. 2007. "Afterimages." *Environment and Planning D: Society and Space* 25: 890–910.

Felman, Shoshana, and Dori Laub. 1992. *Testimony: Crises of Witnessing in Literature, Psychoanalysis, and History*. London: Routledge.

Friedländer, Saul. 2008. *The Years of Extermination: Nazi Germany and the Jews, 1939–1945*. New York: HarperCollins.

Giaccaria, Paolo, and Claudio Minca. 2011. "Topographies/Topologies of the Camp: Auschwitz as a Spatial Threshold." *Political Geography* 30: 3–12.

Harrison, Paul. 2007. "'How Shall I Say It?' Relating the Nonrelational." *Environment and Planning A* 39: 590–608.

———. 2010. "Testimony and the Truth of the Other." In *Taking-Place: Non-Representational Theories and Geography*, edited by Ben Anderson and Paul Harrison, 161–90. Aldershot: Ashgate.

Hatley, James. 2000. *Suffering Witness: The Quandary of Responsibility after the Irreparable.* Albany: State University of New York Press.

Hones, Sheila. 2010. "Literary Geography and the Short Story: Setting and Narrative Style." *Cultural Geographies* 17: 473–85.

Kerner, Aaron. 2011. *Film and the Holocaust: New Perspectives on Drama, Documentaries, and Experimental Films.* London: Continuum Books.

LaCapra, Dominick. 1997. "Lanzmann's *Shoah*: 'Here There Is No Why.'" *Critical Inquiry* 23 (2): 231–69.

———. 1998. *History and Memory after Auschwitz.* Ithaca, NY: Cornell University Press.

Lanzmann, Claude. 2007. "From the Holocaust to 'Holocaust.'" In *Claude Lanzmann's Shoah: Key Essays*, edited by Stuart Liebman, 27–36. Oxford: Oxford University Press.

Lanzmann, Claude, Ruth Larson, and David Rodowick. 1991. "Seminar with Claude Lanzmann 11 April 1990." *Yale French Studies* 79: 82–99.

Levi, Primo. 2004. *The Drowned and the Saved.* London: Abacus.

Liebman, Stuart. 2007. Introduction to *Claude Lanzmann's Shoah: Key Essays*, edited by Stuart Liebman, 3–24. Oxford: Oxford University Press.

Lyotard, Jean-François. 1988. *The Differend: Phrases in Dispute.* Minneapolis: University of Minnesota Press.

Massey, Doreen. 2005. *For Space.* London: Sage.

Olin, Margaret. 1997. "Lanzmann's *Shoah* and the Topography of the Holocaust Film." *Representations* 57: 1–23.

Rothberg, Michael. 2000. *Traumatic Realism: The Demands of Holocaust Representation.* Minneapolis: University of Minnesota Press.

Saxton, Libby. 2007. "Fragile Faces: Levinas and Lanzmann." *Film-Philosophy* 11 (2): 1–14.

Skloot, Robert. 2012. "Lanzmann's *Shoah* after Twenty-Five Years: An Overview and a Further View." *Holocaust and Genocide Studies* 26 (2): 261–75.

Stoicea, Gabriela. 2006. "The Difficulties of Verbalizing Trauma: Translation and the Economy of Loss in Claude Lanzmann's *Shoah*." *Journal of the Midwest Modern Language Association* 39 (2): 43–53.

Waxman, Zoë. 2012. "Transcending History? Methodological Problems in Holocaust Testimony." In *The Holocaust and Historical Methodology*, edited by Dan Stone, 143–57. New York: Berghahn Books.

Wieviorka, Annette. 2006. *The Era of the Witness.* Ithaca, NY: Cornell University Press.

16

A Mobile Holocaust? Rethinking Testimony with Cultural Geography

SIMONE GIGLIOTTI

Introduction

The tellability of past experiences of trauma, particularly those of a Holocaust survivor, is rarely a simple task. Writing about the quest for authenticity in her celebrated memoir, *Still Alive: A Holocaust Girlhood Remembered*, Ruth Klüger argues that: "It won't do to pretend that we can evoke the physical reality of the camps as they were when they functioned. Nevertheless, I want my timescapes. Evocations of places at a time that has passed" (Klüger 2001, 68). Klüger's words are instructive. Admitting the impossibility of articulating the "physical reality" of the camp, Klüger is nevertheless committed to the timescape of her memory. Is it not the work of survivor recountings to return to times past, to present readers, listeners, and viewers with evocations of "timescapes" of confinement, of mobilities that are beyond the reach of nonsurvivors? How do historians and other scholars, long charged with the responsibility of interpreting memoirs, video testimonies, and other first-person accounts including diaries, chronicles, and trial testimony, create knowledge from the ostensibly unknowable? In what ways do representations of time and place in these testimonies, memoirs, and other first-person reports reveal emotional geographies of shared and private traumas?

Historians, as with anthropologists and philosophers, have, in the main, identified the Holocaust's principal trauma sites as fixed entities such as ghettos, camps, and prisons, where victims spent days, nights, weeks, months, and sometimes years living in confined, intolerable, and diseased spaces and conditions. In doing so, they followed the victims' hermeneutic model of site-specific contemporaneous chronicle of life and communities under siege, evident, for example, in a number of well-known ghetto chronicles that have been hailed as much for their ethnography as for their moral commitment to witness (Adelson and Lapides 1989; Kruk 2002; Kassow 2007). Geographers have continued this direction with studies of the ghettos and camps, albeit with a mandate to rethink the structures of confinement as fixed and unchanging, applying the so-called spatial turn in the humanities that began in the early 1990s to the interpretation of the Holocaust. The Holocaust is overdue for such an examination from quantitative and qualitative methods. The emergence of qualitative GIS, geovisualization, deep mapping, and new digital media tools are transforming notions of objectivity and representation. For example, Alberto Giordano and Tim Cole's recent study of the Budapest Ghetto uses GIS and social network analysis to provide a microgeography of everyday mobility and exchange (Giordano and Cole 2011), while Auschwitz's intended building potential and actual uses have been examined in terms of spatial planning, Nazi architectural aesthetics (Jaskot et al. 2014), and as a radical geography of "exception and terror" (Giaccaria and Minca 2011). Recent interdisciplinary collaborations comprising historians and geographers have resulted in studies of the Holocaust's multiscaled impacts in Eastern, Central, and Southern Europe (Knowles et al. 2014), with insights underwritten by data compiled in the United States Holocaust Memorial Museum's multivolume *Encyclopedia of Camps and Ghettos* (Megargee 2009).

Apart from these approaches, the spatial turn also has a deep connection with geography, history, and time, and these latter concepts are often linked to themes of region, diaspora, colonial territory, contact zones, and rubrics such as "border" and "boundary" (Bodenhamer 2010, 16). The spatial turn is also evident in efforts to interpret the colonial and imperial aspirations of Germany in terms of contested "political territory" (Szejnmann and Umbach 2012), inserting the Nazi regime's most actively violated areas and peoples, Eastern Europe, back into the historiography and memory of National Socialism and the Holocaust. The resurgence in literature on "the East," its bureaucrats, lands, and killers (Epstein 2010; Snyder 2012; Lower 2013), memory spaces (Himka and Michlic 2013), and communal violence (Bartov 2008), attempts to reclaim the places and experiences that are authentically indicative of the historical record of Nazi genocide: most victims died in Eastern Europe outside of concentration camps. While not the focus of extended analysis in this chapter, the flourishing literature on colonial and imperial imaginaries in Nazi racial thought and as expressed in Germanization, visions of *Heimat*, *judenrein* Europe, ethnic cleansing (Conrad and Ther 2011), war, and the displacement of populations, awaits its interpreters, and detractors.

This brief overview of recent literature indicates the potential of reading the Holocaust and revisiting its planning, implementation, structures, and experiences in geographical terms. The focus of this chapter is to analyze at the ground level experiences of mobility, and consider how they fit into, or do not, this burgeoning literature on Nazism's spatial imaginaries and Holocaust geographies. Indeed, the very demarcation between Nazism and the Holocaust suggests methodological and rhetorical tensions that persist in the interpretation of Nazi genocide. Much of the aforementioned literature on the spatial turn has focused on the perpetrators' intentions, actions, and controlled or delineated territories—the rhetorical and the physical. An examination of the mobility experiences of victims to and from Auschwitz reveals the received experience of this intention, that is, both rhetorical and physical, and also rethinks the symbolic geography of the Holocaust as *necessarily* linked to fixed and visible sites of confinement. An examination of victims' memories of forced movements of deportation and evacuations in January 1945 gestures to Klüger's evocation of "timescapes." The timescape is a physical place imbued with sensory meaning, an evocation of insecurity from persistent homelessness. How did witnesses, without the often-demeaning rituals and routines that regulated life and labor in ghettos and camps, narrate time and place without the advantage of "fixity"? How did being "mobile," on the move, whether in trains or on foot, produce new orientations to time and place? How can insights from cultural geography prompt new directions in interpreting testimonies of mobility during the Holocaust?

In answering these questions, this chapter focuses on seven women's Holocaust testimonies of movements to and from Auschwitz (the women's camps and subcamps). These testimonies are a mix of oral reports, published memoirs, and interviews, and each contains descriptions of deportation in trains to the camp between 1941 and 1944, and evacuation from the same camp complex on foot and by trains in mid-January 1945.

The displacements of deportation and evacuation occurred in shifting geographies of the "Final Solution": during deportations, the Nazi regime targeted victims in countries of occupation, collaboration, and influence and transported them hundreds of kilometers across borders, in civilian view, and in often intolerable conditions (Gigliotti 2009). During the evacuation phase that began in mid-January 1945, an estimated 55,000 surviving camp prisoners were relocated on foot and by train to camps in Germany and Austria. This relocation increased their exposure to the Soviet offensive in the region and ultimately connected the remaining survivors of the Final Solution to a perennially mobile destiny of wartime traffic, uncertainty, and forced labor (Gigliotti, Masurovsky, and Steiner 2014). These January evacuations were a precursor to the murderous death marches of April and May 1945 (Blatman 2011).

The organization of deportation train journeys and evacuations were guided by temporal and geographical orientations: the timetable and the map. Timetables for deportations were devised by German Transport Ministry officials in cooperation with the Deutsche Reichsbahn. The improvised route maps of SS leaders and officials who planned the hasty relocations from Auschwitz in January 1945 may have signified intention and ultimate destination, but their actual course was evolutionary, climate sensitive, and dependent on sudden enemy attacks. The physical environment of landscapes and climates in which victims moved through during deportations and evacuations were formidable agents of fear and potential, often marking boundaries between free and unfree spaces and peoples.

Orienting Time

An examination of what I call "journey time" in train carriages, and the dominant spatial configuration of the crowd during the evacuations from the camps, offers a rethinking of time as a normative and organizing principle of scholarly inquiry into the Holocaust. Time might be the historians' mantle that anchors documentary empiricism, but what is time and place to and for survivors when they are outside the ghettos and camps and without the provisions of authorship (letter and diary writing) and routine?

The perceived disappearance of time in survivor testimonies is related to the destruction of personal, if not political, space. The particular torment of time's absence has an intellectual forebear in writings by several philosophers. In *Time: An Essay*, Norbert Elias advocated a spatial notion of time as a "people in nature" concept. Elias used the phrase "temporal conscience" to describe the civilizatory conversion of external compulsion by the social institution of time (Elias 1992, 11). The "temporal conscience," he contends, is an innate peculiarity of human consciousness that forces each human being to experience everything as taking place within the flow of time. Elias was a critical interpreter of time's capacity to produce self-control and self-constraint:

> What still remains by and large unclear today is the ontological status of time. People reflect on it without really knowing what kind of phenomenon they are dealing with. Is time a natural phenomenon? Is it a cultural object? Or does the substantive form of the word "time" perhaps create only an illusion of an object? (Elias 1992, 12)

Elias's reflection on the ostensible naturalness of time finds a relative in Siegfried Kracauer's critique of the modern conception of history as occurring in an immanent and continuous process in secular time (Kracauer 1966). The mythological metanarrative of secular time has also been addressed by scholars of postmodernism, poststructuralism, space-time relationships, and memory. In several ways they have worked to undo time's pretensions to a narrative mastery of events and assert the unconscious history of time as embodied memory. Bill Schwarz, an English scholar of postcolonial history and memory, writes that "at some point, any theory of temporality has to confront memory, as the subjective dimension of time itself" (Schwarz 2003). Schwarz identifies an opposition between history and memory that underwrites one of the key debates regarding the resistance of survivors' experiences to temporal, linear classification: "Subjective time, memory included, cannot be accommodated within a theory of historical time that is based only on a conceptualization of time as external and objective" (ibid., 141).

Arguments for the renovation of empirical time and the mapping of the past were explored in a recent issue of the journal *Rethinking History* by a range of urban historians and cultural geographers. But what form was the renovation to assume? Philip Ethington argued that history is not only a temporal phenomenon but also a "change through space. Knowledge of the past . . . is literally cartographic: a mapping of the places of history indexed to the coordinates of spacetime" (Ethington 2007, 466). Ethington, influenced by the work of Georg Simmel, argued for an undoing of time's present incarnation as transcendent in its universality (ibid., 469). Ethington insists that

> mapping is the form of interpretation that historians practice. Their hermeneutic operation is intrinsically cartographic, or possibly choreographic, for all life is movement. . . . However daunting may seem the prospect of "mapping" such intangible topoi as love, greed, faith, ambition, racism, justice (and other forms of cultural cognition), the task is unavoidable given that all human actions inscribe topoi, and every topos is simultaneously locatable and meaningful. (Ethington 2007, 487)

Historians were not eager to relocate time as a core historical orientation. In his reply to Ethington, Thomas Bender critiqued the absence of experience in constructions of place. History was not necessarily directional and progressive, but rather a rhetoric of motion. He commented that: "History is more than presence; more than time and being. It is about being and doing" (Bender 2007, 497). David Carr argued that space and place need to be acknowledged as central to experience and made a plea to retire the "rarified intertextual, semiotic world of the structuralists and post-structuralists" and return to the "lived space-time of our ordinary experience, action and interaction" (Carr 2007, 503). Carr appeared to validate the mapping method of "humanist cartography," as suggested by Daniel Dorling and David Fairbairn (1997, 142). This lived cartography is premised on the notion that the mapping of people should primarily depict movement over the short and long term, using data such as occupational concentration, population density, and everyday movements. Critical to this enterprise is the availability of such data of tracking motion and mobility, concentration, and everyday spatial and place-based encounters.

The challenge of how to map time, space, and place history that was so engagingly explored in the *Rethinking History* issue raises several questions that bear directly on how survivors narrate their orientations to time and place during forced relocations. Scholars of Holocaust victims' experiences have been attempting to formulate their own version of a humanist cartography that responds to the qualitative, subjective traces of personal memory rather than privileging temporal validation and authenticity asked by historical positivism. Lawrence Langer's *Holocaust Testimonies: The Ruins of Memory* (1991) addressed the undoing of the temporal conscience. Continuity and chronology are disrupted in a survivors' spoken and visual recollection, revealing layers that range from common to deep memory. Common memory focuses on the everyday, routinized recollection of encounters of the self in ghetto, camp, or other environments, while deep memory is the individualized embodied return of traumatic encounters sparked by unexpected, sensory identification in the present. "Wounded time" is Langer's closest allusion to the self's temporal undoing by trauma: "As memory plunges into the past to rescue the details of the Holocaust experience, it discovers that cessation plays a more prominent role than continuity" (ibid., 75).

Time, or rather its undoing, features in Gabrielle Spiegel's analysis of the Holocaust as representing a phenomenological rupture of the self in extremity, with testimony ostensibly unavailable to history because of its profoundly abject nature. She distinguishes between the chronological time of the historian and "durational" time of the present, an experience of past time endlessly *re*-lived:

> Durational time resists precisely the closure—the putting an end to the past that chronological/historical time necessarily effects; durational time persists as a past that will not pass, hence a past always present. Thus, while Holocaust testimonies may impart the illusion of chronology, in effect their durational structure defeats any attempt at historicization; they are therefore, *structurally* unavailable as history. (Spiegel 2002, 159)

The concepts of "wounded time" and "durational time" are destabilizers of the temporal conscience. By focusing on the representations of "journey time," I suggest that temporal consciousness could not adhere to conventional empiric measures. In survivor narratives, days, dates, and places are sometimes recalled, often unreliably, as an escort to the more dominant enclosure of the crowd. The traumatic embodiment of duration is, I believe, more suggestive of how the loss of Elias's "temporal conscience" becomes reconfigured as temporal consciousness in the frustrated placement of the self in individual and group scenes of stillness and motion among the crowds of deportees in train carriages and masses of evacuees en route from camps. In this sense, the uncovered fundamental self that Henri Bergson described as an irregular, unrecognizable image may be of help: "our perceptions, sensations, emotions, and ideas occur under two aspects: the one clear and precise, but impersonal; the other confused, ever changed, and inexpressible, because language cannot get hold of it without arresting its mobility or fit it into common-place forms without making it into public property" (Bergson 2002, 72).

The Data of Mobility: Witness, Emotion, Landscape

In relation to deportation, temporal metaphors of the past were depicted as accrued confinement experiences. In her interview with David Boder in Paris on August 7, 1946, Henja Frydman depicted the camp as a boundary marker between liberty and confinement, a temporal experience of three months that "passed quickly," unlike her confinement in the cattle car to Auschwitz that lasted "three frightful days" in which there was "no air, and no toilet" and where everyone had to "take care of himself" (Frydman 1946). Frydman's Auschwitz timescape is told as "two years" of imprisonment, when, in reality, it was eighteen months from her arrival as a nineteen-year-old Resistance member to liberation by the Red Army on January 27, 1945. Still, the impossibility of qualitative description about the embodied endurance of that timescape reigns when she tells Boder:

> Now begins the long chapter of two years imprisonment in Auschwitz. It will not be easy for me, indeed, as all of us say when we are asked to tell. To tell! That is the hardest thing for us. Why? Because there are no words. No way of expressing it, that can describe what happened from the day when we disembarked from the train in the lager Auschwitz until the day of our liberation. (Frydman, 1946)

Ruth Klüger's temporal consciousness is confined by space, for her feelings of entrapment are constantly recalled in embodied reminiscences of her train journey from the Theresienstadt Ghetto to Auschwitz. Temporality is expressed in the parallel journeys to corporeal decline and destination: "on the road to Auschwitz, we were trapped like rats" (Klüger 2001, 91). Durational time is experienced as entrapment not only in her sensory memory but also in the affective embodiment of distance between locations, the symbolic death suspended between stillness and motion:

> If I look at a map today, I see the distance from Theresienstadt to Auschwitz is not very great. Yet it was the longest trip I have ever taken. The train stood around, it was summer, the temperature rose. The still air smelled of sweat, urine, excrement. A whiff of panic trembled in the air. It's from this experience that I think I have an idea of what is must have been like in the gas chambers. (Klüger 2001, 92)

Consciousness of temporality also features in Olga Lengyel's memoir, *Five Chimneys* (1972). She is transported late in the war, in 1944, from the capital of Romania to Auschwitz. The temporal consciousness she exhibits is anchored not only in the physical enclosure of the train carriage but also in the erosion of social convention that mass confinement dictates: public displays of going to the toilet and management of privacy. She writes that "the hours slipped away," while the "seventh day of travel" implies a durational embodiment of innumerable torments. Lengyel is a group witness to moments of quarrelling, physical discomfort, and the ultimate corporeal passage into death. The living space of the carriage had its own intervals of interruption with glimpses of landscapes passed. Her temporal consciousness was resocialized according to the internal and external space of transit:

> From time to time in the course of this infernal trip, I tried to forget reality, the dead, the dying, the stench, and the horrors. I stood on several suitcases and peered out of the little window. I gazed at the enchanting countryside of the Tatras, the magnificent forests of fir trees, the green meadows, the peaceful pastures, and the charming little houses. It was all like a scene advertising Swiss chocolates. How unreal it seemed! (Lengyel 1972, 20)

Such displacements do not characterize Sima Vaisman's *A Jewish Doctor in Auschwitz* (2005), an account published soon after the war that details the author's incarceration in Auschwitz and departure in January 1945. It is a narrative with a strong temporal consciousness, a memoir that historians welcomed for its detail and specialized knowledge of the camp's functioning. Vaisman worked as a camp doctor in a subcamp of Brzezinka (a subcamp of Auschwitz), built for the liquidation of the Jews of Hungary, from mid-May 1944. She had heard as early as September 1944 that a "liquidation" of the camp was unfolding, and to her liquidation meant killing. Dates and times are stated with ease: "On the morning of January 18, there's a rumor that the camp is going to be evacuated. The kitchen is no longer functioning; our soup wasn't brought to us at noon. Feverishly, the S.S. are burning the official documents, the account books of the different centers, our registration slops, etc. A large bonfire where they burn empty suitcases" (ibid., 65). The day builds in anticipation and intensity, for at 5:00 p.m. they were told to line up in preparation for departure, and "At six o'clock, our column gets underway and starts marching. We don't dare believe we are leaving Brezinki [*sic*], that they didn't gas all of us, that we are leaving alive. And, in weather that's 15 below, we walk cheerfully, quickly, we leave behind us this cursed land where millions of people lost their lives" (ibid.). When she departs from Auschwitz, her temporal awareness shifts to the camp's remaining population. Crowds of inmates are extruded from their concealed suffering, assembled into a moving captivity in an open, compliant form: "the endless columns of evacuees from Auschwitz, men and women, people in rags, scarcely clothed, sick people, beaten out of hospitals and infirmaries with clubs at the last minute" (ibid., 67).

A temporal consciousness emerges in descriptions of the march as having stages or intervals of fatigue and episodic physical respite or recovery. Vaisman's testimony is unusually descriptive, and therefore very useful, in constructing a corporeal geography of time and place. In temperatures of minus 26 degrees Fahrenheit, she (and her column) had walked a punishing "fourteen hours without stopping, without having a minute of rest. We have only covered 25 miles. The next morning, we have our first pause at 8 in the morning at a public square, in a Polish village, in front of a school. Many of us are missing. Shot along the road, escaped into the woods? We would like to think so, but we have heard too many gunshots in the night" (ibid., 68). With roads ridden with corpses, Vaisman's spatial cognition exemplifies Yi-Fu Tuan's discussion of spatial experience: "The organization of human space is uniquely dependent on sight. Other senses expand and enrich visual space. Thus sound enlarges one's spatial awareness to include areas behind the head that cannot be seen. More important, sound dramatizes spatial experience" (Tuan 1977, 16). In Vaisman's case, the sonic dimension confirmed her unfree mobility.

After a brief rest for frozen bread and margarine, their group set off again, the columns brought to manageable surveillance: "After two hours, we again have to line up in rows of five and press on. We are walking poorly, we drag ourselves lamentably along the road and, a few miles further on, they bring us to a stable. A few cows are still there, it smells bad but there is a little straw. We throw ourselves on this straw that seems to us as soft as the finest beds, we pile on top of each other, for there are 2,000 of us in one single stable, and we sleep a dreamless sleep" (Vaisman 2005, 69).

Regina Laks Gelb's oral testimony of her departure from Auschwitz in January 1945 contains an inherent chaos (Laks Gelb 2001). Unlike the temporal awareness of Vaisman's testimony, Gelb's transit experiences are recalled without temporal references. Many issues remain unclear from reading it:

temporal indicators such as when she was evacuated, an estimate of the distance walked, and her stopping points en route. While ostensibly not useful to historians looking for reliable empirical evidence, her testimony evokes Langer's notion of "wounded time." She is an afflicted teller, her abjection confirmed in her ability to offer little more than visual cues about her march.

Her testimony about time and place demonstrates the problems of placing the self in a historically usable framework, continuing the characterization of the otherworldliness of Auschwitz, and relocating it to the fields and roads of Silesia. Temporality is not "time-measured" but rather embodied in place-based moments of disassociation and disorientation. She begins:

> and of course, January rolled around, and Auschwitz was being evacuated, and we with it. Luckily enough, we were dressed not in rags, but we had clothing on that could keep us warm. And we were driven on a march toward trains, that were somewhere. There were stations somewhere, but I really don't know. Because at that point much as I was trying to be very independent, not a burden to my sisters, I always knew they were really looking out for my benefit. But right here, I could not overcome my physical person. Meaning we were marched without stopping, and without sleeping, in the snow. And I completely and totally separated myself in sleep, from the reality, the prevailing reality, the march. . . . I think that lasted for almost a week. It was day and night. And I remembered the hallucination, that is the only thing that I really remember. There was a white road, and there were trees on both ends, both sides of the road. And it seemed to me like I was in an enclosure, because it was white and dark. And that I was in some kind of enclosure, that was a total fantasy. But I actually lived that fantasy. (Laks Gelb 2001)

In addition to episodic mentions of train journeys and evacuations, some deportees provide spatial trajectories of confinement. Hadassah Marcus, a David Boder interviewee (Rosen 2010), provides a number of locations in her spatial trajectory of victimization: the Warsaw Ghetto, transport to Majdanek, then to Auschwitz, relocation from that camp on foot on January 18, 1945, and then by train to Ravensbrück and Neustadt. The protracted and punishing evacuation transport is clearly the most indelible timescape of her itinerary, combining mixed transport methods, resulting in a relentless and potentially deadly fate on the roads of Silesia:

> In the year 1945 . . . it was also on the 18th, 1st month, i.e., the 18th of January, when the Russians were already standing behind our backs. We were aroused from sleep in the middle of the night, and we had to march on foot. In that night we covered thirty kilometers, without any rest, running, not walking. We weren't given anything for the road. We were told to take along our Kotzen, but we had to throw it away in the middle of the road, because the

> [our] strength had given out. If one wanted to rest a moment on the road, if he only gave a bow with the head, he was taken out and immediately shot. Very many men fell at that time, more than women. After covering that night thirty kilometers, we were given in the morning fifteen minutes to rest. After that we walked a whole day, again a whole night, and were loaded into RR-cars. He who couldn't make it so fast into the RR-cars was also taken away and shot. In RR-cars we traveled five days. (Marcus 1946)

David Boder also interviewed Helen "Zippi" Tichauer about her experiences of deportation from Slovakia, her work in the women's camp in Auschwitz I, and numerous mobility experiences (Tichauer 1946).

A graphic artist by trade, Tichauer is an especially important witness whose wartime experiences were analyzed in a recent book (Matthaeus 2009). While the book focused on her work in Auschwitz and the silences and contradictions in her 1946 interview with Boder, it did not examine at length the advantages Tichauer derived from working in the Political Department (New Arrivals) at Auschwitz. The following exchange with Boder depicts the environmental and physical conditions of the Auschwitz death march and their ostensible directionless destination:

HELEN TICHAUER: The trip proceeded day and night in a way here and there. Just like the march on foot, so the trip by train, without provisions, in a storm, without toilets, with nothing.

DAVID BODER: Still, what kind of guard did you have?

HELEN TICHAUER: Guarded by the SS. They traveled with us. We traveled. One did not know exactly where to.

She later reveals that because of her work in the political office in Auschwitz, she knew that prisoners were transferred from Auschwitz to Ravensbrück, and such knowledge enabled her to accurately predict the transport's destination: "We always calculated where what is located. I knew exactly. If the road leads to Oranienburg, then the road takes us to Ravensbrück. And Ravensbrück was a famous German concentration camp for women." After further journeys between camps and escapes, Tichauer eventually arrived back in Slovakia on May 28, 1945, armed with one particularly indispensable orienting item: "I came home without shoes [?]. I came home without clothes. The only thing that I brought with me from Germany was a map" (Tichauer 1946).

Tichauer's map, her only possession from the war, guided her survival in those early postwar days and in her journey back home to Slovakia. Her testimony, like the other women's accounts examined, reveals variations in the possibility to witness depending on one's work in the camp, mobility within designated areas, degree of trauma and abuse endured during deportations and in the camp environment, the condition of one's physical and mental health prior to evacuation from the camp, and the presence of family or support networks during evacuations.

The disorientations of time, place, and the body to deportation and evacuations can benefit from geographical approaches to place and displacement. These approaches contribute to the earlier discussion on the spatial turn in Holocaust studies but also expand it, applying such phrases as emotional and corporeal geographies to describe witnesses' encounters with landscapes after the fixed confinement of ghettos and camps.

Cultural Geographies: Place and Affect

Testimonies of deportation and evacuation are mobility narratives that undermine finding rootedness or place, with the victims themselves the preeminent targets of the sedentarist mentality that dominated Nazi Germany's need to control the mobile populations of "wandering Jews" and nomadic Sinti and Roma peoples. Indeed, this construction of sedentarism has a long-established discourse, a normative presumption that privileges the need to be in place, as Tim Cresswell noted in his brief musings on the topic: "Thinking of the world as rooted and bounded is reflected in language and social practice. Such thoughts actively territorialize identities in property, in region, in nation—in place. They simultaneously produce discourse and practice that treats mobility and displacement as pathological" (Cresswell 2006, 42). The solution to the dangers of mobility of unwanted peoples in Nazi Germany, and their urban spaces of dwelling, was anchored to the *Heimat* philosophy: "In Nazi Germany, Gypsies, Jews, and gay people were murdered by the millions. Behind this genocide lay a well-developed ideology of sedentarism. Gypsies, Jews, and gays were described as rootless in order to legitimate the Holocaust. Nazi academics and writers built up a German mythology based on deep soil, the forest, and roots" (ibid.).

The advent of the mobility turn, which has recently entered the field of migration studies and its territorially dispossessed groups—refugees and asylum seekers (Faist 2013)—was supposed to addresses the gaps of the so-called spatial turn, which to Mimi Sheller and John Urry was also "a-mobile" and "sedentarist." They note that: "It is not a question of privileging a 'mobile subjectivity,' but rather of tracking the power of discourses and practices of mobility in creating both movement and stasis. A new mobilities paradigm delineates the context in which both sedentary and nomadic accounts of the social world operate, and it questions how that context is itself mobilised, or performed" (Sheller and Urry 2006, 211).

A relevant aspect of mobilities research relates to emotional embodiments of space and place, the "recentring of the corporeal body as an affective vehicle through which we sense place and movement, and construct emotional geographies" (Sheller and Urry 2006, 216). Historians have not been ignorant to the evidence of emotional behavior or corporeality in rethinking histories of everyday life, power, and class. Writing in 1985, Peter and Carol Stearns noted that historians and anthropologists tended to confuse "thinking about emotion with the history of emotion" and that history can be used to understand "our collective emotional past" (Stearns and Stearns 1985, 815). Joanna Bourke, a historian of warfare and killing, notes in her writing about fear, that to study emotions, they have to be made visible, and be approached as a language game in terms of generic and narrative conventions to explore the grammar and physiology of emotions (Bourke 2003, 113). Barbara Rosenwein also makes a plea for an intervention with emotions: "The new narrative will recognize various emotional styles, emotional communities, emotional outlets, and emotional restraints in *every* period, and it will consider how and why these have changed over time" (Rosenwein 2002, 845). While these historians advocate for the relevance emotions have as a historical truth, they do not place them in a context, or attend to their variability in place, as explained by Davidson and Milligan who suggest that witnessing is a multisensorial event: "We can, perhaps, usefully speak of an emotio-spatial hermeneutic: emotions are understandable—'sensible'—only in the context of particular places. Likewise, place must be *felt* to make sense. This leads to our feeling that meaningful senses of space emerge only via movements *between* people and places" (Davidson and Milligan 2004, 524).

The literature on emotional geographies and its relevance for discussions of mobility is further complicated by the intrusion of affect, so long a key signifier in psychoanalytic studies of Holocaust trauma that it seems to have made a rather controversial entry into cultural geography (Pile 2010; Hadi Curti et al. 2011; also Colls 2004; Davidson and Milligan 2004; Bondi 2005; Bissel 2008). Although the theme of silence as an ethical response to the Holocaust is long-standing, it is somewhat marginalized now. Pile seems to import the notion of unrepresentable silence to describe the difference between emotional and affectual geographies, almost to an unworkable premise: "Affectual geography . . . provides an ontological model of the mind and body that presumes that it is always already partitioned, with discrete spaces being sectioned off from one another. These partitions, consequently, are universal and immutable: emotions and affects have their place, and they do not move" (Pile

2010, 10). It is difficult for survivors to actually recall those shared communities of emotions that undergo transference from an individual to the perceptual environment of the group as the "interactions between bodies and (the manipulation of) flows of affect" (ibid., 11). Rather, radical abjection is privatized in the recollection of Regina Laks Gelb: "But right here, I could not overcome my physical person. Meaning we were marched without stopping, and without sleeping, in the snow. And I completely and totally separated myself in sleep, from the reality, the prevailing reality, the march" (Laks Gelb 2001).

Testimonies of deportation and evacuation actively recall embodied memories and places to evoke Klüger's "timescapes," but of what use are they to geographers in thinking about emotional and corporeal placescapes? Geographers such as Owain Jones have criticized the forgetting of memory in cultural geography (Jones 2011). A key feature of Jones's critique is his attack on presentism. He challenges geographers to engage with memory's role in the affective, performative practices of everyday life, in ways that can address the complex ecologies of memory (and forgetting) that interlink through "individual practicing bodies, texts, materialities, past/present/future timespaces to make the present time deep/complex rather than flat/pure" (ibid., 876). And yet the content of Holocaust testimonies of deportation and evacuation, with their focus on presentism, demonstrate how ecologies of memory and interdependence work on the ground. They provide placemarks of crime, care, and threat in an affective geography and violent mutations of spatiotemporal markers. As indicated in the testimonies excerpted, not only were time and place reoriented to suit the conditions of mobility, for example in the rerouting of evacuation paths to avoid enemy capture, which made the evacuation legs more arduous. Mobility was also redefined to suit the aggressive character and violence of degraded and inhumane transports, whether in trains or on foot. Violence, murder, and attempted escapes were common and sustained in both types of movement.

To be sure, scholarly studies of Holocaust mobility in its violent and aggressive forms are not extensive in number, and yet there is room for a serious investigation of violent and affective mobilities that complicate unspecific and rather sanitized descriptions of place-making and forced movement. I agree with the following premise that, "mobile methodologies . . . concerns not only physical movement, but also potential movement, blocked movement, immobilization, forms of dwelling, and place-making" (Fincham, McGuinness, and Murray 2010, viii), but an understanding of specific examples of affective displacement, such as that initiated by Tim Cole's study of microgeographies of return to Auschwitz (Cole 2013), asks for a reception in cultural as much as physical geography's attention to landscape, tourism, and violence.

The forced movement of deportations and evacuations reveals that place, like time, had multiple meanings, and incarnations, in survivors' recollections. Some testimonies excerpted in this chapter reveal the importance of linking experience to a physical place, such as Tichauer's, and yet others also reveal the impossibility of place, especially when moving through foreign, strange, or hostile regions, to provide security or knowledge. Time, too, was endured, embodied and violated—deportees and evacuees were locked into a suffocating presentism of movement, both of the oppressor's making. Everyday civilian environments of roads, fields, towns, barns, and schools became microgeographies of fatigue, risk, intervention, and death. No longer did ghettos and camps provide the focal places of memory, but nor were the behaviors that those experiences of confinements generated forgotten. The perceived loss of time to regulate individual behavior and interaction demonstrated the power of what philosopher Edith Wyschogrod (1985, 16) called the "death world," the emergence of a new regime of living in which the scaffolding of experience reigns. The sense of closure from the past and future as expressed in the "death-world" magnified the immediacy and meaning of action in a surreal present seething with chaos. The rupture with the continuity of the "life-world" was initiated when "the mode of temporalisation in the death world closes off the future of its inhabitants" and becomes "enforced at the vital level of existence by the system of compulsory enclosure which removed the individual from familiar surroundings and reduces mobility" (ibid., 18). The endurance of the "death world" that Wyschogrod identified in the fixed regime of the camps was rehearsed in the captivity of train journeys and repeated, if not actively controlled, in the open walking columns that left Auschwitz during January 1945. Although time's loss was inscribed in testimonies as the rhetoric of decline, time itself was not so much destroyed as was the reliance on it to orient and navigate the modern subject. The temporal conscience was exposed as illusory. Its mourned disappearance was experienced as panic, disorder, and regression. The decivilizing process of closed mobility in train carriages and open mobility during evacuation columns secured the temporal unmaking of the individual. It also produced a resensitization of the self to space and place.

Conclusion

This chapter has used insights from cultural geography to interpret memories of mobility during the Holocaust. It took a small number of testimonies of women survivors of Auschwitz and commented on their allusions to time and place during two experiences of forced movement: deportation and evacuation. It used Auschwitz as the focal point, intending to rethink that location as the symbolic high point of the Holocaust and offer its value as a site that people arrived at, moved through, and left. It concluded that without the advantage of fixed sites of ghettos and camps, the familiar organizing structures of life and death, deportees and evacuees struggled to make meaning of their experiences and their place in the oppressor's world. A focus on mobility, on the "in-between" of Holocaust places such as ghettos and camps, affirms that cultural geography can provide a valid alternative to the prevalent and persistent sedentary approach to histories of the Holocaust's places of ghettos and camps. The approach of mobility is also acutely responsive to the actual historical and spatial conditions that defined victims' life worlds regardless of their location. Through examining experiences outside of ghettos and camps, where conditions of documentation were very often difficult, memories of forced movement reveal that emotionality and corporeality are the overlooked evidence of displacement and survival. To endure and survive mobility, victims formed what Barbara Rosenwein (2002) has termed "emotional communities"—affective bonds and emotional behaviors—that underwrote survival and concealed displays of excessive emotions to avoid being seen as fallible, vulnerable, and weak.

The approach of mobility applied in this chapter has aimed to enrich the spatial turn in Holocaust studies and return it to the ground level and to the victims. As much as this mobility indicated that the Final Solution was now on the move, it also revealed that mobility itself was remade, stained with violence and death.

References

Adelson, Alan, and Robert Lapides, eds. 1989. *Łódź Ghetto: Inside a Community under Siege.* New York: Viking.

Bartov, Omer. 2008. "Eastern Europe as the Site of Genocide." *Journal of Modern History* 80 (September): 557–93.

Bender, Thomas. 2007. "Theory, Experience, and the Motion of History." *Rethinking History* 11 (4): 495–500.

Bergson, Henri. 2002. "The Idea of Duration." In *Henri Bergson: Key Writings*, edited by Keuth Ansell Pearson and John Mullarkey, 49–77. London: Continuum.

Bissell, David. 2008. "Comfortable Bodies: Sedentary Affects." *Environment and Planning A* 40: 1697–712.

Blatman, Daniel. 2011. *The Death Marches: The Final Phase of Nazi Genocide*. Cambridge, MA: Belknap Press of Harvard University Press.

Bodenhamer, David J. 2010. "The Potential of the Spatial Humanities." In *The Spatial Humanities: GIS and the Future of Humanities Scholarship*, edited by David J. Bodenhamer, John Corrigan, and Trevor M. Harris, 14–29. Bloomington: Indiana University Press.

Bondi, Liz. 2005. "Making Connections and Thinking through Emotions: Between Geography and Psychotherapy." *Transactions of the Institute of British Geographers* 30 (4): 433–48.

Bourke, Joanna. 2003. "Fear and Anxiety: Writing about Emotion in Modern History." *History Workshop Journal* 55 (1): 111–33.

Carr, David. 2007. "Commentary on 'Placing the Past: "Groundwork" for a Spatial Theory of History.'" *Rethinking History* 11 (4): 501–5.

Cole, Tim. 2013. "Crematoria, Barracks, Gateway: Survivors' Return Visits to the Memory Landscapes of Auschwitz." *History and Memory* 25 (2): 102–31.

Colls, Rachel. 2004. "'Looking Alright, Feeling Alright': Emotions, Sizing, and the Geographies of Women's Experiences of Clothing Consumption." *Social and Cultural Geography* 5 (4): 583–614.

Conrad, Sebastian, and Philipp Ther. 2011. "On the Move: Mobility, Migration, and Nation, 1880–1948." In *The Oxford Handbook of Modern German History*, edited by Helmut Walser Smith, 573–90. Oxford: Oxford University Press.

Cresswell, Tim. 2006. *On the Move: Mobility in the Modern Western World*. London: Routledge.

Davidson, Joyce, and Christine Milligan. 2004. "Embodying Emotion Sensing Space: Introducing Emotional Geographies." *Social and Cultural Geography* 5 (4): 523–32.

Dorling, Daniel, and David Fairbairn. 1997. *Mapping: Ways of Representing the World*. Essex: Pearson Education.

Elias, Norbert. 1992. *Time: An Essay*. Oxford: Blackwell.

Epstein, Catherine. 2010. *Model Nazi: Arthur Greiser and the Occupation of Western Poland*. Oxford: Oxford University Press.

Ethington, Philip J. 2007. "Placing the Past: 'Groundwork' for a Spatial Theory of History." *Rethinking History* 11 (4): 465–93.

Faist, Thomas. 2013. "The Mobility Turn: A New Paradigm for the Social Sciences?" *Ethnic and Racial Studies* 36 (11): 1637–46.

Fincham, Ben, Mark McGuinness, and Lesley Murray, eds. 2010. *Mobile Methodologies*. Basingstoke: Palgrave Macmillan.

Frydman, Henja. 1946. "Oral History Interview." In *Voices of the Holocaust: Illinois Institute of Technology*, August 7, edited by David Boder. Paris, France.

Giaccaria, Paolo, and Claudio Minca. 2011. "Topographies/Topologies of the Camp: Auschwitz as a Spatial Threshold." *Political Geography* 30 (1): 3–12.

Gigliotti, Simone. 2009. *The Train Journey: Transit, Captivity, and Witnessing in the Holocaust*. New York: Berghahn.

Gigliotti, Simone, Marc J. Masurovsky, and Erik Steiner. 2014. "From the Camp to the Road: Representing the Evacuations from Auschwitz, January 1945." In *Geographies of the Holocaust*, edited by Anne Kelly Knowles, Tim Cole, and Alberto Giordano. Bloomington: Indiana University Press.

Giordano, Alberto, and Tim Cole. 2011. "On Place and Space: Calculating Social and Spatial Networks in the Budapest Ghetto." *Transactions in GIS* 15 (1): 143–70.

Hadi Curti, Giorgio, Stuart C. Aitken, Fernando J. Bosco, and Denise Dixon Goerisch. 2011. "For Not Limiting Emotional and Affectual Geographies: A Collective Critique of Steve Pile's 'Emotions and Affect in Recent Human Geography.'" *Transactions of the Institute of British Geographers* 36: 590–94.

Himka, John Paul, and Joanna B. Michlic, eds. 2013. *Bringing the Dark Past to Light: The Reception of the Holocaust in Post-Communist Europe*. Lincoln: University of Nebraska Press.

Jaskot, Paul B., Anne Kelly Knowles, Chester Harvey, and Benjamin Perry Blackshear. 2014. "Visualizing the Archive: Building at Auschwitz as a Geographic Problem." In *Geographies of the Holocaust*, edited by Anne Kelly Knowles, Tim Cole, and Alberto Giordano. Bloomington: Indiana University Press.

Jones, Owain. 2011. "Geography, Memory, and Non-Representational Geographies." *Geography Compass* 5 (12): 875–85.

Kassow, Samuel D. 2007. *Who Will Write Our History? Emanuel Ringelblum, the Warsaw Ghetto, and the Oyneg Shabes Archive*. Bloomington: Indiana University Press.

Klüger, Ruth. 2001. *Still Alive: A Holocaust Girlhood Remembered*. New York: Feminist Press at the City University.

Knowles, Anne Kelly, Tim Cole, and Alberto Giordano, eds. 2014. *Geographies of the Holocaust*. Bloomington: Indiana University Press.

Kracauer, Siegfried. 1966. "Time and History." *History and Theory* 6: 65–78.

Kruk, Herman. 2002. *The Last Days of the Jerusalem of Lithuania: Chronicles from the Vilna Ghetto and the Camps, 1939-1944*. Translated by Barbara Harshav. New Haven, CT: YIVO Institute for Jewish Research.

Laks Gelb, Regina. 2001. "Oral History Interview." In *United States Holocaust Memorial Museum*.

Langer, Lawrence L. 1991. *Holocaust Testimonies: The Ruins of Memory*. New Haven, CT: Yale University Press.

Lengyel, Olga. 1972. *Five Chimneys: The Story of Auschwitz*. St. Albans: Mayflower.

Lower, Wendy. 2013. *Hitler's Furies: German Women in the Nazi Killing Fields*. New York: Houghton Mifflin Harcourt.

Marcus, Hadassah. 1946. "Oral History Interview." In *Voices of the Holocaust: Illinois Institute of Technology*, September 13, edited by David Boder.

Matthäus, Jürgen, ed. 2009. *Approaching an Auschwitz Survivor: Holocaust Testimony and Its Transformations*. Oxford: Oxford University Press.

Megargee, Geoffrey P., ed. 2009. *The United States Holocaust Memorial Museum Encyclopedia of Camps and Ghettos, 1933–1945*. Vol. 1 (Parts A and B). Bloomington: Indiana University Press.

Pile, Steve. 2010. "Emotions and Affect in Recent Human Geography." *Transactions of the Institute of British Geographers* 35: 5–20.

Rosen, Alan. 2010. *The Wonder of Their Voices: The 1946 Holocaust Interviews of David Boder*. Oxford: Oxford University Press.

Rosenwein, Barbara. H. 2002. "Worrying about Emotions in History." *American Historical Review* 107 (3): 821–45.

Schwarz, Bill. 2003. "Already the Past: Memory and Historical Time." In *Regimes of Memory*, edited by Susannah Radstone and Katharine Hodgkin, 135–51. London: Routledge.

Sheller, Mimi, and John Urry. 2006. "The New Mobilities Paradigm." *Environment and Planning A* 38 (2): 207–26.

Snyder, Timothy. 2012. *Bloodlands: Europe between Hitler and Stalin*. New York: Basic Books.

Spiegel, Gabrielle M. 2002. "Memory and History: Liturgical Time and Historical Time." *History and Theory* 41 (May): 149–62.

Stearns, Peter N., and Carol Z. Stearns. 1985. "Emotionology: Clarifying the History of Emotions and Emotional Standards." *American Historical Review* 90 (4): 813–36.

Szejnmann, Claus-Christian, and Maiken Umbach, eds. 2012. *Heimat, Region, and Empire: Spatial Identities under National Socialism*. Basingstoke: Palgrave Macmillan.

Tichauer, Helen. 1946. "Oral History Interview." In *Voices of the Holocaust: Illinois Institute of Technology*, September 23, edited by David Boder. Feldafing, Germany.

Tuan, Yi-Fu. 1977. *Space and Place: The Perspective of Experience.* Minneapolis: University of Minnesota Press.

Vaisman, Sima. 2005. *A Jewish Doctor in Auschwitz: The Testimony of Sima Vaisman.* Hoboken, NJ: Melville House Publishing.

Wyschogrod, Edith. 1985. *Spirit in Ashes: Hegel, Heidegger, and Man-Made Mass Death.* New Haven, CT: Yale University Press.

17

What Remains? Sites of Deportation in Contemporary European Daily Life: The Case of Drancy

KATHERINE FLEMING

I

Countless locations around the world have been the setting for tragedies of one sort or another: massacres, deportations, enslavements, suicides, murders. What should become of such sites, after the blood has been washed away and the physical marks of what happened in them have been removed? Should they be condemned? Set aside as memorials, converted into museums or places of reflection? Should normal life carry on around and in them, or is it disrespectful or even blasphemous to think that it could? Different answers are given to these questions; they vary according to the power yielded by the victimized party in question (or that party's lobby), to the magnitude of the tragedy, to the political and social considerations at play in the polities in which they took place, and to the value of the real estate they occupy. Far easier to decommission a remote field than to give up a valuable central urban zone.[1] Debates over how best to memorialize sites of past collective trauma are still more complicated when countervailing forces are present—those that for varying reasons hope not to encourage but rather to suppress memory, or where feelings are ambivalent.

A complex web of factors—political, economic, and sentimental—motivates advocates both for remembering and forgetting. In a world in which real estate has become a tremendous driver of national economies, protecting the value of properties from the potential damage inflicted on them by past traumatic events is a concern. In the United States, the National Association of Realtors (NAR) refers to homes in which murders, suicides, or other violent events have taken place as "stigmatized properties." "Stigmatized properties" take far longer to sell than "nonstigmatized properties," and when they do sell, they sell for a lower price (Larsen and Coleman 2001, 1–16). While some US states have disclosure laws, which require sellers to share information about stigmatizing events, many do not—and angry buyers who have unknowingly purchased a home only later to find out that it was the past scene of a gory crime have (largely unsuccessfully) sought recourse in the courts. It seems that it is not generally death or the thought of the dead that keeps buyers away, but specifically violent and traumatic death. For instance, the proximity of a cemetery does not in statistical terms have an adverse effect on the price of single-family homes, and in some instances can even have a positive one (Larsen and Coleman 2010, 33–49). But proximity to—and more potently, being the actual site of—a horrific past violent event can (so to speak) be a real deal killer. Houses where murders have taken place are particularly difficult to sell.

What is it that is so troubling about such places? After all, they bear no mark of the past. They are not themselves the agents of the crimes committed within them; they are simply the mute stages upon which humans have enacted horror. Yet the site itself is a constant "reminder" of what happened—and thus a suggestion, perhaps, that we are, or could be, complicit—or could be a future victim ourselves. Our generalized horror at the unavoidability of the human condition becomes focused on specific places. In avoiding bad places, and marking them off, we aim to protect ourselves as much as we hope to remember anything particular to these places.

The idea that specific places can "trap" the bad "energy" of past horrors is of course culturally widespread. In contemporary Western popular culture, it is a recurrent theme of novels and films. The most famous of the genre is arguably the series of eleven films inspired by the 1977 book, *The Amityville Horror: A True Story*, by Jay Anson, which tells the tale of a series of paranormal hauntings of a Long Island home in which a mass murder had taken place. Its new owners claimed to have been so terrorized that they had to leave the property after living there for less than a month. Whether the book in fact told a "true story" has been hotly debated, but the actual home exists, and is still visited by gawkers who find the place significant. The immense success of the series testifies to the abiding fascination with the notion that sites are forever marked by the events for which they were the stage.

For the owners of such a property, an array of steps can be taken to expiate or expunge the past: exorcism, renovation, address change, and (in the most extreme of circumstances) demolition and rebuilding.[2] Through various expiating measures, what is being aimed at of course is *forgetting*: the owner wants to be able to "forget" what happened in his or her home, and wants others to "forget" it, too. More accurately—since this sort of forgetting can be difficult to accomplish—the owner wants in some way for the past to have been expiated through acknowledgment of it, such that it is acceptable for the banalities of everyday life to continue to play themselves out on a stage that once was the setting for something violent and egregious. Thus, paradoxically, such places must be purged of their past through appropriate remembering of it; only once appropriate remembering has been enacted can the past be set aside or bracketed.

The best-selling Anglo-French novelist Tatiana de Rosnay has long been fascinated with the notion that past atrocities haunt the places in which they took place. Her 2003 novel, *La Mémoire de Murs* (*The Memory of the Walls*) is about a violent murder that later haunts a woman who has unknowingly moved into the apartment in which it had been committed. Clearly, this earlier novel was a warm-up for the much bigger question in which Rosnay was more deeply interested: four years later Rosnay's 2007 *Sarah's Key* further developed the theme through the story of a young Jewish boy whose tragic death during the Holocaust continued to haunt his Paris apartment. *Sarah's Key* is a microcosm for a much bigger set of questions: How can a normal European life continue, on a continent that is bathed in the blood of the Holocaust? How can the quotidian and unexceptional unfold in the same places where the exceptional and horrific have taken place? How is the past to be honored in the present? How can we forget it, and how can we allow ourselves to forget? And how are the present and the future to exist side by side with that past, without on the one hand obliterating or belittling it, yet on the other without being overshadowed by it completely? In France, where the question of the extent of domestic complicity with the Nazi occupiers is hotly debated, these questions are particularly fraught, to the extent of being virtually inarticulable. It is telling that Rosnay, who is bilingual in French and English, first wrote *Sarah's Key* in English: as she commented, "I felt that writing about such a sensitive French subject would somehow be easier for me if I used my 'English' side" (Rosnay 2013).

In addition to political sensitivities, a particular set of problems emerges in cases in which the "stigmatized property" isn't just a single-family residence, or isn't a private property at all. Taken as a metaphor, the real estate concept of the "stigmatized property" can be enlarged and adapted to fit broader circumstances. What if the "stigmatized property" is part of the public space, or otherwise indispensable to the life of the society it occupies? And what if the question of whether what happened there should be forgotten, or remembered, is deeply contested? For of course the greatest horrors aren't those of fantastical movie plots and the paranormal, but are those enacted *by* real people, *upon* real people, often with the complicity not just of other individuals, but of entire polities. The vexing problem of how to expunge a place of its violent past is not just a favorite trope of horror movies, and is not directed just toward forgetting. It has been a major dilemma surrounding many real places, in real life, as well, and is often connected to the question of how best to remember, and honor, a horrific past event, while at the same time allowing for some "normal" quotidian life to resume.[3]

Such questions connect to an extent to "sites of memory," the concept and term made famous by Pierre Nora in his multivolume work on historical memory (Nora 1992–96). In Nora's formulation such sites can be real or symbolic—actual places or conceptual topoi that function as zones for remembering that which otherwise would be forgotten. More precisely, they are zones for the deliberate and symbolic "remembering" of things that *have* been forgotten, or at least, of things that are not "remembered" communally in the true sense of the term. Because they aren't, "sites of memory"—*lieux de mémoire*—have been designated to help us—force us—to "remember," or at least to feel that we do. What we "remember" in these places, of course, are things that we didn't ever know, or "remember," in the first place. Sites of memory are didactic, and deliberately so; as such, they are generally tightly conceived, choreographed, controlled, and monitored by various official entities (governmental groups, or groups of private citizens who have gathered together as a formal entity dedicated to the "memory" of an event in their or their forebears' past). Often, they are dedicated to the memory of the victims of past events, or, more precisely, the memorialization of victims has long been the focal point for the commemoration of past events. Victims serve as the portal into a broader "remembering" of the larger events of which their suffering and victimization was a part. The victims are remembered, but this remembering is forever bound up with the horrors to which they were subject.

Broadly speaking, these sites fall into one of two categories: spaces that do not themselves have any particular inherent significance, but that have been filled with material objects that do (museums are the quintessential example), or that have been designated as holding significance as a place of memorialization (for example, the Vietnam War Memorial). The second broad category consists of spaces which themselves have some particular significance, almost invariably because of a specific event that took place on the site at some historic moment. And some—and here the World Trade Center Memorial would be a good example—are both: a specific location of significance because of events that took place there, with the add-on of both a memorial and a museum.

This essay is concerned with the second type of site—in which something of particular significance took place, and which of consequence has been designated (by some, at least) to be a *lieux de mémoire*—a place of particular significance in the collective "memory" (or memorialization) of the broader events of which it was a part. But the site I consider here, Drancy, the housing complex just outside Paris that served as a Jewish concentration and deportation camp during World War II, while certainly a *lieux de mémoire*, is perhaps better thought of also as a realtor thinks about "stigmatized properties"—a place where the impulse to forget and cover over is potently present, and coexists alongside the will to remember.

Three sets of interrelated tensions quite vividly mark Drancy, in specific, and to varying degrees also mark other such sites in general. First is the tension between remembering/memorializing/honoring the past, on the one hand, and allowing for an ongoing, normalized present, on the other. This tension can be characterized as a gray zone that lurks between remembering and dwelling; in Drancy, a low-income housing area, this "dwelling" is both figurative and quite literal. Second is the tension between different "marginalized groups": the (largely immigrant) Jews who were imprisoned in Drancy and deported from there to Auschwitz, and the largely Muslim urban poor and working class who now inhabit the housing complex. Third is the tension between different times—between a past that is made eternally present through the designation of Drancy as a memorial to the Holocaust, and a present and future that Drancy's residents, and Drancy as a community, are attempting to build for themselves.

Drancy as a "stigmatized property" is a place of past violence that has gone on to have not simply formal significance as a site of memory/memorialization, but that *in addition* also serves an ongoing, quotidian, and banal function, one that threatens to (the reaction varies by observer) cover up, degrade, demean, or erode the significance of the tragedies for which they set the stage. Indeed, in the case of Drancy—a housing complex the residents of which live quite literally in the shadow of the Holocaust—there is a powerful argument to be made that the site's two functions are in direct contradiction to each other: for the residents to have a normal existence, they cannot be expected to think constantly of the Holocaust, yet for one of the most appalling chapters of the Holocaust in France to be remembered, Drancy must not lapse into being simply, or solely, a housing complex.

The significance of such "dual" sites is contested—players as diverse as victims' advocacy groups, developers, governmental agencies, and religious organizations engage in pitched battles over the extent to which the past events marking a given site should be forgotten, or remembered. Such battles are not unique to Drancy; they characterize countless public sites connected to the deportation and roundup of European Jews during World War II that have gone on to a commonplace afterlife. Consider that there is virtually no key train station in continental Europe that is not in some way implicated in the Holocaust. Public squares and parks in countless cities were used as roundup points. For "normal" life to be reestablished after the war, a good deal of forgetting of necessity had to happen. In a continent that was itself a great battlefield, not all sites of horror could be set aside. Instead, certain sites have been preserved and elevated to a symbolic significance that is intended to capture not just the specific horror of their own specific past and place, but transcendently to memorialize the broader horror of the Holocaust, writ large. The paradigmatic example, of course, is Auschwitz. As Edward Said observed: "Think of geographical designations like Auschwitz, think of what power and resonance they have, over and above a particularly specifiable moment in history or a geographical locale like Poland or France" (Said 2000, 180). In certain circles (among French Jews, for instance, or historians of the Holocaust), Drancy verges on such "power and resonance," though to most, the name, if known, is simply a place on the map.

The process of triage—according to which some sites of the Holocaust end up museums, others end up with a plaque or other more modest memorial, and still others end up with nothing at all—is fraught and ongoing to this day. Sites of deportation pose particularly vexing problems. In many instances they mark the last place that deportees saw "at home" before their deportation. They also are among the most public sites of the Holocaust: most are centrally located, and the process of roundup and deportation was one of which neighbors and other spectators were very much aware. But for these very same reasons, sites of deportation are like "stigmatized properties" that must be made saleable again: it is not plausible to set aside significant space in the midst of the urban public sphere, declaring it too sacred or shameful or otherwise unfit for daily life.

Consequently some (like Plateía Eleftherías in Thessaloniki, site of one of the most notorious roundups of Greek Jews during the Holocaust) have gone on to a totally banal afterlife. Plateía Eleftherías used to be a central city site: among other historical events, the Young Turk Revolution of 1908 started in a massive demonstration there, where the new Ottoman Constitution was declared, but is now a parking lot. (Its name—which means "Freedom Square"—carries a now bitterly ironic meaning for those few who remember its role in the Holocaust.) There is a Holocaust memorial at the site—a large black sculpture by Alexios Menexiadis—but the site itself has all the meaninglessness and base utility of a municipal garage. In France, the largest site of roundup and deportation was the Vélodrome d'Hiver, a massive indoor bicycle-racing track on the south side of the Boulevard de Grenelle in Paris, just steps from the river Seine. It has been razed to the ground and in its place stands a sterile office building. A modest plaque at one side of the building declares the history of the place, while a somewhat larger memorial beside the river commemorates France's deported Jews. The synagogue on Via Luigi Carlo Farini in Florence—where that city's deported Jews were collected for deportation and now are commemorated—is one of the most well-fortified buildings in Italy today, with security approximating that of an Israeli airport; while some semblance of an ongoing Jewish daily life continues there, its normality is seriously undercut by the heavily fortified nature of the site—which itself stands as a permanent reminder of the horrors perpetrated upon Europe's Jews during World War II (and a reminder of the sense of vulnerability still felt by many European Jews today).

II

Drancy is a "commune," or suburb, of Paris, six miles northeast of the city center, a stop on the RER (commuter train) line—and now home, largely, to Muslim North African immigrants. The name of the place is now banal (*banalisé*)—and in many ways, it long has been, except to those who have a connection to its past history and place within the World War II. But when historians speak of "Drancy" they mean it to denote something more specific: not to the town of Drancy, but the deportation camp there that between 1941 and 1944 served as the key collection and departure point for Jews from France. Of the nearly 80,000 Jews deported from France during the war, more than three-quarters were deported from Drancy. Virtually all of them were sent from there to Auschwitz-Birkenau.

Unlike many other major topoi of the Holocaust in France—the "carousel" at the École Militaire in Paris's 7th arrondissement; the former Vélodrome d'Hiver (Vel d'Hiv) in the 15th; the Marais (a once Jewish neighborhood in the 3rd and 4th)—Drancy is in a place upon which no one would stumble by mistake. It is a modest suburb, a station on the way to Roissy–Charles de Gaulle airport (and, more recently, a site of major new real estate development, which touts it as "affordable yet just minutes from central Paris"). And within the commune of Drancy, the camp itself—an apartment building—is not something upon which one would stumble. Finally, if one *did* happen upon it, it would be largely unrecognizable as having any particular significance within the broader landscape. A brisk half-hour walk from the Drancy RER stop, the camp looks now from the outside like a modest strip mall, with dry cleaners and key-copy shops and couscous restaurants occupying its street-side ground-floor spaces. No street signs show the way, no plaques mark the building's walls.

Expunge from the imagination any image that the word "camp" might invoke. Instead, think: "housing project." The camp at Drancy was first designed as a modernist experiment in low-cost suburban living, a project of the Office d'habitations à bon marché (the low-cost residences office, or HBM). Started in 1931, construction tapered off in 1935 because of a global financial crisis. A long gray block of residential towers and interspersed open spaces, it remained unfinished when the Nazis confiscated it in 1941; for a year before that it had housed Spanish political prisoners after serving briefly as German barracks (Rajsfus 1996, 31–32; Sabbagh 2002, 5–6; Wieviorka and Laffitte 2012, 11–13). The design is basic: three buildings of five stories each, two long ones on the sides and a shorter one on the base, form a squared-off "U" shape around a dismal open courtyard. While not tall, the buildings are immense: each of the U's sides is about 200 meters long. Capacity was intended to be under 1,000 but during its time as a concentration camp it held up to 7,000 residents or more. As a camp, a double fence of barbed wire surrounded it, with guards living in the tower apartment blocks beside it. Its "apartments" had no furniture—the camp's internees slept on the floor, on straw, and used old tin cans as latrines, so repellent were the collective facilities located at one end of the complex's courtyard. While many of the letters sent by internees to friends and family outside show a false optimism—"we are doing well," "we are staying strong"—between the lines we read the hunger and the cold and the growing sense of desperation; Drancy was a cold and gray and terrifying place (Sabbagh 2002).

The Jews who were sent there during the war were thus among its very first, albeit temporary residents, and spent their internment living in what was in effect a construction site. Some internees (especially those deported early in the war) spent months living there before being deported to the death camps in Poland. Zacharie Mass, a doctor whose letters from Drancy to his wife in Maisons-Alfort compose one of many moving eyewitness accounts that survive, lived within its walls from October 16, 1941, until July 31, 1943. His letters show how totally dependent the camp's residents were on the outside world for their survival. Virtually every one of the hundreds of letters he wrote to his wife (who was not interned) mentions his desperate need for her to send him supplies: food, medicine, clothing—but above all, food (Mass 2012). Perhaps the most commonly occurring word in the entire corpus of letters we have from Drancy is the French *colis*, or package—the packages shipped to internees by family and friends outside, the contents of which they were dependent upon for their every need. The camp's inhabitants were often rounded up with nothing but the clothes on their backs. Residents lived for months without a change of undergarments, with malfunctioning showers and no soap—soap was so scarce that it was one of the most pilfered items in the camp (Wieviorka and Laffitte 2012, 43). There was a desperate need for medicine and warm clothing. Internees suffered from running sores, typhoid, edema, and hypothermia, among other maladies. In spring of 1942, 10 percent of the camp's population was in the camp's hospital (Rajsfus 1996, 173).

Following the infamous "Rafle du Vél d'Hiver" of July 16 and 17, 1942, in which more than 13,000 individuals, many of them children, were rounded up across Paris and dumped for days in the Vélodrome d'Hiver before being transferred to camps, Drancy became less an internment camp than a transit camp. Over the four months that followed, close to 30,000 internees were deported from Drancy, in thirty-one convoys (Rajsfus 1996, 146; Wieviorka and Laffitte 2012, 131). Life for the internees became even more excruciating, both physically and psychologically. Its most tragic inhabitants, about whom even the most hardened person could not read without weeping, were thousands of children, brutally separated from their parents just after the Vélodrome d'Hiver roundups, who spent three weeks in Drancy in late August and early September 1942. Between the ages of two and twelve, "they were dumped in groups of 100–120 in rooms bereft of furniture with repulsively filthy sleeping pallets on the ground. Chamber pots had to be placed on the hallway landings because many were too young to go downstairs alone to use the W.C. in the courtyard. In this period, the standard meal in the camp was cabbage soup. Very quickly, all of the children were struck with diarrhea. They soiled their clothes and the pallets upon which they sat all day long and on which they slept at night." The children were left alone at night in blackout conditions, crying out for their mothers, weeping and screaming. "On occasion all of the children in one room would scream in terror and despair" (Rajsfus 1996, 147–48).

Drancy came to be called *La Cité de la Muette*—the Silent City—ostensibly as a nod to the peaceful yet quasi-urban environment it was designed to evoke. The housing project still carries this same name today, though actually *muette* derived from the word *muete*, a cage for hunting dogs; before urbanization the area was likely a favored hunting zone (Wieviorka and Laffitte 2012, 12). Historically, Drancy was certainly far more of a cage than a place of peace.

"Drancy" today doesn't have the name recognition of, say, Auschwitz. It is entirely plausible to say, "I live in Drancy," or "I am going to Drancy"—indeed, thousands of people say these things every single day—and for it not to sound dissonant or haunting. And indeed, people *do* live in Drancy—and not simply in the "commune" of Drancy, but in the "Drancy" that is *La Cité de la Muette*. Drancy today has been restored to its original purpose and is a low-income housing complex—albeit one encompassing a Holocaust memorial. The horrific scenes described in witness testimony from its years as an internment and transit camp were played out in stairwells and rooms that exist today in more or less exactly the same form as seventy years ago. Drancy's residences, at least to those who know and care, are the ultimate "stigmatized property." The central strangeness of Drancy as it exists today is the juxtaposition of the utterly banal, if somewhat depressing, circumstances of the daily lives lived within its walls with the knowledge of the extraordinary circumstances of life as it was lived within those same walls by those interned there by the Nazis. In this awkward coupling of past and present, it is unclear whether it is the past that encroaches on the present, or the present that degrades the past.

In the spring of 2009, construction workers renovating buildings at Drancy discovered a large cache of graffiti, hitherto largely unseen, hiding on the buildings' walls underneath layers of drywall and other coverings. With the discovery some sixty years after the war of the wrenching last messages its internees wrote upon its walls, the "Silent City" spoke, asserting the memory of what it once had been (*Des noms sur des murs* 2009). We know from testimonials and memoirs that writing graffiti on the walls at Drancy became a ritual act for many inhabitants, who duly recorded their names, their date of arrival, and the date of their ultimate deportation. Others wrote poems, or commentary, or brief biographical sketches. But while the graffiti became the centerpiece of a major exhibit at the Memorial de la Shoah in central Paris, in Drancy itself it is unclear how best to memorialize a past atrocity in the midst of an entirely new population that now inhabits it.

Perhaps the ultimate silence of *La Muette* is the fact that until July 1943 it was under French administration and management, under the umbrella of the French Ministry of the Interior. Like many European countries under Nazi German occupation during the war, France has had a difficult time conceding the extent to which the crimes of the war were not merely Nazi crimes, but were also enabled by local occupied authorities and populations, indeed at times with their enthusiastic participation. Plaques in Paris commemorating the Holocaust now regularly include some wording about the "active complicity" of the French government in the roundup and deportation of Jews in France. But the topic is still sensitive and contested. And not only was Drancy under direct French administration—many roundups of Jews who were taken there were effected by the French. The most notorious, "Le Rafle du Vél d'Hiver" was enacted by French policemen and with the complicity of a number of French civil servants. Memorials to the Holocaust in France thus also testify to the varying extent to which France is willing to acknowledge its place within it. They serve a doubly didactic function: on the one hand, they ostensibly teach the observer about the past, but on the other, more darkly, they have the expiating function of a confession. They state not merely, "this was done," but also, in some instances at least, "we did this." But in the case of Drancy, who is the "we" to whom such a confession would speak?

Drancy today is a memorial to the Holocaust in France. Parked forlornly but prominently at the base of the squared-off U formed by the complex's residences is a boxcar, once used to transport Jews from Drancy to Bobigny, where they were transferred to trains bound for Auschwitz. Markings on the car indicate that it was designed to hold eight horses, or forty men; during the deportations from Drancy it was actually used to transport one hundred people at a time. A Magen David (Star of David) on its side indicates the Jewish nature of its cargo. As if this feature of the contemporary landscape were not grim enough, beside it stands a ghastly memorial, a massive sculptural representation by the artist Shlomo Shelinger of the "doors of death"—in reference to Drancy's grim epithet, "the antechamber of death." The children living in the complex walk past this daily—a large recreation center is among the conglomerate of buildings just beside the monument. Not fifty meters from the monument, in the center of the U, is a dismal abandoned playground, ringed by parking places for the building's residents.

The extent to which the sites of the Holocaust should be remembered is a bigger European question, inextricably linked to debates over who was victim and who was victimizer during the war of which the Holocaust was a part. It is actually a sort of cluster bomb of questions—a densely packed container for many subquestions—about the extent to which Holocaust commemoration is central to the commemoration of World War II as a whole, about the balance between commemoration of the war's Jewish and non-Jewish victims, and, most uncomfortably, about the extent to which some of the war's victims were complicit in the victimization of others. This latter question is particularly acute in those countries that were under German occupation. In the case of France, a long and painful revisionist history has been unfolding arguably starting with the 1972 publication of Robert Paxton's *Vichy France: Old Guard and New Order, 1940–1944*, which began to reveal the uncomfortable truth that the Vichy government's collaboration with the Nazis was willing rather than forced (Paxton 1972). The logical extension has been a consideration of the extent to which French collaboration with the Nazis reached well beyond the official members of the Vichy government. Drancy today is a microcosm of some of the most fraught debates in contemporary French society. Predominant among them are questions relating to the memory of the Holocaust in contemporary France, on the one hand, and the place of (largely Muslim) immigrants, on the other. In many ways, these debates are two different historical instantiations of the same "problem," the "problem" of the outsider. In Drancy, they collide.

France, like much of the rest of Europe, is now struggling with the "Muslim Question"—arguably analogous to the "Jewish Question" of the first half of the twentieth century. To read the European press, one of the most pressing "concerns" today is the "question" of Islam in Europe. There are now sixteen million Muslims living in EU member states; virtually all are immigrants or first- or second-generation Europeans. Because of dropping birth rates among the white European population, even with the slowing of Muslim immigration to Europe, the percentage of the European population made up by Muslims will continue to grow. The Pew Forum on Religion and Public Life projects that by 2030, there will be about sixty million Muslims in Europe, about 8 percent to 9 percent of the total European population. And half of them—thirty million—will be in today's EU. France will see close to 10 percent of its population made up of Muslims.

These demographics have created huge anxiety, which in France revolves around the question of "integration"—a notion based in the belief in an identifiable shared French corpus of historical, cultural, and social content. Central to that corpus is the experience of World War II—a pivotal moment in France's modern history, and paradoxically a topos of great nostalgia, in that the war marked a historical break between "traditional" French society and the rapidly escalating forces of Europeanization and globalization that came in its wake. The tenth of France's population that is Muslim has no connection to that sense of nostalgia, or to the effects of the war as it was played out on the European continent. What is the proper place of the "remembering" of the Holocaust in their acculturation and "integration" as French residents? And when monuments proclaim that "we" were complicit, are they to consider themselves part of that "we"?

III

It is not clear if it is possible at once to honor the past and to normalize the present. At stake is the awkward interplay between stigmatization and memorialization. Stigmatization suggests shame, while memorialization suggests its opposite: honor. The victims are honorable; the perpetrators and their acts the locus of shame. These dueling connotations are at the very heart of debates over how best to remember the Holocaust. But in the case of Drancy, a past-marginalized French population (immigrant Jews) is peculiarly being memorialized for a new marginalized French population (immigrant Muslims), in which the latter (Muslims) are the literal inhabitants of the property in which the former (Jews) were imprisoned. And the confessional function—the statement not just of *what* was done, and *to whom*, but also the declaration that in some way "we"—that is, the French people—did it, is being made to a people who have no connection to that "we" at all.

Drancy is a troubling case of a property that is at once stigmatized, and not stigmatized enough. The looming monument and squat boxcar stand as testament to its grim past, while its current state—dilapidated and gloomy, as most public housing projects are; home to a largely disenfranchised population, as most public housing projects also are—is at once an insult to the past and an insult to its present. The whole building could not be set aside as a monument, after all; it occupies a large piece of land in France's most densely populated urban zone. But at the same time, it could not be set aside because of the ludicrousness of that proposition in light of what it is: a housing block. It would be theater of the absurd to empty a massive building of inhabitants to commemorate its former inhabitants. And thus we have the camp itself now serving as home to hundreds of families, whose children play beside a monument testifying to a past of which they were not a part, commemorating other children the memory of whom is almost unbearable.

Meanwhile, Drancy—the commune, that is, rather than the camp—is rapidly growing in desirability as a place to live. Walking from the RER to *La Muette*, one is confronted by a series of billboards, advertising new "green" construction projects going up in the area. Immediately across from Drancy (the former camp) a large advertisement beside a construction site proclaims: "Your new address in Drancy." On the billboard, a happy family sits on a leafy balcony, enjoying life in Drancy. Silent—mute—just beside it is the remnant of *Cité de la Muette*—by implication, an *old* address in Drancy; one so old that it is now virtually forgotten. "Imagine yourself in Drancy," beckon the ads, while right next door is a Drancy, still standing, in which it is all but impossible to imagine oneself.

In Drancy, two solutions to the problem of destigmatization live side by side: memorialization and forgetting. The two populations they mark, like the different populations to whom they aim to speak, scarcely cross paths. The present Drancy and the past Drancy cannot hear each other. The peculiar missed message and the dual tracks along which Drancy operates are in many ways representative of the problem of the geographies of the Holocaust. How is it that places of past significance are to have meaning in their contemporary contexts? Indeed, one of the great challenges of the commemoration of the Holocaust is to find ways to make it universal enough that in theory it is part of *everyone*'s past. And another is the problem posed by Drancy: how to resolve the tension between the need to destigmatize a place's past and the need to keep it stigmatized—and thus, remembered, if uninhabitable—forever. Edward Said has written of what he calls "the right to a remembered presence" of a place's past inhabitants (Said 2000, 184). It is important that there not be a conflict between this right and the actual daily presence of those who now inhabit it—yet there always is. How is that conflict to be adjudicated such that both the past and the present are paid their due?

There is a vast variation in the degree to which the public sites of the Holocaust have been banalized, sacralized, or otherwise set apart from or integrated into the ongoing daily life of the cities in which they are found. Integration is not invariably a sign of forgetting, or, worse, of denial. Are sites of atrocity and suffering better marked as sacral spaces, or given new life as something connected both to past and present? While it is impossible to give definitive answers to these questions, it is increasingly important—as we hurtle ever further away from the Holocaust itself, and lose any true "memory" of it that we ever had—that we pause to consider them, just as we might pause, however briefly, to read a commemorative plaque, or glance at a statue, as we cross through a busy public square to run an errand or catch a bus.

Notes

1. Compare the size of the Flight 93 National Memorial in Shanksville, Pennsylvania (over a thousand acres) with the National September 11 Memorial and Museum at the World Trade Center, in lower Manhattan on the site of the former World Trade Center "twin towers" (eight acres). The memorialization and remembering—of even the most extreme and universally acknowledged national tragedies—can be assigned a price tag, just like any other material transaction.

2. The *Amityville Horror* home used to be 112 Ocean Avenue; subsequent owners had the address changed to 108 Ocean Avenue.

3. A recent, and particularly tragic, example is the Sandy Hook Elementary School in Newtown, Connecticut, where in December 2012, a twenty-year-old gunman shot and killed twenty children and six school staff members. It seems likely that the building will be torn down and the lot left vacant.

References

Larsen, James E., and Joseph W. Coleman. 2001. "Psychologically Impacted Houses: Broker Disclosure Behavior and Perceived Market Effects in an Unregulated Environment." *Journal of Real Estate Practice and Education* 4 (1): 1–16.

———. 2010. "Cemetery Proximity and Single-Family House Price." *Appraisal Journal* 78 (1): 33–49.

Mass, Zacharie. 2012. *Passeport pour Auschwitz: Correspondance d'un médecin du camp de Drancy*. Paris: Editions le manuscrit.

Des noms sur des murs: Les graffiti du camp de Drancy. 2009. Exhibit, Mémorial de la Shoah, Paris.

Nora, Pierre. 1992–96. *Les lieux de mémoire*. 3 vols. Paris: Gallimard.

Paxton, Robert. 1972. *Vichy France: Old Guard and New Order, 1940–1944*. New York: Columbia University Press.

Rajsfus, Maurice. 1996. *Drancy: Un camp de concentration très ordinaire, 1941–1944*. Paris: Le cherche midi éditeur.

Rosnay, Tatiana de. 2003. *La mémoire des murs*. Paris: Le livre de poche.

———. 2013. "Frequently Asked Questions (FAQ)." Accessed June 10. http://www.tatianaderosnay.com/index.php/faq.

Sabbagh, Antoine. 2002. *Lettres de Drancy*. Paris: Tallandier.

Said, Edward. 2000. "Invention, Memory, and Place." *Critical Inquiry* 26 (2): 175–92.

Wieviorka, Annette, and Michel Laffitte. 2012. *À l'intérieur du camp de Drancy*. Paris: Perrin.

Acknowledgments

A book like *Hitler's Geographies* is always the result of a collective effort. For this reason, we wish to appropriately acknowledge all the people who contributed with their expertise and their commitment to its realization. We would thus like to start by thanking the authors for their much-appreciated contributions. We are also extremely grateful to John O'Loughlin, Rory Rowan, James Sidaway, and Dan Stone for reading and commenting on parts of this manuscript, and to the referees for their detailed and extremely helpful comments and suggestions.

We would also like to thank our publisher, and in particular our editors: Abby Collier for supporting this project in its earliest stages, and Mary Laur for guiding us through the several revisions and the process of completing the book. We are sincerely grateful also to Logan Ryan Smith, Dawn Hall, and Michael Koplow for managing many key editorial aspects, and to Maartje Roelofsen, Trista Chih-Chen Lin, and Alexandra Rijke for acting as our "in-house" copyeditors in various stages of the crafting of this collection. It is important to acknowledge that during the long period of preparation of this volume we have both received substantial institutional support, in terms of making time available for its completion, but also in terms of funding. We are therefore sincerely thankful for the support received from the Department of Geography at Royal Holloway, University of London, the Environmental Sciences Department at Wageningen University, and the University of Torino. All of these institutions made reciprocal visits and the attendance at numerous conferences possible. Claudio has also benefited from the research leave spent in 2009 as a visiting professor at the Department of Geography, University of British Columbia in Vancouver. Paolo is particularly grateful to the Remarque Institute and the Global Research Initiative at New York University for the hospitality and support during his visitorship in 2014.

Contributor Biographies

Trevor Barnes is Professor and Distinguished University Scholar at the University of British Columbia, Vancouver, where he has taught since 1983. His main research interests are in the history of twentieth-century geography, especially during periods of war, both hot and cold. He has published over 150 journal articles and book chapters. His most recent books are *Reading Economic Geography* (Blackwell, 2003), *Politics and Practices in Economic Geography* (Sage, 2007), and *The Wiley-Blackwell Companion to Economic Geography* (ed., with J. Peck and E. Sheppard, Blackwell, 2012).

Mark Bassin is Research Professor of the History of Ideas, in the Center for Baltic and East European Studies at Södertörn University in Stockholm. His research focuses broadly on problems of space, ideology, and identity in Russia and Germany. He has held teaching appointments at the University of Birmingham, University College London, the University of Wisconsin–Madison, and UCLA, and has been Visiting Professor at the University of Chicago, Copenhagen University, and Pau University (France). He is the author of *Imperial Visions: Nationalist Imagination and Geographical Expansion in the Russian Far East, 1840–1865* (Cambridge University Press, 1999), and has coedited the collections *Space, Place, and Power in Modern Russia: Essays in the New Spatial History* (Northern Illinois University Press, 2010) and *Soviet and Post-Soviet Identities* (Cambridge University Press, 2012).

Richard Carter-White completed his PhD at the University of Exeter and is currently an Assistant Professor at the University of Tokyo. His research is concerned with the spatialities of genocide, and particularly the spatial historiographies generated by literary, cinematic, and testimonial representations of the Holocaust.

Andrew Charlesworth was the first geographer to begin a comprehensive study of the geography of the Holocaust, looking at landscapes of memorialization in particular. He was the first to innovate a field study program on the Holocaust for undergraduates. In 2008 he held a Charles H. Revson Foundation Fellowship at the US Holocaust Memorial Museum. It was there that he became disenchanted with certain forms of academic study of the Holocaust and embarked on a novel centered on the murder of a high school teacher by three of his students in the Permanent Exhibition.

Tim Cole completed his PhD in Geography at the University of Cambridge. He is now Professor of Social History at the University of Bristol. He is the author of *Images of the Holocaust / Selling the Holocaust* (Routledge, 1999), *Holocaust City* (Routledge, 2003), *Traces of the*

Holocaust (Continuum, 2011), and is a coeditor of *Militarized Landscapes* (Continuum, 2010) and of *Holocaust Geographies* (2012).

Clemens Driessen is a philosopher and social scientist studying the lives of animals in our technological culture. Currently he is a Lecturer in the Cultural Geography chair group at Wageningen University. Previously he was a postdoctoral researcher in the Geography Department of King's College London, where he worked on an ESRC-funded research project into contemporary and historical attempts at "back breeding" of wild herbivores for use in "rewilding," a mode of nature conservation that attempts to instill natural processes to (re)produce ecosystems by the introduction of keystone species. Earlier at Wageningen University, he researched ethical concerns in livestock farming, combining philosophical, ethnographic, and historical approaches to investigate our ambivalent and changing relations with animals.

Stuart Elden is Professor of Political Geography at Warwick University. He is the author of books on Heidegger, Foucault, Lefebvre, and territory, including *Speaking against Number: Heidegger, Language, and the Politics of Calculation* (Edinburgh University Press, 2006) and *Terror and Territory: The Spatial Extent of Sovereignty* (University of Minnesota Press, 2009). His most recent monograph, *The Birth of Territory*, was published in 2013.

Katherine Fleming is the Alexander S. Onassis Professor of Hellenic Culture and Civilization in the Department of History at New York University (NYU). She also serves as Vice Chancellor and Senior Vice Provost, and as Director of the Remarque Institute. A specialist on religion and nationalism in the Mediterranean, she is the author, among other books, of *Greece: A Jewish History* (Princeton University Press, 2007), which won numerous awards, among them the National Jewish Book Award. Fleming holds a BA (Barnard/Columbia) and an MA (Chicago) in comparative religion, and a PhD (UC Berkeley) in History.

Michael Fleming has taught human geography at Jesus College and Pembroke College, Oxford University and has been a visiting researcher at the Pułtusk School of Humanities and at the Institute of History, Polish Academy of Sciences in Warsaw. He is currently a Professor at the Academy of Humanities and Economics in Łódź, Poland, and at the Polish University Abroad, London. His current research examines Polish intelligence on the Holocaust. He is the author of *Communism, Nationalism, and Ethnicity in Poland, 1944–1950* (Routledge, 2010).

Paolo Giaccaria is an Assistant Professor in Political and Economic Geography at the University of Turin. He received his PhD from the London School of Economics. His work focuses on geographical theories, in particular ANT and complexity theory. Currently, his main research topic is the relationship between spatiality, modernity, and biopolitics, with special attention to the geographies of the Third Reich and of the Holocaust. A second focus of research lies with the modern—colonial and orientalist—making of the Mediterranean space, in particular the practices of cosmopolitan citizenship in imperial Mediterranean cities. He is the coauthor of *Local Development and Competitiveness* (Springer, 2001), *Torino nella competizione europea* (Rosenberg and Sellier, 2002), and *Geografia del sistema manifatturiero piemontese* (Carocci, 2010).

Simone Gigliotti is a Senior Lecturer in the History Programme at Victoria University, Wellington, New Zealand. She is the author of *The Train Journey: Transit, Captivity, and Witnessing in the Holocaust* (Berghahn Books, 2010), among other publications on the Holocaust and genocide.

Joshua Hagen is Professor of Geography at Marshall University. He has published widely on issues related to urban planning, historical preservation, and nationalism in Germany, including in *Journal of Historical Geography*, *Environment and Planning D: Society and Space*, *Cultural Geographies*, *Journal of Urban History*, and *Annals of the Association of American*

Geographers. He has also coauthored work on geopolitics and border studies, including *Borderlines and Borderlands: Political Oddities at the Edge of the Nation-State* (Rowman and Littlefield, 2010) and *Borders: A Very Short Introduction* (Oxford University Press, 2012).

Jamie Lorimer is Associate Professor in the School of Geography and Environment at the University of Oxford. His research explores the politics of Nature, with specific reference to wildlife conservation. The research that informs his chapter emerges from an ESRC-funded project on the history and politics of European rewilding. His current research explores the microbiome as a new space for science studies and geographical enquiry. He is the author of *Wildlife in the Anthropocene: Conservation after Nature* (University of Minnesota Press, 2015).

Claudio Minca is Professor and Head of Cultural Geography at Wageningen University. His current research centers on three major themes: tourism and travel theories of modernity; the spatialization of (bio)politics; and the relationship between modern knowledge, space, and landscape in postcolonial geography. His most recent books are *Travels in Paradox* (with T. Oakes, Rowman and Littlefield, 2006), *Social Capital and Urban Networks of Trust* (with J. Hakli, Ashgate, 2009), *Real Tourism* (with T. Oakes, Routledge, 2011), *On Schmitt and Space* (with R. Rowan, Routledge, 2015), and *Moroccan Dreams* (with L. Wagner, I. B. Tauris, forthcoming).

Mechtild Rössler is a historian of geography and cultural landscape specialist. She has received an MA in cultural geography from Freiburg University (Germany) with a focus on urban geography and landscape studies and a PhD from the Faculty for Earth Sciences, University of Hamburg, on the history of planning. She is teaching regularly in World Heritage studies at different universities (Turin, University College Dublin, and BTU Cottbus) and is currently carrying out a research project on the History of the World Heritage Convention led by Professor Cameron (Canada Research Chair on Built Heritage), University of Montreal. She has published seven books, more than one hundred articles, mostly on World Heritage, landscape issues, history of geography and planning as well as nature/culture interaction, and contributed to the editorial board of three international journals.

Dan Stone is Professor of Modern History at Royal Holloway, University of London. He is the author or editor of more than a dozen books, including, most recently, *Histories of the Holocaust* (Oxford University Press, 2010), *The Holocaust and Historical Methodology* (ed., Berghahn, 2012), *The Oxford Handbook of Postwar European History* (ed., Oxford University Press, 2012), *Holocaust, Fascism, and Memory* (Palgrave, 2013), and *The Liberation of the Camps: The End of the Holocaust and Its Aftermath* (Yale University Press, 2015).

Gerhard Wolf completed his PhD at Humboldt University Berlin. He is a DAAD-Lecturer for Modern German Studies at Sussex University (since 2008), after having worked at the History Department at the Free University Berlin (August 2006), and at the Memorial Site House of the Wannsee Conference (August 2001). His most recent publication is *Ideologie und Herrschaftsrationalität—Nationalsozialistische Germanisierungspolitik im annektierten Polen* (Hamburger Edition, 2012).

Jürgen Zimmerer is Professor of Modern History at the University of Hamburg. His research interests include German colonialism, comparative genocide, and environmental violence. Recent publications include *Von Windhuk nach Auschwitz? Beiträge zum Verhältnis von Kolonialismus und Holocaust* (LIT, 2011), *Kein Platz an der Sonne: Erinnerungsorte der deutschen Kolonialgeschichte* (Campus, 2013), and *Climate Change and Genocide*, a special issue of the *International Journal of Human Rights* (2014, 18/3).

Index